INGENIOUS MECHANISMS
FOR DESIGNERS AND INVENTORS

VOLUME I

INGENIOUS MECHANISMS

FOR DESIGNERS AND INVENTORS

VOLUME I

Mechanisms and Mechanical Movements Selected
from Automatic Machines and Various Other Forms
of Mechanical Apparatus as Outstanding Examples
of Ingenious Design Embodying Ideas or Principles
Applicable in Designing Machines or Devices Requiring Automatic Features or Mechanical Control

Edited by

FRANKLIN D. JONES

INDUSTRIAL PRESS INC.
200 MADISON AVENUE, NEW YORK 10016

Industrial Press Inc.
200 Madison Avenue
New York, New York 10016-4078

INGENIOUS MECHANISMS
FOR DESIGNERS AND INVENTORS—VOLUME I

33 35 37 39 41 43 45 44 42 40 38 36 34 32

PREFACE

When the designer or inventor begins to originate or develop some form of automatic machine or other mechanical device, he is confronted by two important problems: The first one is purely mechanical and relates to the design of a mechanism that will *function* properly. The second problem is a commercial one and pertains to designing with reference to the cost of manufacture.

In order to solve the mechanical part of the problem, especially when an intricate motion or automatic control is required, a wide knowledge of the principles underlying those mechanical movements which have proved to be successful, is very helpful, even to the designer who has had extensive experience. The purpose of this treatise is to place before inventors and designers concise, illustrated descriptions of many of the most ingenious mechanical movements ever devised. These mechanisms have been selected not only because they are regarded as particularly ingenious, but also because they have stood the test of actual practice. Many of these mechanisms embody principles which can be applied to various classes of mechanisms, and a study of such mechanical movements is particularly important to the designer and student of designing practice owing to the increasing use of automatic machines in almost every branch of manufacture.

The second problem mentioned, that of cost, is directly related to the design itself which should be reduced to the simplest form consistent with successful operation. Many mechanical movements *are* ingenious because they are simple in design. Simplified designs usually are not only less costly but more durable. Almost any action or result can be obtained mechanically if there are no restrictions as to the number of parts used and as to manufacturing cost, but it is evident that a design should pass the commercial as well as the purely mechanical test. In this connection it is advisable for the designer to study carefully mechanical movements which actually have been applied to commercial machines. Practically all of the mechanisms shown in this treatise have been utilized on automatic machines of various classes.

CONTENTS

CHAPTER I

CAMS AND THEIR APPLICATIONS

A STUDY of the various mechanical movements and automatic regulating devices used on automatic and semi-automatic machines of different types, will show that mechanical movements based on the same general principles are often applied to machines which differ widely as to type and purpose. For instance, a mechanism for obtaining an intermittent motion may possibly be utilized in connection with almost any mechanical device requiring such motion, after certain changes have been made. Frequently these necessary changes will alter the form and perhaps the entire arrangement without changing the underlying principle governing the operation. This explains why designers of automatic machinery find that a knowledge of mechanical movements of all kinds is valuable, because an understanding of one design often suggests an entirely different application.

Many of the most ingenious mechanical movements and regulating devices ever devised will be found in this treatise. Some of these with more or less modification have been used so generally on different classes of machinery that they may be considered standard, whereas many others are more special and are not so generally known. All of these mechanisms, however, are believed to embody some mechanical principle that is likely to be useful to designers and inventors. These various mechanisms have been grouped in chapters according to the general types or classes to which they belong, partly to show different modifications of a given type and also to assist

1

users of this book in finding a mechanical movement suitable for a particular application.

This first chapter deals with cams because they are widely used in the design of automatic machines of practically every type. In fact, by the use of some form of cam it is possible to obtain practically an endless variety of movements and irregular motions, many of which could not possibly be derived by other mechanical means. Even though some other type of mechanism might be substituted, the cam provides the simplest and cheapest method of obtaining most of the special unusual actions required in automatic machine design.

General Classes of Cams. — The name "cam" is applied to various forms of revolving, oscillating, or sliding machine members which have edges or grooves so shaped as to impart to a follower a motion which is usually variable and, in many cases, quite complex. Cams are generally used to obtain a motion which could not be derived from any other form of mechanism. Most cams revolve and the follower or driven member may have either a rectilinear or oscillating motion. The acting surface of the cam is in direct contact either with the follower or with a roller attached to the follower to reduce friction. The exact movement derived from any cam depends upon the shape of its operating groove or edge which may be designed according to the motion required.

Cams may be classified according to the relative movements of the cam and follower and also according to the motion of the follower itself. In one general class may be included those cams which move or revolve either in the same plane as the follower or a parallel plane, and in a second general class, those cams which cause the follower to move in a different plane which ordinarily is perpendicular to the plane of the motion of the cam. The follower of a cam belonging to either class may either move in a straight line or receive a swinging motion about a shaft or bearing. The follower may also have either a uniform motion or a uniformly accelerated motion. The working edge or groove of a uniform motion cam is so shaped that the follower moves at the same velocity from the beginning to the end of the stroke. Such cams are only adapted to comparatively slow speeds, owing to the shock resulting from the sudden movement of the follower at the beginning of the stroke and the abrupt way in which the motion is stopped at the end of the stroke. If the cam is to rotate quite rapidly, the speed of the follower should be slow at first and be accelerated at a uniform rate until the maximum speed is attained, after which the motion of the follower should be uniformly decreased until motion ceases, or a reversal takes place; such cams are known as "uniformly accelerated motion cams."

Plate Cam. — Several different forms of cams are shown in Fig. 1. The form illustrated at *A* is commonly called a "plate cam," because the body of the cam is in the form of a narrow plate, the edge of which is shaped to give the required motion to the follower. This follower may be mounted in suitable guides and have a reciprocating motion (as indicated in the illustration) or it may be in the form of an arm or lever which oscillates as the cam revolves. When the follower is in a vertical position as shown, it may be held in contact with the cam either by the action of gravity alone or a spring may be used to increase the contact pressure, especially if there are rather abrupt changes in the profile of the cam and the speed is comparatively fast.

Positive Motion Cam. — The cam illustrated by diagram *B*, Fig. 1, is similar to the type just described, except that the roller of the follower engages a groove instead of merely resting against the periphery. Cams of this general form are known as "face cams" and their distinctive feature is that the follower is given a positive motion in both directions, instead of relying upon a spring or the action of gravity to return the follower. The follower, in this particular case, is in the form of a bellcrank lever and is given an oscillating motion. One of the defects of the face cam is that the outer edge of the cam groove tends to rotate the roller in one direction and the inner edge tends to rotate it in the opposite direction. A certain amount of clearance must be pro-

vided in the groove and, as the roll changes its contact from
the inner edge to the outer edge, there is an instantaneous
reversal of rotation which is resisted, due to the inertia of

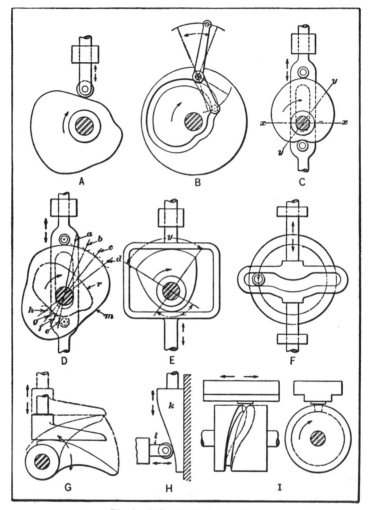

Fig. 1.   Different Types of Cams

the rapidly revolving roll; the resulting friction tends to wear
both the cam and the roll.   This wearing action, however,
may not be serious when the cam rotates at a slow speed.   If

the speed is high, there is also more or less shock each time the follower is reversed, owing to the clearance between the roller and the cam groove.

Plate Cam Arranged for Positive Motion. — In order to avoid the defects referred to in connection with the face cam, the follower of a plate cam is sometimes equipped with two rollers which operate on opposite sides of the cam, as shown at *C,* Fig. 1. With such an arrangement, the curve of the cam for moving the follower in one direction must be complementary to the curve of the remaining half of the cam, since the distance between the rollers remains constant. In other words, this cam may be designed to give any motion throughout 180 degrees of its movement, but the curvature of the remaining half of the cam must be a uniform distance from that of the first, at all points diametrically opposite. Then the distances measured along any center line, as at *xx* or *yy,* are constant and equal the distance between the follower rollers. For this reason, the term *constant diameter cam* is sometimes applied to this class which is adapted for heavier work than the grooved face cam illustrated at *B.* The follower or driven member is slotted to receive the camshaft, and this slot acts as a guide and keeps the rollers in alignment with the center of the cam.

Return Cam for Follower. — When the curvature of one half of a cam is not complementary to the curvature of the other half, a special return cam is necessary, if the follower is equipped with two rollers in order to secure a positive drive. A main and return cam is illustrated at *D,* Fig. 1. The main cam may be laid out to give any required motion for a complete revolution of 360 degrees, and the return cam has a curvature which corresponds to the motion of the return roller on the follower. After the main cam is laid out to give whatever motion is required, points as at *a, b, c, d,* etc., are located on the path followed by the center of the roller, and, with these points as centers, the points *e, f, g,* and *h* are located diametrically opposite, and at a distance equal to the center-to-center distance between the rollers. These latter points lie in the path followed by the center of the return roller, and by striking arcs from them having a radius equal to the roller radius, the curvature or working surface of the return cam may be laid out. One method of arranging these two cams is to place the follower between them and attach the rollers on opposite sides of the followers. The camshaft, in some cases, carries a square block which is fitted to the elongated slot in the follower to serve as a guide and a bearing surface.

Yoke Type of Follower. — Another form of positive motion cam is shown at E, Fig. 1. In this case, the follower has a surface which is straight or tangential to the curvature of the cam. With a follower of this kind, there is a limitation to the motion which can be imparted to it, because, when the contact surface is flat or plane, it is evident that no part of the cam can be concave since a concave surface could not become tangent to the straight face of the follower, and even though the follower is curved or convex any concave part of the cam must have a radius which is at least as great as the radius of any part of the follower. The type of cam shown at E, like the one illustrated at C, can only be laid out for a motion representing 180 degrees of cam rotation; the curvature of the remaining half of the cam must be complementary to the first half or correspond to it. The follower of the cam shown at E has a dwell or period of rest at each end of its stroke, the parts x and y being concentric with the axis of the camshaft. This general type of cam has been used for operating light mechanisms and also to actuate the valves of engines in stern-wheel river steamers.

Inverse Cams. — On all of the cams previously referred to, the curved surface for controlling the motion has been on the driving member. With a cam of the inverse type, such as is shown at F (Fig. 1) the cam groove is in the follower and the roller which engages this groove is attached to the driving member. The motion of this cam can be laid out for only 180 degrees of movement. The inverse type of cam is used chiefly on light mechanisms, the particular cam illustrated at

F being designed to operate a reciprocating bar or slide. The curved part of the slot in the follower has the same radius as the path of the driving roller, and serves to arrest the motion of the slide momentarily. The well-known Scotch yoke or slotted cross-head is similar to an inverse cam having a straight slot that is perpendicular to the center line of the follower. (The motion obtained with the Scotch yoke and its practical application is referred to in Chapter IX.)

Wiper and Involute Cams. — The form of cam shown at *G,* Fig. 1, is simply a lever which has a curved surface and operates with an oscillating movement through an arc great enough to give the required lift to the follower. A cam of this kind is called a "lifting toe" or a "wiper" cam, and has been employed on river and harbor steamboats for operating the engine valves. Many involute cams are somewhat similar in form to the type illustrated at *G,* and they are so named because the cam curve is of involute form. Such cams are used on the ore crushers in stamp mills. Several cams are placed on one shaft and as they revolve the rods carrying the stamps are raised throughout part of the cam revolution. Disengagement of the cam and follower then causes the latter to drop.

Cams having Rectilinear Motion. — Some cams instead of rotating are simply given a rectilinear or straight-line motion. The principle upon which such cams operate is shown by diagram *H,* Fig. 1. The cam or block *k* is given a reciprocating motion in some form of guide, and one edge is shaped so as to impart the required motion to the follower *l.* An automatic screw machine of the multiple-spindle type is equipped with a cam of this general type for operating side-working tools, the tool-slide receiving its motion from the cam which, in turn, is actuated by the turret-slide. This type of cam is also applied to an automatic lathe for operating the radial arm or tool-holder.

Cams for Motion Perpendicular to Plane of Cam. — The cams previously referred to all impart motion to a follower which moves in a plane which either coincides with or is parallel to the plane of the motion of the cam. The second general class of cams previously referred to, which cause the follower to move in a plane usually perpendicular to the plane of the motion of the cam, is illustrated by the design shown at *I*, Fig. 1. This form is known as a "cylinder" or "barrel" cam. There are two general methods of making cams of this type. In one case, a continuous groove of the required shape is milled in the cam body, as shown in the illustration, and this groove is engaged by a roller attached to the follower. Another very common method of constructing cylinder cams, especially for use on automatic screw machines, is to attach plates to the body of the cam, which have edges shaped to impart the required motion to the follower. When a groove is formed in the cam body, it should have tapering sides and be engaged by a tapering roller, rather than by one of cylindrical shape, in order to reduce the friction and wear.

Automatic Variation of Cam Motion. — Ordinarily the motion derived from a cam is always the same, the cam being designed and constructed especially for a given movement. It is possible, however, to vary the motion, and this may be done by changing the relative positions of the driving and driven members by some auxiliary device. This variation may be in the extent or magnitude of the movement or a change in the kind of motion derived from the cam. The cam mechanism shown at *A* in Fig. 2 is so arranged that every other movement of each of the two followers is varied. The bellcrank levers *a* and *b*, which are the followers, have cam surfaces on the lower ends, and they are given a swinging motion by rolls *d* and *e* pivoted to arm *c* which revolves with the shaft *h* seen in the center of the arm.

The requirements are that each lever have first a uniform motion and then a variable motion; it is also necessary to have a change in the variable stroke until twelve strokes have been completed, when the cycle of variable motions is repeated. For instance, every other vibration of each lever is through a certain angle, and for twelve alternate vibrations the stroke is changed from a maximum to a minimum, and *vice versa,* the angle of the uniform vibration being the mean or average movement for the variable strokes. The uniform vibration is obtained when roll *d* engages the cam surface on either lever *a* or *b*, and the variable movement is derived from roll *e* on the opposite end. This roll is mounted eccentrically on bushing *f* which is rotated in its seat by star-wheel *g*, one-twelfth revolution for each revolution of arm *c*; consequently, the roll is moved either toward or away from the axis of shaft *h*, thus varying the angle of vibration accordingly.

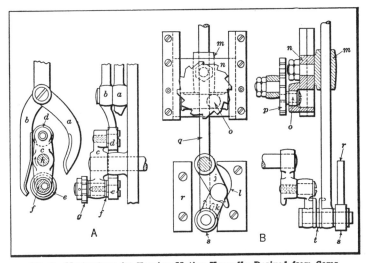

Fig. 2. Mechanisms for Varying Motion Normally Derived from Cams

Another mechanism which serves to vary the motion derived from a cam surface is shown at *B* in Fig. 2. This mechanism is used in conjunction with one previously described. A motion represented by the curvature *l* of a plate cam is reproduced by the upper end of the rod or lever *q*. One movement of the rod end is an exact duplicate of the cam curvature, and this movement represents the mean of a cycle of twelve movements, each of which is a reproduction of the curvature on an increasing or diminishing scale from maximum to minimum, or *vice versa*. The lever returns to the starting position with a rectilinear or straight-line motion.

The lever is given a reciprocating movement by crank *j* and connecting link *k*. The roll *s* at the lower end of the lever is kept in contact with cam surface *l* by spring *t*. The lever *q* is fulcrumed and slides in the oscillating bearing *m* which is supported by the slotted cross-head *n*. This cross-head is operated by roll *o* which is carried by a crankpin on a twelve-tooth ratchet wheel *p*. When the mechanism is in action, the crank *j* throws connecting link *k* out of line with lever *q* and the resulting tension on spring *t* causes roll *s* to follow the outline or curvature *l* of the cam until the upper end of the

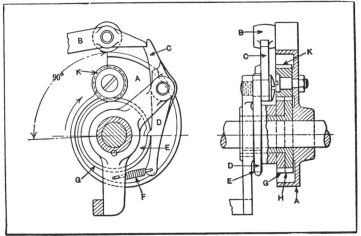

Fig. 3.   Arrangement for Varying Dwell of Cam Follower

travel is reached; then the connecting link *k* is thrown out of line with lever *q* in the opposite direction, which causes spring *t* to force roll *s* against the straight return guide *r*. For each revolution of the crank, a pawl turns the ratchet wheel *p* one tooth, so that the slotted cross-head *n* and the bearing *m* are gradually raised and then lowered. As the result of this upward and downward movement of bearing *m*, which is the fulcrum for lever *q*, the motion is increased and then diminished the desired amount.

Varying Dwell of Cam Follower. — The mechanism illustrated in Fig. 3 is for varying the dwell of a cam follower or the length of time it remains stationary. The cam *A* lifts lever B during three-fourths of a revolution, and during the dwell the follower B is held up by the latch C. This latch is controlled by pawl D, cam E, and spring F. The cam E has ratchet-shaped notches in its edge and is made integral or in one piece with a twenty-four-tooth gear G. The ratchet and gear are revolved upon the hub of a twenty-five-tooth stationary gear H, by the planetary pinion K, once for every twenty-four revolutions of cam A. With this particular mechanism, the lever B is given a dwell of 90 degrees for the first revolution; thereafter the dwell increases 360 degrees after each

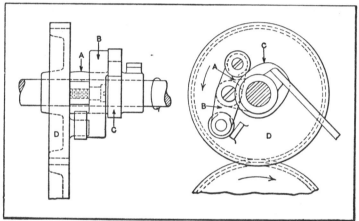

Fig. 4. Application of Cam for Varying Rotary Motion

rise of the follower, until the fourth period (which gives 1530 degrees dwell) when the dwell decreases until it is again 90 degrees; that is, during the fourth period the rise occurs while the cam makes three-fourths revolution, and then there is a dwell equivalent to $4\frac{1}{4}$ revolutions. Twenty-four revolutions are required to complete a cycle of movements. When milling the teeth in cam E, the index-head was arranged for twenty-four divisions, but teeth were cut only at the following divisions: 1-2-4-7-11-16-20-23. When the mechanism is in use, latch C is disengaged whenever pawl D enters a notch in cam E, thus allowing lever B to drop suddenly.

Variable Rotary Motion derived from Cam. — An unusual application of a cam is illustrated in Fig 4. In this case, a cam is used to impart a variable angular velocity to a gear which makes the same number of revolutions as its driving shaft. The driving shaft carries a casting *A* to which is fulcrumed the lever *B* which, in turn, has a roll on each end. One roll engages a cam *C* which is supported upon the shaft but does not revolve with it. The other roll bears upon a lug on the side of gear *D* which is also free upon the shaft, but is constrained to revolve with it either faster or slower, accord-

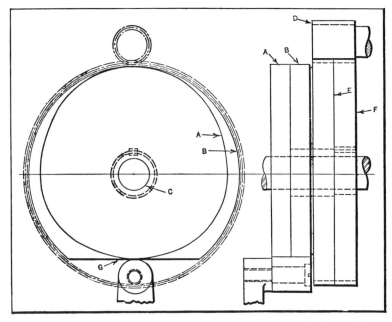

Fig. 5. Two-part Cam which Alternately Increases and Decreases Stroke of Follower

ing to the relative positions of lever *B* and cam *C*.

Cam which Alternately Increases and Decreases Motion of Follower. — An automatic paper-tube rolling machine has a driven member that must dwell during three-fourths of a revolution of shaft *C* (Fig. 5) and then be given a stroke that varies gradually in length for successive cam revolutions. This variation is obtained by using a cam having two sections *A* and *B*. These two sections are both driven by pinion *D* through gears *E* and *F*. Gear *E* is integral with cam *B* and has 105 teeth, whereas gear F is keyed to the hub of cam A and has 104 teeth. Both gears have the same outside diameter, and the difference in tooth numbers provides a differential movement between the cams, so that one cam is continually changing its position relative to the other.

The dwell is obtained when the roll of the follower is in contact with the concentric part of cam B. When cam A is in the position shown, the maximum stroke occurs as the follower traverses across the flat edge G of cam B. The stroke of the follower is gradually reduced as A turns relative to B, thus filling the segment-shaped space at G, so that finally the cam is nearly concentric all around. The motion of the follower is somewhat irregular, but for this particular application, the irregularity is immaterial, as the essential requirement is to have the follower, after the 364th revolution of the pinion, at a distance from the center of shaft C equal to the dwelling position.

Automatic Variation of Cam Rise and Drop According to Pressure Changes. — The special design of cam illustrated in Fig. 6 normally has a 120-degree rise, a 60-degree dwell, a 90-degree drop, and a 90-degree dwell. In the operation of the machine to which this cam was applied, however, it was necessary to vary the motion derived from the cam in accordance with the pressure exerted upon a certain part of the machine; for instance, if the pressure exceeds a given limit during a dwell, the rise must take place in 90 degrees instead of 120 degrees; whereas, if the pressure decreases below the desired amount, the drop must be lengthened to 120 degrees. The mechanism for automatically varying the cam motion is comparatively simple, as the illustration indicates.

The main cam A carries two auxiliary cams B and C. These cams are driven by pins, which pass through them as shown by the sectional view, and they are free to slide upon these pins and the shaft, parallel to the axis of the shaft. Cam B carries a roller K and cam C, a roller L. Adjacent to these movable cams, there is a disk D having two sets of ratchet teeth and two side cams M and N. (The end view of this

disk is shown at the lower part of the illustration.) A pawl
F rests upon the block G until the increase or decrease of
pressure interferes with the balance of the spring shown and
causes pawl F to drop into engagement with a ratchet tooth.
As soon as this engagement occurs, disk D stops rotating and
cams M and N come into engagement with rollers K and L
and force cams B and C over toward cam A, so that they engage the wide cam-roller on the follower, and give it the re-

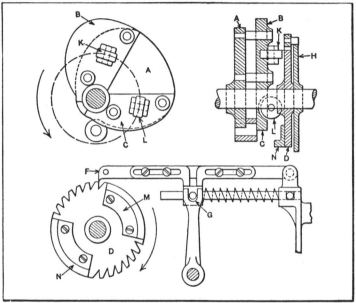

Fig. 6. Cam equipped with Mechanism for Varying Rise and Drop
According to Predetermined Pressure on Another Part of the Machine

quired variation of movement. The cam H returns pawl F
to the neutral position.

Sectional Interchangeable Cams for Varying Motion.— A
flexible cam system was required that made it possible to
vary the motion relative to the complete cycle of movements
by substituting one interchangeable cam section for another,
instead of using a large single cam for each variation. Two
distinct methods of obtaining practically identical results were
successfully evolved. One mechanism was a rotary type and

the other involved the use of rectilinear motion for the cam sections. Both mechanisms might properly be called "magazine" cams, because the cam sections are continually placed in action and then replaced by others in successive order. The rotary design is illustrated in Fig. 7. The cam sections shown at A are semi-circular. The continuity of the cam surface is obtained by making each semi-circular section in the form of a half turn of a spiral with close-fitting joints, the complete cam appearing like a worm. The sections are fed longitudinally along the shaft and successively under the

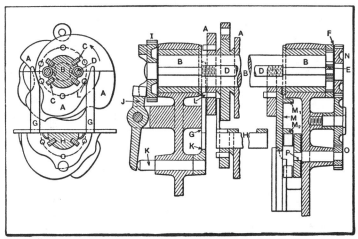

Fig. 7. Cam Mechanism Provided with Interchangeable Sections for
Varying Motion of Follower

lever roller at a rate of advance equaling the lead of the spiral. Four feathers C are provided to guide and retain the cams. The two screws D producing the longitudinal movement are driven by pinions E meshing with an internal gear F, which is fastened to the bearing. As the feathers extend only to within the width of one cam from the left bearing, two sections drop from the shaft at every revolution, the dropping sections being guided by the guides G. The double cam upon the driving gear I, the lever J, and the carrier-slide K provide the means for hanging the semi-circular cams upon the magazine bar H. The slide K catches each piece by the pins L and, by pushing one, causes the further one to slide onto the lifting slide M which engages its grooved hub. The gears N and O, in the ratio of 1 to 2, and disk P operate a slide for returning the cams to their shaft. The rollers on P successively engage the steps M_1 and M_2, thus raising the slide which drops back automatically.

To facilitate engagement between the cam threads and the screws, the square threads of the latter are V-shaped at the entering ends, and, to insure locking the cams to the shaft quickly, the ends of the feathers recede into pockets and fly out by the action of springs. Any part of the system may be

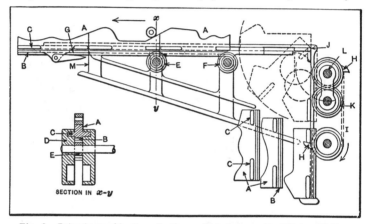

Fig. 8. Interchangeable Cam Sections which have a Rectilinear Motion

changed by placing the desired section in a holder and introducing it between the slide K and the magazine bar. The cam to be removed—the dropping cam—comes out upon an inclined runway of the holder.

The alternate design is the rectilinear cam system shown in Fig. 8. The mechanism consists of the cam sections A, provided with rack teeth at B. (See also detail sectional view.) Each section has four lugs C which act as guides in the ways D. A pinion E feeds the sections along beneath the lever roller, and the frictionally driven pinion F assembles them. When any section has passed beneath the roller, it is automatically hung upon the magazine chute. The forward lugs C are made slightly longer than the rear ones, to span the gap G; but the rear lugs enter the gap just as the forward lugs clear the ways. The sections are taken from the lower part of the ways in the magazine by spring-controlled forks H upon the chain I which engage the lugs and lift the cams until the smaller lugs strike at the corner J. The linked gear K meanwhile engages the rack, and as it swings about the center L, it lifts the cam up against the ways; here the resistance offered to further motion of the links causes K to rotate about its own center and slide the cam into place.

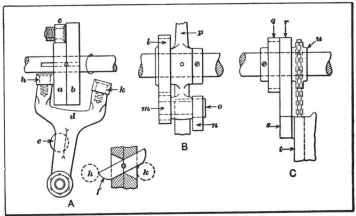

Fig. 9.   (A) Double-shifting Cam; (B) Lever Vibrated from Shaft on which it is Fulcrumed; (C) Shaft Oscillated by Cam Located on it

Substitute sections are introduced at M, and the replaced sections are lifted from the ways.

Double Two-revolution Cam of Shifting Type. — The cam mechanism illustrated at A in Fig. 9 is so arranged that two revolutions of a double cam are necessary in order to give the required motion to a follower. One revolution is required for the rise or upward movement of the follower and a second revolution for the "dwell," during which the follower remains stationary. The cam sections a and b are fastened together and are free to slide upon their shafts a distance equal to the face width of one section. The two cam sections are driven by means of a spline. Roll c is attached to the follower or driven member and, in the illustration, is shown in contact with the spiral cam *a*, from which the upward movement is derived. The cam *b* is simply a circular disk mounted concentric with the shaft. The lever *d* for shifting the double cam is operated by a "load-and-fire" mechanism having a spring plunger at *e*. (The load-and-fire principle is explained in Chapter VI on "Reversing Mechanisms.")

When the mechanism is in operation, cam *a* lifts roll *c* to its highest position, when lever *d* shifts the double cam along the shaft, leaving roll *c* upon cam *b*, where it remains during a dwell of one revolution; the cam is then immediately shifted in the opposite direction, thus allowing roller *c* and the driven member to drop instantaneously upon cam section *a*. The movement of shifting lever *d* is derived from the double-ended lever *f* (see detailed view) which extends through a slot in the cams. This lever is pivoted at the center and is free to swing in one direction or the other, until it rests against the sides of the opening. With the double cam in the position shown in the illustration, end *f* engages roll *h* and forces it to the left until spring plunger *e* comes into action and suddenly throws the lever over the full distance. The opposite end of lever *f* swings far enough to clear roll *k* before this roll is thrown over.

Lever Vibrated from Shaft on which it is Fulcrumed.— A cam which is used for vibrating a lever twice for each revolution of a shaft on which it is fulcrumed is illustrated at *B* in Fig. 9. A gear *l* attached to the shaft drives a pinion *m* which is one-half the size of the gear. This pinion revolves cam *n*, and the shaft for the pinion and cam has a bearing in the end of lever *p*. The cam revolves in contact with a stationary roll *o* which causes the lever to vibrate about the shaft as a center twice for every revolution.

Shaft Oscillated by Cam located on it. — Fig 9 shows, at *C*, how a shaft can be given an oscillating or rocking movement by a cam which is mounted on the shaft. The cam *r* is attached to gear *q* which is driven from an outside source. As the cam revolves in contact with roll *s*, a reciprocating motion is imparted to slide *t*. A chain attached to this slide passes over a sprocket *u* which is fast to the shaft. The other end of the chain is fastened to a tension spring beneath the slide, which serves to hold the roll *s* into engagement with the cam.

Double-track Cam. — A cam that provides the required motion and "dwells" for a slide on a special flat-wire form-

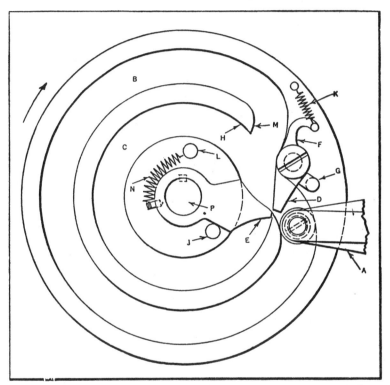

Fig. 10. Cam having Two Concentric Grooves which are Engaged Alternately by a Roller on the Driven Slide

ing machine (See Fig. 10) is so designed that the follower *A* has a dwell at each end of its stroke. The cam has two concentric grooves *B* and *C*, and as it rotates, the roller on follower *A* is transferred alternately from one groove to the other by means of the switching levers *D* and *E*. The roll in the illustration is about to come into contact with lever *D*, which will swing around until the lug *F* engages pin *G*;

then edge D will be flush with edge H and follower A will pass into groove C. The path thus formed to guide the roll into groove C is positive, as lever D is against stop G and lever E is against J.

As soon as the roll passes the end of lever D, the latter snaps back to the position shown, through the action of spring K. The follower now dwells at one end of its stroke, as groove C is concentric. When the cam has revolved far enough to bring the roll into contact with lever E, the latter swings around until it strikes stop-pin L, and then the edge E is flush with end M. A path has now been formed which leads the roll into the outer groove B, after which lever E snaps back to the position shown, through the action of coil spring N. The cam rotation then continues until the roll is again in the position illustrated, when the cycle just described is repeated.

The follower is rigidly connected to a slide (not shown) which operates a mandrel for forming the stock. The cam receives its motion from shaft P. The required movement could have been obtained by the use of an ordinary cam, thus reducing the speed of shaft P one-half, but because of difficulties due to conflicting machine speeds, it was considered advisable to employ the special cam described.

Spiral Cam for Reciprocating Motion. — A positive spiral cam drive for imparting a reciprocating motion to a slide is shown in Fig. 11. The cam C, which has a spiral groove, revolves continuously in the direction indicated by the arrow, and transmits motion to slide D through engaging rollers A and B which are connected by rocker arm E, and are arranged to engage the cam alternately. If roller A is in the inner position or at the inner end of the cam groove as shown, it will be traversed to the outer end of the groove while the cam makes $1\frac{1}{2}$ revolutions; as this roller approaches the outer end of the groove, it engages a cam insert F (see also detail sectional view) placed in the groove; consequently, roller A rides up the inclined surface of this cam insert, which causes rocker E to force the other roller B down into engagement

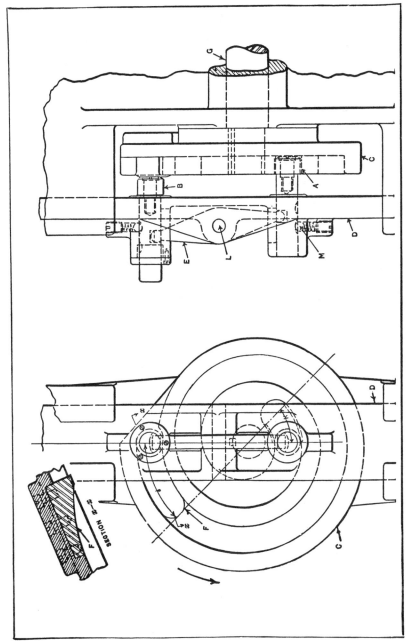

Fig. 11. Spiral Cam which is Engaged Alternately by Two Rollers on Rocker Arm of Driven Slide

with the inner part of the cam groove; then the return stroke of the slide begins as the cam continues to revolve, and when roller *B* has reached the outer position, thus completing one cycle, the action is reversed, roller *A* being again forced into engagement at the inner position of the cam groove. It will be seen then that three cam revolutions are required for the forward and return strokes of the slide, and the rollers successively traverse from the inner to the outer positions.

At the beginning and end of the spiral, the groove is milled concentric with driving shaft *G* (as indicated by the arrows

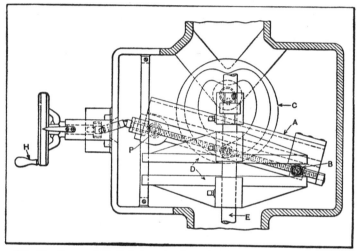

Fig. 12. Cam and Slotted Cross-head Combination with Adjustment for Varying Stroke

J and *K*) which provides a dwell equivalent to one-eighth revolution of the cam at each end of the stroke. The concentric sections *J* and *K* also permit the rolls to enter and leave the groove freely. The spiral groove advances uniformly so that a uniform motion is imparted to the driven slide. The rocker *E*, which swings on pin *L*, has rounded ends that engage grooves cut in the roller plungers. Pawl *M* which is backed by a spring, drops into either of two half round grooves in the plunger for locating it in the upper and lower positions. The other plunger has the same arrangement. The cam insert F is of hardened tool steel and the rolls are beveled at the bottoms to correspond with the curve of the insert.

Cam-stroke Adjustment without Stopping Machine. — The mechanism shown in Fig. 12 is for traversing the table of a grinding machine along the bed. This machine, which is of a comparatively small size, is intended for internal and external grinding operations; thus it is necessary to provide means for readily changing the stroke of the table. With the mechanism illustrated, any variation in stroke can be obtained from zero to the maximum while the machine is operating. The motion for the table is derived from a heart-shaped cam C mounted on a vertical shaft which is driven through a speed-changing mechanism. This cam engages a roll attached to the lower side of an oscillating arm A having on its upper side another roll B which can be adjusted relative to the pivot P about which the arm oscillates. This upper roll operates between the parallel faces of yoke D, and the latter is attached to a rod E located beneath the table of the machine. On the under side of the table and extending throughout its entire length is a dovetailed slide-way in which is fitted a block that is attached to and moves with the reciprocating rod E. By means of a suitable lever, this block, which fits into the dovetailed slide-way, can be clamped in various positions for changing the location of the table. The action of the mechanism is as follows: When the cam C is rotating, arm A oscillates about pivot P and, through roller B, transmits a rectilinear motion to yoke D, rod E, and the table The length of this movement or stroke is governed by the position of roll B relative to pivot P, which may be varied by means of a screw that is connected through a universal joint with a shaft upon which handwheel H is mounted. When roll B is moved inward until it is directly over pivot P, no movement will be imparted to yoke D or the table.

Crank and Cam-lever Combination. — An interesting form of mechanism is illustrated in Fig. 13. This mechanism is used on moving picture cameras and also for feeding films through printing machines. It is commonly referred to as a "claw" mechanism or movement. The claw or hook A is double and engages evenly spaced perforations that are along each edge of the film. When this device is applied to a moving picture camera, the film is drawn, from a roll in the film box, down in front of the lens and then passes to a reel in the receiving box. The film remains stationary during each exposure and is drawn downward between successive exposures which are made at the rate of sixteen a second. The hook A, which engages the film and moves it along intermittently and with such rapidity, receives its motion from a crank and cam-lever combination. The two intermeshing gears B and C

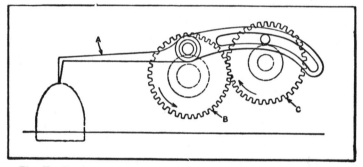

Fig. 13. Crank and Cam Combination for Operating Claw Mechanism of Moving Picture Camera

revolve in opposite directions. Gear B has a crankpin upon which the hook is pivoted. An extension of this hook has a curved cam slot that engages a pin on gear C. As the two gears revolve, the hook is given a movement corresponding approximately to the D-shaped path indicated by the dotted lines. While this mechanism is shown in a horizontal position in the illustration, it would normally be vertical with the hook uppermost, when in operation. Some of the other claw mechanisms in use differ from the one shown in regard to the arrangement of the operating crank and the cam or curved slot for modifying the crank motion. For instance, the cam, in some cases, is a separate part that is placed between the crank and the film hook, a pin on the hook lever engaging the cam slot. Another type of claw mechanism derives both the downward motion for moving the film and the in and out movements of the film hook from separate cam surfaces.

Group of Cams engaged Successively. — The mechanism to be described was designed to engage with the driving shaft first one and then another of the cams in a group of five mounted upon the same shaft. It was necessary to have these cams operate their respective levers successively back and

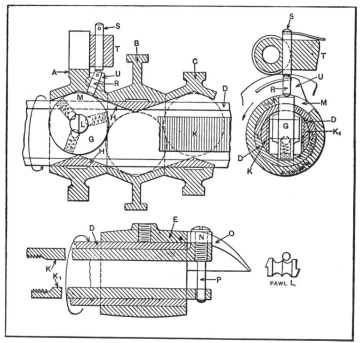

Fig. 14. Cams in a Group Engaged Successively

forth from one end of the group to the other, and while any one cam was in action the others must remain stationary with their lever rolls on a 90-degree dwell. Eight revolutions of the shaft were required to complete one cycle of movements. The device for controlling the action of these cams is shown in Fig. 14. The cams A, B, C, etc., are mounted upon a hollow shaft D carried in bearings E. The engagement of successive cams with the hollow shaft is effected by a roll-key G

which is caused to move inside of the shaft from end to end. This motion of the roll-key is obtained from ratchets K and K_1. (See longitudinal section at lower part of illustration, which is taken at an agle of 90 degrees to upper section in order to show more clearly the construction.) As the roll-key is moved along, it follows the inclined surfaces H which bring it into engagement with the respective cam keyways, as at M. Within the roll-key there is a double-ended pawl L (see also detail view) which is held into engagement with either ratchet K or K_1 by balls and springs. The ratchets are cut oppositely and are given a reciprocating movement by cam O, roll N, and roll screw P which causes both ratchets to reciprocate together. A similar equipment on the opposite end of the ratchet makes the motion positive. When the roll-key has engaged the last cam in one direction, the return of the ratchet causes the pawl L to rise onto a higher surface, thereby throwing it into mesh with the other ratchet and effecting the reversal.

Obtaining Resultant Motion of Several Cams. — A driven member or follower is given a motion corresponding to the resultant motion of four other cam-operated followers by the mechanism to be described. These followers are in the form of levers, which are equally spaced and fulcrumed upon one bar. Four of the levers are operated independently by four positive-motion cams. The fifth lever, which is in the center of the group, receives the resultant motion of all the others; that is, the forces acting upon the other four levers are automatically resolved and their resultant in magnitude and direction is transmitted positively to the fifth lever. It is not necessary to show the cams or levers to illustrate the principle involved, but the ingenious apparatus by means of which the resultant motion is obtained is shown in horizontal section in Fig. 15. Each of the four levers is connected by a knuckle joint to one of the racks A, B, C, and D. These racks are free to slide up and down independently and are arranged in two pairs. One pair meshes with pinion E and the other pair with pinion F. As the arrangement of the mechanism is symmetrical it will only be necessary to describe the action of one side. Any movements of the levers connecting with racks A and B will be transmitted to pinions E and G, which are mounted on one stud and rotate together. A stationary rack H and a sliding rack J engage pinion G. The sliding rack J carries a pinion K which, in turn, engages a stationary rack L and a sliding rack M. Pinion N is located on sliding bar P to which is attached the fifth lever previously referred to.

In order to illustrate the action of this mechanism, assume that rack A lifts one inch, rack B drops one-half inch, rack C is stationary, and rack D lifts one-quarter inch. The resultant is a three-quarter inch rise. In analyzing the motion, it should be remembered that a pinion moving along a stationary rack will cause a movable rack on the opposite side to travel with

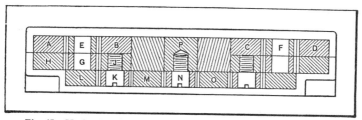

Fig. 15. Mechanism for Obtaining Resultant Motion of Several Cams

twice the pitch-line velocity of the pinion, which fact and its converse are here applied. The racks A and B acting upon pinion E will cause it to rise $\frac{1}{2} \times (1 - \frac{1}{2}) = \frac{1}{4}$ inch. This movement is doubled in the sliding rack J which, therefore, travels one-half inch, and it is again doubled in sliding rack M which as a movement of one inch. Rack M, in turn, moves pin N and the fifth lever slide P one-half inch. If the action of racks C and D is analyzed in a similar manner, it will be found that rack O has a movement of one-half inch, and rack N, one-quarter inch, which gives a total rise of the lever attached to slide P of three-fourths inch. To further illustrate the action, if all of the cam levers should drop one inch simultaneously, the result would be a drop of four inches for the middle lever attached to slide P.

CHAPTER II

INTERMITTENT MOTIONS FROM RATCHET GEARING

IT is frequently necessary for machine parts to operate intermittently instead of continuously, and there are various forms of mechanisms for obtaining these intermittent motions. A tool-slide which is given a feeding movement at regular intervals is an example of a part requiring an intermittent movement. Automatic indexing mechanisms which serve to rotate some member, periodically, a definite part of a revolution, after the machine completes a cycle of operations, represent other applications of intermittent movements. The usual requirements of an intermittent motion, when automatic in its action, are that the motion be properly timed relative to the movement of parts operating continuously and that the member receiving the intermittent motion be traversed a predetermined amount each time it is moved. The movement may be uniform or it may vary periodically. When the machine part which is traversed intermittently must be located in a certain position with considerable accuracy, some auxiliary locating device may be utilized in conjunction with the mechanism from which the intermittent motion is obtained. For example, the spindle carriers of multiple-spindle automatic screw machines are so arranged that the carrier is first rotated to approximately the required position by an intermittent motion, and then it is accurately aligned with the cutting tools by some form of locating device.

Ratchet Gearing. — One of the simplest and most common methods of obtaining intermittent movements is by means of ratchet gearing. This type of gearing is arranged in various

ways, as indicated by the diagrams in Fig. 1. In its simplest
form, it consists of a ratchet wheel *a* (see diagram *A*), a pawl
b, and an arm or lever *c* to which the pawl is attached. The
arm *c* swings about the center of the ratchet wheel, through
a fractional part of a revolution, as indicated by the full
and dotted lines which represent its extreme positions. When
the movement is toward the left, the pawl engages the teeth

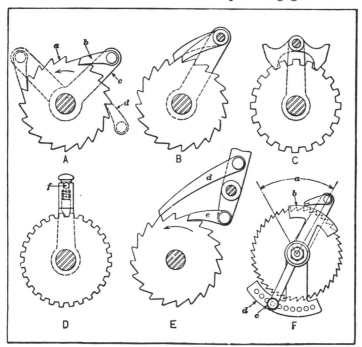

Fig. 1. Different Arrangements of Ratchet Gearing

of the ratchet wheel so that the latter turns with the arm.
When the arm swings in the opposite direction, the pawl
simply lifts and slides over the points of the teeth without
transmitting motion to the ratchet wheel. If a load must
be sustained by the ratchet gearing, a fixed pawl located at
some point, as indicated at *d*, is used to prevent any backward rotation of the ratchet wheel.

With gearing of this general type, the faces of the ratchet
teeth against which the end of the pawl bears should be so

formed that the pawl will not tend to fly out of mesh when a load is applied. In order to prevent such disengagement, the teeth should be so inclined that a line at right angles to the face of the tooth in contact with the pawl will pass between the center of the ratchet wheel and the pivot of the pawl. If the face of this tooth should incline at such an angle that a line at right angles to it were above the pawl pivot, pressure against the end of the pawl would tend to force it upward out of engagement with the ratchet wheel.

Multiple Pawls for Ratchets. — When a single pawl is used as shown at *A*, Fig. 1, the arm which carries it must swing through an arc equal to at least one tooth of the ratchet wheel; hence the pitch of the teeth represents the minimum movement for the wheel. If two or more pawls are used, a relatively small motion of the arm will enable successive teeth to be engaged without decreasing the pitch of the ratchet wheel. The principle is illustrated by diagram *B* which shows two pawls in position instead of one. As will be seen, one pawl is longer than the other by an amount equal to one-half the pitch of the ratchet teeth. With this arrangement, the movement of the arm may equal only one-half the pitch, if desired, the effect being the same as though a single pawl were applied to a wheel having teeth reduced one-half in pitch. By using three pawls, each varying in length by an amount equal to one-third of the tooth pitch, a still finer feeding movement could be obtained without actually decreasing the pitch of the teeth and thus weakening them.

Reversal of Motion with Ratchet Gearing. — A simple method of obtaining a reversal of motion is illustrated by diagram *C*, Fig. 1. A double-ended pawl is used and, in order to reverse the motion of the ratchet wheel, this pawl is simply swung from one side of the arm to the other, as indicated by the full and dotted lines. Reversible ratchet wheels must have teeth with bearing faces for the pawl on each side.

Another method of obtaining a reversal of motion is shown at *D*. The pawl, in this case, is in the form of a small plunger which is backed up by a spiral spring. One side of the pawl is beveled so that the pawl merely slides over the teeth on the backward movement of the arm. When a reversal of movement is required, the pawl is lifted and turned half way around, or until the small pin f drops into the cross-slot provided for it, thus reversing the position of the working face of the pawl.

Frictional Ratchet Mechanisms. — The types of ratchet gearing previously referred to all operate by a positive engagement of the pawl with the teeth of the ratchet wheel. Some ratchet mechanisms are constructed on a different principle in that motion is transmitted from the driving to the driven member by frictional contact. For instance, with one form, the driving member encircles the driven part which has cam surfaces that are engaged by rollers. When the outer driving member is revolved in one direction, the rollers move along the inclined cam surfaces until they are wedged tightly enough to lock the driven part and cause it to turn with the operating lever. When the driver is moved in the opposite direction, the backward motion of the rollers releases them. This general principle has been applied in various ways.

Double-action Ratchet Gearing. — It is sometimes desirable to impart a motion to the ratchet wheel during both the forward and backward motions of the ratchet arm or lever. This result may be obtained by using two pawls arranged as illustrated by diagram E, Fig. 1. These pawls are so located relative to the pivot of the arm that, while one pawl is advancing the ratchet wheel, the other is returning for engage· ment with the next successive tooth.

Variable Motion from Ratchet Gearing. — Ratchet gearing, especially when applied to machine tools for imparting feeding movements to tool-slides, must be so arranged that the feeding motion can be varied. A common method of obtaining such variations is by changing the swinging movement of the arm that carries the operating pawl. In many cases the link which operates the pawl arm receives its motion either from a crank or a vibrating lever, which is so arranged that the pivot for the rod can be adjusted relative to the center of rotation for changing the movement of the operating pawl and the rate of feed.

One method of adjusting the motion irrespective of the movement of the operating pawl is illustrated at F in Fig. 1. The pawl oscillates constantly through an arc a, and this angle represents the maximum movement for the ratchet wheel. When a reduction of motion is desired, the shield b is moved around so that the pawl is lifted out of engagement with the ratchet wheel and simply slides over it during part of the

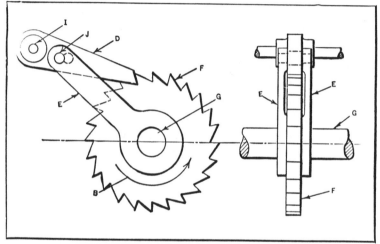

Fig. 2. Ratchet Mechanism to Prevent Reversal of Rotation and Arranged to Lift Pawl and Eliminate Noise when Ratchet Wheel is Rotating Clockwise

stroke. Thus, when the shield covers three of the teeth as shown in the illustration the motion of the ratchet wheel is reduced the same as though the swinging action of the pawl lever had been diminished an amount corresponding to three of the teeth. With the particular arrangement illustrated, the shield is held in any position by means of a small spring plunger c that engages holes in a stationary plate d.

Ratchets having Lifting Pawls to Prevent Noise. — Fig. 2 shows the construction of a ratchet mechanism that was designed for use on machines in which the noise of the pawl passing over the teeth of the ratchet is objectionable. Incidentally, the continuous wear on the ratchet teeth and the end of the pawl is eliminated by the arrangement shown.

The ratchet wheel *F* revolves with the shaft *G*. The pawl *D* swings freely on the pivot *I*, which is held in the stationary part of the machine. The connecting links *E* are free on the shaft *G* and are held together at their upper ends by the rivet *J* which has a shoulder on both sides. This permits the links to be tightly fastened together and still be free to swing on

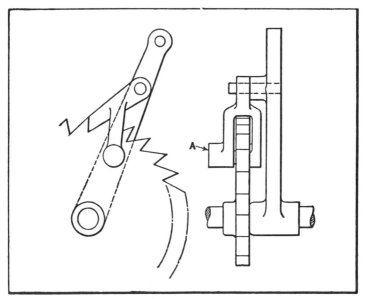

Fig. 3. Ratchet Mechanism with Silencing Device

the pawl *D*. The links *E* are sprung together, or toward each other at the lower end, so that they have a slight friction bearing on the sides of the ratchet wheel *F*. There is an elongated hole in pawl *D* through which rivet *J* passes.

The action of the mechanism is very simple but effective. When shaft *G* is turned clockwise, ratchet wheel *F* turns with the shaft. The friction on the sides of ratchet wheel *F* has a tendency to revolve links *E* with the wheel. The tendency to revolve, however, is prevented by rivet *J* which passes through pawl *D*. As rivet *J* shifts to the right-hand end of the slot in D this action results in lifting pawl D out of contact with wheel F and holding it out of contact as long as shaft G is turned in a clockwise direction. The height that pawl D is lifted above the ratchet wheel is controlled by the length of the slot through which rivet J passes. As soon as shaft G revolves in the opposite direction, as indicated by arrow B, links E tend to revolve with the ratchet, and this results in bringing pawl D downward into contact with the teeth of the ratchet wheel, as shown in the illustration.

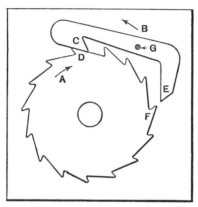

Fig. 4. Ratchet Mechanism having Double-ended Pawl

Another ratchet equipped with a silencing device is illustrated in Fig. 3. Boss A contains a spring plunger provided with a fiber tip. ,This plunger produces a slight friction on the ratchet sides and so causes the pawl to be lifted from the ratchet teeth on the idle stroke and kept from contact with the teeth until the working stroke. Many modifications of this principle are possible.

Silent Ratchet having Double-ended Pawl. — The ratchet mechanism shown in Fig. 4 has a double-ended pawl which operates silently. When the ratchet wheel turns in the direction indicated by arrow A, or when the pawl rotates in the direction indicated by arrow B, the end C of the pawl is raised by tooth D, thus bringing the end E into position to be engaged by tooth F. The engaging faces of the teeth are sloped so that the pawl will slide to the root and obtain a full contact. No spring is attached to the pawl.

When used as a feeding device, a frictional resistance, such as a friction washer placed on the fulcrum pin G, must be provided to eliminate rattle and insure the proper functioning of the pawl. When used simply to prevent the reversal of either member, no frictional resistance is necessary. In laying out a ratchet of this type it should be borne in mind that one of the pawls is just on the point of passing the tip of one of the teeth when the other pawl is midway between the tips of two teeth. It should also be noted that this type of ratchet, when used as a feeding mechanism, provides for feeding or indexing in multiples of one-half of a tooth space.

Silent Ratchet of Ball or Roller Type. — The design of ratchet shown in Fig. 5 should not be confused with the friction type. Power is transmitted by gripping the balls or

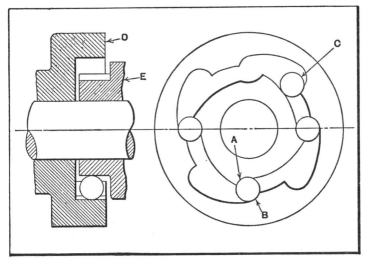

Fig. 5. Design that Transmits Power by Gripping Balls or Rollers
between the Driving and Driven Members

rollers between surfaces A and B of the driving and driven members, not between cam surfaces. No springs are employed to bring the balls into place, gravity alone being relied on. Usually only three balls are in action; in the illustration it will be observed that ball C is not in engagement. Either member D or E may serve as the driver. When this mechanism is used in a drive where the movement need not be accurate, it is not necessary to machine the engaging surfaces, and iron castings serve well unless the strain is severe.

Ratchet Designed to "Dwell" Automatically. — When a feed-shaft or other driven member requires a "dwell" after

every partial revolution, this may be obtained by a double
ratchet mechanism arranged like the one shown in Fig. 6.
This particular mechanism is designed to give a dwell equivalent to 3 teeth, or 3/16 revolution of the ratchet wheel, after
every movement equal to 13 teeth, or 13/16 revolution.

Ratchet wheel B has the idle period or dwell, and ratchet
wheel A carries a shield or guard F which prevents the pawl
E of wheel B from operating during the dwell. Ratchet wheel
B is keyed to shaft D, and the auxiliary ratchet wheel A is

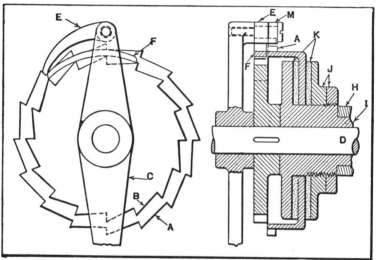

Fig. 6. Double Ratchet Having Shield which Prevents One Pawl from
Engaging Wheel During Dwell

confined between two leather disks K, the pressure required
being obtained from check-nuts J. Pawl E engages wheel B,
as mentioned, and pawl M engages wheel A. These two pawls
are pivoted to and operated by lever C, which gives them a
movement that is slightly greater than three ratchet wheel
teeth.

The function of the auxiliary ratchet A is merely to carry
shield F around so as to prevent E from engaging wheel B
during the idle period. The illustration represents the beginning of the dwell, which will continue until pawl M has moved
A around so that shield F does not interfere with the action

of pawl E. Shaft D is a running fit in sleeve I, which is a
force fit in part H of the machine.

Automatic Variation in Ratchet Feed Motion. — A special
attachment on a wood-turning machine requires a comparatively heavy feed at the beginning, followed by a finer feed
for finishing. This alternate retarding and accelerating feed
motion is obtained automatically from a ratchet mechanism

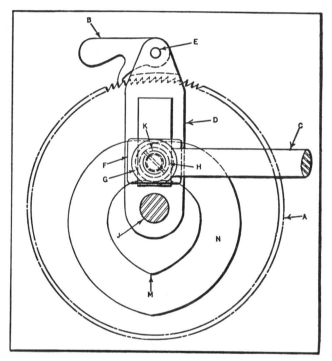

Fig. 7. Ratchet Feed Movement which is Increased and Decreased
Alternately as Cam Varies Radial Position of Crankpin

(see Fig. 7) which is so designed that the radial position of
the ratchet lever crankpin is continuously increased and decreased by a cam. The ratchet wheel A is secured to the feed-
screw shaft and the cam groove N is cut in one side. Ratchet
lever D is free to turn on shaft J, and it carries the feed pawl
B. Lever D is given a swinging or oscillating movement by
link C which connects with stud K. This stud is driven into

the slide or cross-head F, and it has a projection on the other side carrying the cam roll G which engages cam groove N.

It is evident that as ratchet A is intermittently rotated, the cam will first increase the radial position of pin K until point M is passed, and then will return pin K to the minimum radial position shown by the illustration. This increase and decrease between the centers of shaft J and pin K will, of course, have a corresponding effect upon the arc through which lever D swings and the resulting movement imparted to ratchet wheel A and the feed-screw.

Automatic Reduction of Intermittent Movement. — The mechanism to be described is applied to a chucking grinder for automatically reducing the cross-feeding movement and depth of cut, as the diameter of the part being ground approaches the finished size. The head which carries the grinding wheels (three or four wheels are used on this machine) is given a reciprocating motion on the bed of the machine, and the work-spindle head is mounted on a bracket that can be set at an angle relative to the motion of the wheel-carrying slide for taper grinding. The shaft which transmits motion to the cross-feed mechanism shown in Fig. 8 derives its motion from a cam surface on a swinging member of the wheel-head reversing mechanism, which is of the bevel gear and clutch type controlled by a load-and-fire shifting device. The universally jointed telescopic shaft F_2 transmits motion to the cross-feed mechanism at whatever angle the swiveling bracket and work-spindle may be set. The cross-feed screw M_2 has mounted on it a handwheel K_2 and a spur gear N_2. This spur gear is connected with ratchet wheel H_2 by a tumbler gear arrangement controlled by lever J_2, which thus provides for reversing and disengaging the feeding movement. The ratchet wheel is operated by a pawl O_2, pivoted to lever G_2 which, in turn, receives its movement from rockshaft F_2. This movement is positive in the direction which operates the ratchet wheel H_2, and through it the cross feed. In the other direction, motion is derived from a spring R_2 until the point of plunger S_2 brings up against the adjustable stop T_2. As the position of T_2 governs the extent of the movement of the swinging of lever G_2, a greater or less cross feed is effected at each stroke.

The position of stop T_2, and the amount of feed, is governed by two things. In the first place, the knurled nut U_2

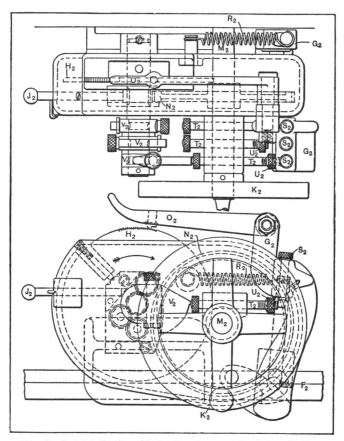

Fig. 8. Ratchet Feeding Mechanism Arranged to Automatically Diminish the Feeding Movement

furnishes a check to its backward movement, and thus regulates the rate of cross feed. Screwing this nut out increases the feed — screwing it back decreases it. In the second place, the feed is controlled by cam V_2, which is adjustably clamped on the shaft of ratchet wheel H_2, and revolves with it in the direction of the arrow. As the feeding progresses, the lower edge of V_2 comes into contact with the left-hand end of stop T_2, gradually limiting its movement from that permitted by the adjustment of U_2 until finally, in the position shown, the swinging of lever G_2 is stopped altogether, thus stopping the cross feed. The diminishing depth of cut thus provided for, as the desired finished diameter is approached, tends to improve the work in regard to accuracy and finish. It will be noted in the plan view that there are three stop cams V_2, three stops T_2, and three feed adjusting nuts U_2 and plungers S_2.

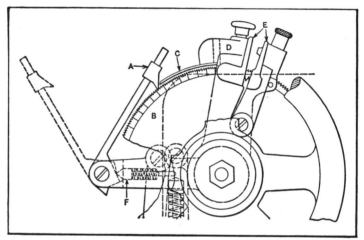

Fig. 9. Ratchet Gearing Arranged to Disengage Automatically after a Predetermined Movement

Any one of these three latter may be pressed down into working position, thus giving a separate cross-feed stop and rate of feed for each of three operations.

Automatic Disengagement of Ratchet Gearing at a Predetermined Point. — The action of ratchet gearing can be stopped automatically after the ratchet wheel has been turned a predetermined amount, by equipping the wheel with an adjustable shield which serves to disengage the pawl after the required motion has been completed. This form of disengaging device, as applied to the cross-feeding mechanism of a cylindrical grinding machine, is shown in Fig. 9. This mechanism is used to automatically feed the grinding wheel in toward the work for taking successive cuts, and it is essential to have the mechanism so arranged that it can be set to stop the feeding movement when the diameter of the work has been reduced a predetermined amount. When the pawl *A* is in mesh with the ratchet wheel *B*, the grinding wheel is fed forward an amount depending upon the position of screws (not shown) which control the stroke of pawl *A*. The automatic feeding movement continues at each reversal of the machine table, until the shield *C*, which is attached to head *D*, intercepts the pawl and prevents it from engaging with the ratchet wheel, thus stopping the feeding movement. The arc through which the ratchet wheel is turned before the pawl is disengaged from it, or the extent of the inward feeding movement of the grinding wheel, depends upon the distance between the tooth of the pawl and the end of the disengaging shield. With the particular mechanism illustrated, a movement of one tooth represents a diameter reduction of 0.00025 inch, so that the amount that the wheel moves inward before the feeding motion is automatically disengaged can be changed by simply varying the distance between the shield and the pawl. To facilitate setting the shield, a thumb-latch *E* is provided. Each time this thumb-latch is pressed, the shield moves a distance equal to one tooth on the ratchet wheel. For instance, if the shield is at the point of disengagement and the latch is pressed sixteen times, the shield will move a distance equal to sixteen teeth. As each tooth represents 0.00025 inch, a feeding movement of 0.004 inch will be obtained before the pawl is automatically disengaged. This mechanism prevents grinding parts below the required size, and makes it unnecessary for the operator to be continually measuring the diameter of the work. It is located back of a handwheel (which is partly shown in the illustration) that is used for hand adjustment. The pawl is kept in contact with the ratchet wheel and is held in the disengaged position by a small spring-operated plunger *F*.

Non-stop Feed Ratchet Adjustment—1. — Ordinarily, the feeding movement obtained with a ratchet feeding mechanism is varied by changing the radial position of the operating crankpin, but this is not readily accomplished without stopping the machine. The variable ratchet feeding mechanism shown in Fig. 10 may be adjusted while operating. It consists of a fixed crankpin *A* mounted on a crank disk *B,* which, in

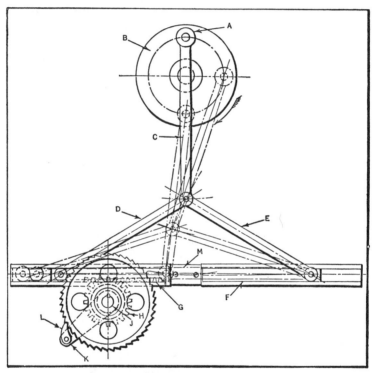

Fig. 10.   Ratchet Feed Mechanism which may be Adjusted without
Stopping the Machine or Driving Crank

turn, is mounted on a main drive shaft. One end of the pitman *C* is connected to the crankpin, and the other end is connected to links *D* and *E.* Fastened to link *E* is a sliding bar *F,* while link *D* is fastened to the rack *G.* Rack *G* meshes with pinion *H,* which is free on feed-shaft *J* but is connected to an arm *K* carrying pawl *L.* This pawl meshes with the ratchet wheel which is keyed to the feed-shaft.

Thus arranged, the driving motion of the main shaft is transmitted to the pawl arm with the sliding bar F abutting against the block M; and if the sliding bar is held against M during the downward stroke of the crankpin, all of this motion will be imparted to the pawl arm, swinging the pawl the maximum distance back around the ratchet wheel. On the upward movement of the crankpin, the pawl arm will be pulled upward and the pawl, engaging with the ratchet wheel, will turn the wheel through a maximum degree of rotation.

However, if the sliding bar is allowed to have some movement away from block M and the movement of the rack and pawl arm is impeded, part of the downward motion will be transmitted to the sliding bar and subtracted from that of the pawl arm, depending upon how far the sliding bar, on the one hand, and the rack and connected elements, on the other hand, are allowed to move. For example, if no limit is placed on the movement of the sliding bar, and no movement of the rack and connected parts is allowed during the downward stroke of the crankpin, all of the movement will be imparted to the sliding bar, and on the upward stroke of the crankpin, the sliding bar will merely come back to M, and there will have been no feeding motion imparted to the ratchet wheel and feed-shaft.

It will thus be seen that by regulating the proportion of movements of the sliding bar and the rack, any desired degree of rotation of the ratchet wheel and the feed-shaft may be effected. Suitable arrangements may be provided for setting the position of the sliding bar for each stroke of the machine. This can be easily done without stopping the machine. This mechanism is especially suitable for feeding structural steel, etc., through a punching machine, where the spacing of the holes is variable. The main drive shaft then represents the main shaft of the punching machine.

Non-stop Feed Ratchet Adjustment —2.— Many machines have feed rolls in one form or another, the two roll shafts being geared together, as for example, by gears A and B (see Fig. 11). The lower shaft of this particular design has a

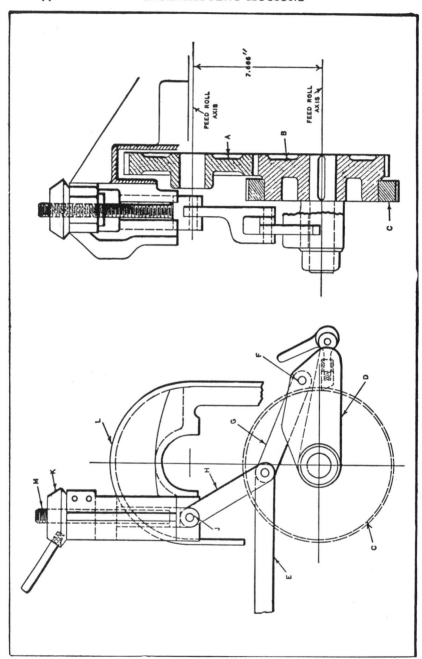

Fig. 11. Another Mechanism for Adjusting Ratchet Feeding Movement without Stopping the Machine

ratchet wheel C operated by a pawl carried by lever D, which, according to the former arrangement, was given a reciprocating movement by the direct action of a connecting-rod E pivoted to it, the motion imparted to E depending upon the radial position of a crankpin. When connecting-rod E is directly connected to lever D, changes in the feeding movement requires stopping the machine in order to increase or decrease the radial position of the crankpin (not shown) to which connecting-rod E is attached. Moreover, such trial adjustments may have to be repeated before the desired feeding movement is obtained. In operating these machines, which run at 200 revolutions per minute, it is important to feed the correct amount because the feeding is against a fixed stop; consequently, if the feeding movement is too long, the ratchet teeth and feed rolls wear out very rapidly, and too short a movement would cause scrap blanks.

To avoid wasting time in adjusting the feed mechanism and also secure more accurate adjustments, a feed mechanism was designed which permits varying the feed a very slight amount if desired and while the machine is running at full speed. This improved design is so arranged that the motion is transmitted from connecting-rod E to pawl lever D through link G and a link H which swings about a pivot J that may be adjusted vertically by turning nut K. As pivot J is moved upward less motion is transmitted to pawl lever D, whereas a downward adjustment of the pivot has, of course, the opposite effect, the feed being increased.

The casting L, which supports this adjusting mechanism, has a vertical slot in which the rectangular head of bolt M slides as nut K is turned. This nut is prevented from moving vertically by the retaining collar seen in the side view. Link H is pivoted at J to the forked end of bolt M. The parts should be so proportioned that link H will not swing back beyond the vertical line. Links H and G of this mechanism measure 5 11/16 inches from center to center of the holes, and connecting-rod E is about 4 feet long. The crankpin to which E connects is first adjusted roughly to give sufficient motion, and then after the machine is started pivot *J* is adjusted vertically until the feeding movement is exactly what is required. This mechanism works perfectly, and not only saves time, but has lengthened the life of both ratchets and feed rolls. Although some form of friction ratchet would make it possible to obtain a theoretically perfect adjustment, a positive action is desired for this mechanism. A triple form of pawl now is used. The distance between these pawls is equal to one-third of the pitch of the ratchet wheel teeth; hence, in effect, the ratchet wheel has three times the actual

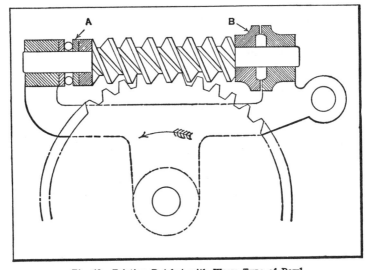

Fig. 12. Friction Ratchet with Worm Type of Pawl

number of teeth, so that comparatively fine adjustments are possible without eliminating the positive drive.

Friction Ratchet of Worm-pawl Type. — In Fig. 12 is shown an ingenious ratchet consisting of a worm-wheel and a worm mounted on a forked lever. When this design is enclosed in an oil-tight case and the parts are properly made, it functions with a high degree of accuracy. However, it is a comparatively expensive arrangement. Either the worm or the worm-wheel may be used for driving the mechanism. It will be seen that the worm-shaft is equipped with a ball thrust bearing at A and is provided at B with a bearing that sets up an appreciable friction when subjected to a load. Bearing B and one race of bearing A are keyed to the shaft.

When the forked lever is operated in the direction indicated by the arrow, the thrust is against the plain bearing B, and the frictional resistance prevents the worm from revolving; hence the worm acts as a ratchet and turns the worm-wheel. On the contrary, when the forked lever is operated in the opposite direction, the worm revolves, because it overcomes the relatively slight friction of bearing A; consequently, the

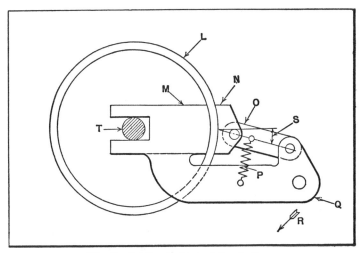

Fig. 13. Friction Ratchet of Toggle Type

worm-wheel remains stationary while the worm is swung backward.

By regulating the amount that the forked lever is moved in the counter-clockwise direction when the worm is the driving member, the movement of the worm-wheel can be varied. This device should not be regarded as a worm and worm-wheel mechanism in the ordinary sense, as no rubbing action takes place between the worm and the worm-wheel teeth when the two members are under a load. Added advantages of this construction are taking up the load without shock, and silent operation. An important point to be observed in designing a ratchet of this type is to make the helix angle of the worm such that the worm will just revolve when the load is on bearing A, and will remain stationary when the load is on bearing B.

Friction Ratchet of Toggle Type. — The friction ratchet shown in Fig. 13 is so arranged that the flange L of the friction wheel is gripped between the body M and shoe N. The friction surfaces are kept in contact by means of a light spring P. When the arm Q moves in the direction indicated by the arrow R, body M and shoe N merely slide over flange L, but when it moves in the opposite direction, the mechanism is

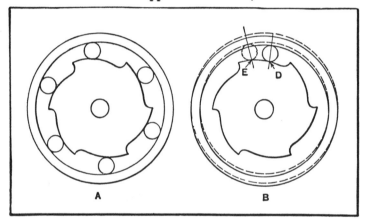

Fig. 14. Two Ratchet Designs that Function by Jamming Balls between the Driving and Driven Members

friction-locked, so that arm Q, body M, shoe N, and link O all move together as a solid piece with the wheel.

The advantage of this construction is that the sliding or friction surfaces have a comparatively large area and therefore great durability. This has the disadvantage, however, that, should a heavy film of oil accumulate on flange L, the ratchet will slip unless the angle S is made not larger than 7 degrees. A thin layer of oil absorbed on flange L will not cause the ratchet to fail. In designing this type of ratchet, care should be taken to see that body M is made very rigid. Provision should also be made for a radial movement of body M; this may be done by slotting the end to fit the central shaft T. Flange L and body M may be made of cast iron, and shoe N of soft steel.

Friction Ratchets of Cam Type. — Two styles of friction ratchets are illustrated in Fig. 14. In the one at A the driving rollers are jammed between an internal circular surface of the outer member and cam surfaces on the inner member. With a ratchet of this type, springs are often utilized to force the rollers into contact with the jamming surfaces. One

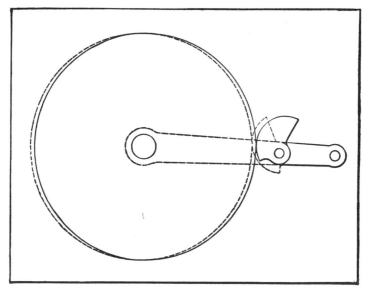

Fig. 15. Friction Ratchet Consisting of a Disk, Segment, and Lever

disadvantage of this design is that the rollers always bear at the same point on the inner member with the result that in time a depression is formed on each curved surface. When this occurs, the device can no longer be relied upon, as a jamming action is impossible because of the shoulders that are formed. The ratchet shown at B was designed to overcome this disadvantage. The outer member is placed slightly eccentric relative to the center about which the inner member revolves, as indicated on an enlarged scale by the dotted lines, and so the jamming of the rollers is distributed on each curve between points D and E. By this provision the forming of depressions is obviated.

A type of friction ratchet, which is fairly reliable in the ordinary form, when the disk is concentric, is shown in Fig. 15. However, if the disk is slightly eccentric relative to the fulcrum of the lever to which the engaging segment is attached, much better results will be secured, because the jamming is again distributed over a considerable area of the segment surface, and the forming of a depression on this member is avoided. This will be apparent by reference to the dotted positions of the ratchet disk and segment. Care must be taken to see that the amount of eccentricity is not too great, or the segment will not jam properly on one half of the disk periphery.

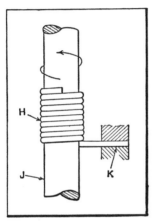

Fig. 16. Ratchet of Coil Spring Type

Friction Ratchets of Coil Spring Type. — The friction r a t c h e t shown in Fig. 16 consists of a close-wound coil spring H which is assembled on the main spring shaft J, and has one end fixed at K. The inside diameter of the spring is slightly less than the diameter of the shaft J so that it normally grips the shaft. A torque in the direction indicated by the arrow tends to open the spring an imperceptible amount, beginning at the free end, until the friction is overcome. The shaft then turns freely. A torque in the opposite direction, however, increases the friction between the spring and shaft, owing to the "belt-like tension," and thus locks the shaft.

A simple method of employing a coil spring to provide a ratchet drive is illustrated in Fig. 17 (see upper illustration). The drum A is fastened to shaft B, and spring C is wound to fit snugly on drum A. One end of the spring is fastened to lever D. If lever D is rotated in the direction opposite that indicated by the arrow, the spring will slide or slip on the drum. If rotated in the direction indicated by the arrow, the spring will tighten on the drum and drive shaft B.

This type of ratchet tends to produce a frictional drag upon the driven member during its idle stroke due to the gripping action of the spring. This drag can be made to serve a useful purpose in some cases, or it can be reduced to a

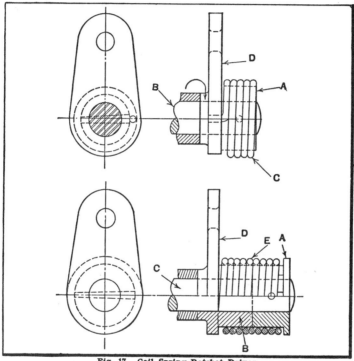

Fig. 17. Coil Spring Ratchet Drives

negligible value by increasing the number of turns and increasing the inside diameter slightly so that the grip on the drum is loosened.

The spring ratchet shown by the lower illustration depends for its drive upon its gripping action on both the driver and the driven member and neither end of the spring is fastened. The drum B, machined on lever D, drives the drum A on shaft C, by the gripping action of spring E.

Intermittent and Continuous Friction Ratchet Feed Mechanism.

— Mechanisms for obtaining both intermittent and continuous feeds that will permit of minute adjustments are sometimes required. A mechanism of this kind which has been successful is shown in Fig. 18. This mechanism was first applied to a horizontal drilling machine used to drill a hole 1 3/32 inches in diameter and 18 inches long in a chrome-nickel steel crankshaft. It was found, upon trial, that a feed of 0.007 inch per revolution was the ideal feed for this work. For some unknown reason, a feed of 0.008 inch made the drill chatter, while a feed of only 0.006 inch caused the work to become glazed so that the drill would not cut after being

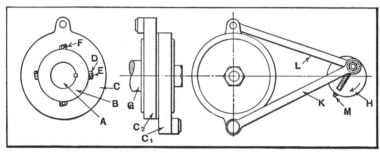

Fig. 18. Adjustable Friction Ratchet Feeding Mechanism which may be Arranged either for Intermittent or Continuous Motion

in use a short time. The drill used was of the oil-tube twist type, with flutes 20 inches long, and was made of high-speed steel.

A feed of the same type as was used for the crankshaft drilling operation was successfully applied to a machine for cutting fabrics. A feeding mechanism of the type described can also be used to advantage in connection with the roll feeds of punch presses. The mechanism used for the drilling operation is of the continuous-feed type, but when used with the fabric cutting machine or in connection with punch press feeding rolls, it is employed to give an intermittent feed.

Referring to the accompanying illustration, the feed or drive shaft A is keyed to a hardened and ground steel collar B around which is placed a hardened and ground ring C. In ring C are suitable recesses D in which are placed hardened rolls E that are backed up by springs F. The action of this unit is as follows: When the outer ring C is rotated counter-clockwise, there is practically no resistance to its movement, but if it is rotated clockwise, the hardened rolls E are forced against collar B as they ride up the inclined surface of recess D. The rolls finally become wedged, and thus drive collar B which is keyed to shaft A.

In order to obtain an intermittent feed, it is merely necessary to use one of the friction units which is only capable

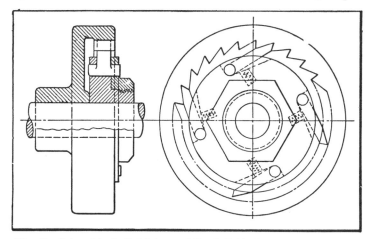

Fig. 19. Internal-tooth Ratchet especially Suitable for High-speed Drives

of driving shaft A in one direction. This unit may be connected to a rotating shaft on the machine by means of an adjustable crank, so arranged that its throw can be easily adjusted to obtain a fine or a coarse feed. For a continuous feed (of varying velocity due to the crank action) it is necessary to employ two units, as indicated in the two views at the right-hand side of the illustration. In this mechanism two outer rings C_1 and C_2, and one collar B are required. If both rings are arranged to drive when turning clockwise, then link L and ring C_2 will be driving while the pin of crank disk H is moving through the upper half of its circle, or from left to right, and link K and ring C_1 will be driving while the crankpin is moving through the lower half of its circle, or from right to left. If the two rings are arranged to drive while rotating counterclockwise, the action just described will be reversed, ring C_2 being the driver when the crankpin is moving from right to left, and C_1 driving when the pin is moving from left to right.

The crank disk H may be placed on the main drive shaft or any other shaft that may be conveniently located for the purpose. The screw at M provides a means of regulating the throw of the crank and may be equipped with a dial reading to hundredths or thousandths of an inch to facilitate the accurate adjustment of the rate of feed.

As there is no perceptible interval between the release of one ring and the engagement of the other, a practically continuous feed is obtained. If desired, it is permissible to make the recess for the friction rolls in collar B instead of in the rings. This construction has the advantage of giving a slightly better contact for the friction rolls owing to the fact that the contact surface of the outer ring is formed to a larger radius.

Internal Ratchet for High-speed Drive.— A ratchet having internal teeth is the best design to employ in a high-speed drive, because centrifugal force tends to assist the action of the pawl; this is particularly true when the pawl is mounted on the driving member. Fig. 19 shows an internal-tooth ratchet in which power is transmitted to the driven member through the medium of four pawls, each of which is assisted in its engagement with the ratchet teeth by a small coil spring.

Cam-operated Ratchet Pawl.— A ratchet feeding mechanism employed to revolve a heavy cylinder is shown in Fig. 20. This device is designed to start the rotation of the cylinder with a sudden movement, stop it quickly, and hold it stationary for a predetermined period, and then repeat the operation. The view at the left-hand side of the illustration shows the mechanism in the position it occupies when the cylinder is being held stationary. The view at the right-hand side shows the mechanism in the position occupied just before the indexing movement begins.

It will be noted that there are but four principal parts to the mechanism, namely, the ratchet wheel A, the pawl B, the crank C, and the cam D. Cam D rotates continually in the direction indicated by the arrow. When the roller E on crank C reaches the highest point on cam D, the ratchet A will have been completely indexed and be locked by the tooth f on the crank C from further rotation in the direction of the arrow, while the pawl B will prevent it from being rotated in the reverse direction.

Ratchet Mechanism for 90-degree Indexing Movement. — A 90-degree indexing mechanism which has no idle return

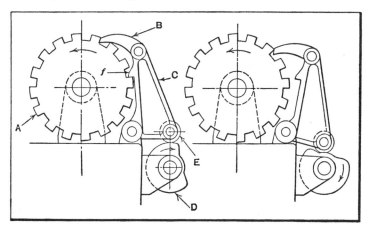

Fig. 20. Cam-operated Ratchet Feeding Mechanism

throw and which can be operated by a short stroke of the indexing member or lever is shown in Fig. 21. The screw or stud A is connected to the part of the machine that produces the indexing motion. The links B and C are connected to the arms D and E which carry the pawls F and G. Arms D and E are free to revolve on the indexing shaft H to which the index-wheel J is fastened.

To impart an indexing movement of 90 degrees to wheel J, pawls F and G are given a movement of 45 degrees in one direction and then returned to their normal position. A downward movement of stud A of the right distance, as indicated by dimension K, will draw levers D and E down so that they are rotated through an angle of 45 degrees. This downward movement brings pawl F backward, so that its point will coincide with the center line X-X, and pawl G will move forward until its point also coincides with center line X-X. Thus pawl G indexes wheel J through an angle of 45

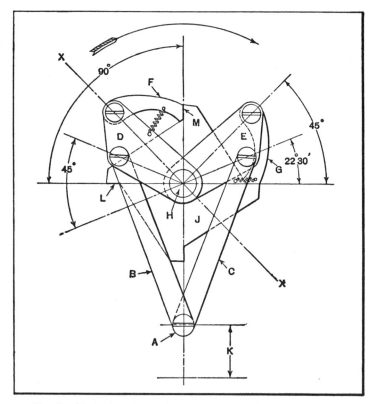

Fig. 21. Mechanism for Producing a 90-Degree Indexing Movement

degrees so that pawl F will catch on the tooth face L of the index-wheel. The return movement of stud A causes pawl F to move forward and to index wheel J the remaining 45 degrees, while the pawl G moves backward to its former position into contact with the face M of the succeeding tooth ready for the next indexing movement.

Index Mechanism for Ratchet Dial Feed. — A limited quantity of small parts, of a design that could be made most economically on a dial feed press, was required. As the comparatively small quantities to be produced did not warrant the purchasing of a new machine, the dial feed shown in

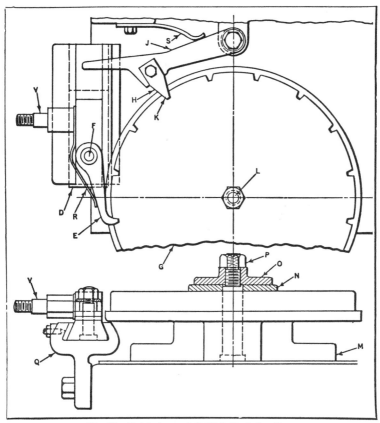

Fig. 22. Ratchet-operated Dial Feed for Press

Fig. 22 was designed and applied to an old press. On the last part of the up stroke of the press ram, a crank-operated rocker arm (not shown) having an elongated slot for receiving stud V, causes the ratchet slide D to be advanced. The pawl E, pivoted on the stud F in slide D, engages one of the notches in the dial-plate G and indexes this member to the next station. At the completion of the cycle, point H on lever J drops into the notch on plate G, locking this member during the down stroke of the press ram and the first part of the upward stroke. The dial-plate G is therefore locked while the dies are in operation.

As the slide D approaches the end of its rearward movement, it comes in contact with the end of lever J and lifts the locking point H from the notch in plate G. While the point H is still held away from plate G the point of pawl E drops into the indexing notch. When slide D has moved forward far enough to permit point H to come in contact with plate G, the indexing notch at K has moved from under point H, which then rides on the periphery of the plate until pawl E has completed the indexing movement, at which time it drops into the next notch.

The dial-plate G is shown blank, without the work stations cut in it, in order to eliminate unnecessary details. The dial-plate is made a good running fit on the stud L, which is driven into the die-bed M. The die-bed is provided with a flange which supports the dial-plate. A friction washer N, made from wood, is held in contact with the upper face of the dial-plate by the washer O and nut P. The bracket Q in which slide D operates is provided with an adjustable gib to compensate for wear on the dovetail faces of the bracket and slide.

The flat spring R serves to keep the pawl E in contact with the dial-plate. The locking point H is made from tool steel, and is rigidly secured to the arm J. The flat spring S serves to hold the locking point in contact with the plate. The stroke of slide D may be changed by adjusting the crank which operates the rocker arm.

Escapements for Clock Mechanisms. — An escapement may be considered as a form of ratchet mechanism having an oscillating double-ended pawl for controlling the motion of the ratchet wheel by engaging successive teeth. Escapements are designed to allow intermittent motion to occur at regular intervals of time. The escapement of a clock is illustrated in Fig. 23. As applied to a pendulum clock the escapement serves two purposes, in that it governs the movement of the scape wheel for each swing of the pendulum and also gives the pendulum an impulse each time a tooth of the scape wheel is released. An escapement should be so arranged that the pendulum will receive an impulse for a short period at the lowest part of its swing and then be left free until the next impulse occurs. The time required for a pendulum to swing through small arcs is practically independent of the length

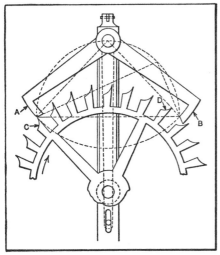

of the arc. For instance, if a stationary pendulum receives an impulse, the time necessary for its outward and r e t u r n movement will be approximately constant regardless of the impulse and arc of swing, within ordinary limits. Thus, if the impulse is of considerable magnitude, the pendulum starts with a relatively high velocity, but the distance that it travels counteracts the increase of speed so that

Fig. 23. Escapement for Controlling Action of Clockwork

the time remains practically constant for any impulse or arc of swing. A pendulum that is swinging freely will adapt the length of its swing to the impulse it receives, and any interference which might be caused by the locking or unlocking of the escapement will affect the regularity of movement less if it occurs at the center of the swing rather than at the ends. As the arc of swing increases, there is a very slight increase in the time required for the movement, and, therefore, it is desirable that the impulses given to a pendulum should always be equal.

One of the earlier forms of escapements was known as

the "anchor" or "recoil" escapement. With this type, the pendulum was never free, but was controlled by the escapement throughout the swing. To avoid this effect, the Graham "dead-beat" escapement, illustrated in Fig. 23, was designed and has been extensively used. When the escapement is in action, the pallets A and B alternately engage the teeth of the scape wheel, which revolves intermittently in the direction indicated by the arrow. With the mechanism in the position illustrated, the point of tooth C is about to slide across the inclined "impulse face" or end of the pallet A, thus giving the pendulum an impulse as it swings to the left. When tooth D strikes the "dead face" of pallet B, the motion of the scape wheel will be arrested until the pendulum reverses its movement and swings far enough to the right to release tooth D; as the point of D slides past the inclined end of B, the pendulum receives another impulse, and this intermittent action continues indefinitely or until the force propelling the scape wheel around, which may be from a spring or weight, is no longer great enough to operate the mechanism. In designing an escapement of this type, the pallets are so located as to embrace about one-third of the circumference of the scape wheel. One of the features of the dead-beat escapement is the effect which friction has on its operation. During each swing of the pendulum, there is a rubbing action between the points of the scape wheel teeth and the surfaces of the pallets, so that the pendulum is retarded constantly by a slight amount of friction. This friction, however, instead of being a defect, is a decided advantage, because, if the driving force of the clock is increased so that the impulse on the pallets becomes greater, the velocity of the pendulum tends to increase, but this effect is counteracted by the frictional retardation caused by a greater pressure of the teeth of the scape wheel on the faces of the pallet. If the driving force be increased, the frictional retardation increases relatively in a greater proportion than the driving effect and, up to a certain point, the time of vibration of the pendulum diminishes. If the force or weight propelling the clock mechanism is continually increased, a neutral point is finally reached, beyond which a greater force causes the time of vibration to increase instead of to diminish. In the design of clock mechanisms, it is desirable to have a driving power of such magnitude that it neither accelerates nor retards the motion of the pendulum. Many modifications of the escapement previously referred to have been devised to meet special requirements. The escapements of watches and of some clocks and portable timekeeping devices have a balance wheel instead of a pendulum to regulate the period of the intermittent action, but all of these escapements operate on the same general principle.

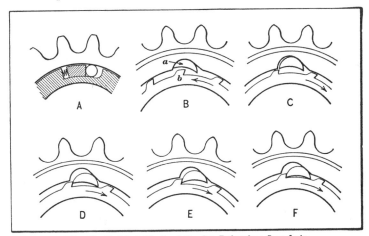

Fig. 24. Ratchet Mechanisms for Releasing Sprockets

Ratchet Mechanisms for Releasing Sprockets. — Some ingenious ratchet mechanisms have been applied to the sprocket wheels of bicycles to permit the pedals to remain stationary while coasting down a grade or hill. A design that has been extensively used is illustrated in principle by the detailed sectional view at A in Fig. 24. The sprocket wheel is not attached directly to the inner member which is shown in section, but motion is transmitted from one part to the other through frictional contact. The inner ring has a series of recesses equally spaced about the circumference. Each of these recesses contains a hardened steel roller or ball, and the bottoms of the recesses are inclined slightly. The rollers are lightly pushed up these inclined surfaces by blocks behind which are small spiral springs. Any relative motion of the inner and outer members of the sprocket causes these steel rollers to either roll up the inclined surfaces and lock the two parts together or to move in the other direction and release the driving and driven members, the action depending upon the direction of the relative movement. For instance, if the outer sprocket is revolved in a clockwise direction, all of the rollers are immediately wedged in their recesses. If the motion of the outer sprocket is suddenly arrested and the inner member continues to revolve, the rollers are immediately released.

An entirely different type of ratchet mechanism designed for use on the sprockets of bicycles is shown by diagrams *B* to *F,* inclusive, which illustrate its method of operation. The exterior sprocket is recessed on the inner side for the reception of a crescent-shaped piece *a,* which acts as the pawl. The depth of the recess and the shape of part *a* are such that the teeth on the inner ring *b* can pass freely when moving in the direction indicated by the arrow at *B;* with motion in this direction, part *a* simply is given a rocking movement in its recess to allow the successive teeth to pass. When the relative motion is in the opposite direction, as indicated by diagram *C,* the teeth on the inner member swing part *a* around in its seat, as shown by the successive diagrams, until it is finally wedged firmly between the two parts as shown at *F.* These so-called "free-wheel" mechanisms were subsequently replaced by an arrangement operating on the same general principle so far as the releasing mechanism was concerned, but so designed that a backward movement of the pedal also applied a brake.

Ratchet-controlled Press Shearing Mechanism. — In manufacturing a certain design of automobile radiator, two fins are cut from the stock by knife blades held in a bracket mounted at the right-hand side of the press. Seven strokes of the ram are required for piercing all the holes in a fin and so these knives function but once at every seven strokes. This intermittent operation is produced by the mechanism illustrated diagrammatically in Fig. 25. On the right-hand side of the ram is mounted a 14-tooth ratchet *A*, which is brought into contact with a spring-actuated finger *B* on the housing of the machine at each upward stroke of the ram, so that the ratchet is indexed one-fourteenth revolution at each stroke.

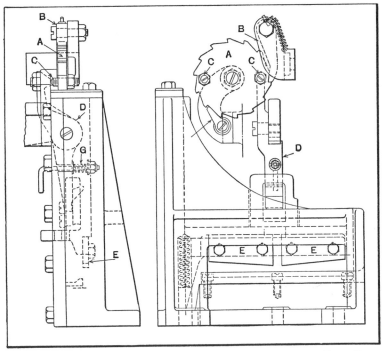

Fig. 25. Mechanism that Operates the Cutting-off Knife Blades in the Power Press at Every Seventh Stroke of the Ram

Diametrically opposite each other on ratchet *A* are two screws *C*, and at each seventh stroke of the ram, one of these contacts with the upper end of lever *D*, swiveling the lower end of the lever to the right so that it engages with the frame containing shear blades *E*. Then, at the next stroke of the ram, lever *D* forces the knife frame down to shear off the fins. The spring-actuated pawl *F* prevents the ratchet *A* from turning backward when contact is made with lever *D*, and spring *G* insures that the latter will disengage from the knife frame when either of screws C has passed the upper end of the lever.

Ratchet Mechanism for Indexing Revolving Chuck. — The chuck shown in Fig. 26 is equipped with two special indexing jaws for gripping a three-way brass fitting and indexing it for machining each of the three open ends without stopping the turret lathe. The indexing is controlled by a lever on the side of the machine, which operates the automatic chuck and bar feed mechanism.

The jaw block A contains the indexing mechanism. The indexing jaw B is integral with the stem C. At D there is an eight-tooth ratchet (see also the plan view). Pawls E and F engage this ratchet, and rotate the jaw. The slides G and H carry the pawls, and are tongued and grooved into the jaw block A.

The forward movement of slide H engages the push-pawl E and rotates the ratchet one-eighth of a revolution, and the return movement engages the hook-pawl F, which completes the one-quarter turn required for each of the three positions. The springs I hold the pawls in the proper position.

The plate J is keyed to the stem C. The under side of plate J is drilled and reamed taper at three places to receive the lock-bolt K for retaining each indexed position. The lock-bolt is operated by a cam made integral with the slide H, as shown by the detail view of the slide (upper right-hand corner of illustration). Referring to the plan view of the jaw, it will be noted that there is considerable clearance between pawl E and ratchet D. This is so that the cam on the jaw H will pass under the roll of the side lever and rock it far enough to extract the lock-bolt from the index-plate before the pawl E starts rotating the ratchet.

Looking at the end view of the chuck, two lugs are attached to the chuck face at L. The lower ends of the yokes M are pivoted to lugs L. The upper ends of the yokes M are pivoted into grooves cut transversely across the slides H and G at N. A projection of the yoke connects through a link P with a plunger Q which slides in the spindle.

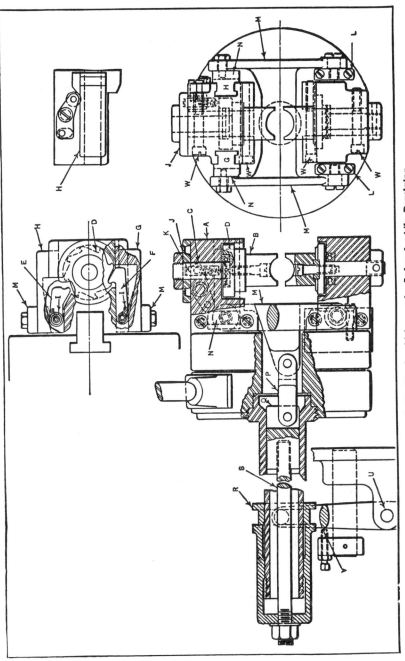

Fig. 26. Turret Lathe Chuck which may be Indexed while Revolving

The sleeve R slides over the rear end of the spindle and replaces the wedge used in the regular automatic chuck mechanism. Sleeve R also connects with the plunger Q through the rod S. A yoke has rolls that engage the groove in the sleeve R. This yoke pivots at U in a bracket on the machine, and its lower end is attached through a link to a rocker arm, swung by a hand-lever which operates an ordinary bar feed mechanism. Stop V limits the rearward movement of the sliding mechanism.

The indexing jaws are both supported on ball thrust bearings. The lower jaw is only a carrier. The jaw blocks are grooved to slip over the master chuck jaw and are bolted entirely through the master jaw, as shown at W in the end view.

These jaw blocks are attached to a two-jaw chuck, but they can be designed for fixtures or faceplates. The material used throughout in the jaws is steel; the yoke and sleeve are of cast iron. Jaws designed as described should not be considered for other than light and medium light work, as the gripping required for heavy work causes a thrust so great that the jaws cannot be revolved by the hand-lever, and to release the jaws slightly for indexing will cause misalignment of the machined surfaces.

In machining the brass parts for which this indexing chuck is used, each of the three diameters is bored to a different depth; hence, three boring heads must be used on the turret. The threads on each of the three diameters, however, are identical and, therefore, could be cut by using only one die-head, which is the method formerly employed. This method required a lot of extra turret indexing in order to swing the die-head back to the work. The present method is to use three die-heads, there being one after each boring position. This arrangement, in conjunction with the indexing mechanism, saves much time, as 665 pieces can be machined daily.

CHAPTER III

INTERMITTENT MOTIONS FROM GEARS AND CAMS

WHEN a shaft which rotates continuously is to transmit motion to another shaft only at predetermined intervals, intermittent gearing is sometimes used. Gearing of this type is made in many different designs, which may be modified to suit the conditions governing their operation, such as the necessity for locking the driven member while idle, the inertia of the driven part, or the speed of rotation. With some forms of intermittent gearing, the driven gear rotates through a fractional part of a revolution once for each revolution of the driver, whereas, with other designs, the driving gear transmits motion to the driven gear two or more times while making a single revolution. The number of times that the driven gear stops before it is turned completely around is varied in each case according to the requirements; the periods of rest may also be uniform or vary considerably.

Gears for Uniform Intermittent Motion. — The design of intermittent gearing illustrated by diagram *A*, Fig. 1, is so arranged that the driving gear, which has only one tooth, revolves fourteen times for each revolution of the driven gear. Each time the tooth of the driver engages one of the tooth spaces in the driven gear, the latter is turned through an arc x. The driven gear is locked against rotation when the driving tooth is not in mesh, because the circular part of the driver fits closely into the concave surfaces between the tooth spaces as they are successively turned to this position. The radius of the driver should be small enough to insure adequate locking surfaces between the tooth spaces, but not so small that sharp weak points will be formed at the edges of the tooth spaces. Counting mechanisms are often equipped with gearing of this general type. In order to vary the relative movements of the driving and driven gears, the meshing teeth may be arranged in various ways. For instance, if a second tooth were added to the driver on the opposite side as indicated by the dottted lines, the driven gear would receive motion for each half revolution of the driver. The diagram at *B* illustrates another modification. In this case, the driven gear has a smaller number of rest periods, and it is turned farther for each revolution of the driver, as the latter has three successive teeth.

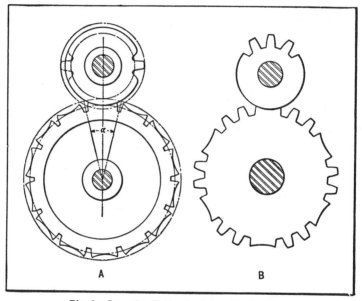

Fig. 1. Gears for Uniformly Intermittent Motion

Variable Intermittent Motion from Gearing. — With some forms of intermittent gearing, the driven gear does not move the same amount each time it is engaged by the driver, the motion being variable instead of uniform or equal. The diagram *A,* Fig. 2, shows an example of the variable motion intermittent type. The driving gear has four driving points around its circumference with numbers of teeth at each place varying from one to four. The tooth spaces on the driven gear are laid out to correspond so that the motion received by the driven gear is either increased or decreased progressively depending upon the direction of rotation of the driver. Gearing of this general type may be arranged in many different ways and is designed to suit the particular mechanism of which it forms a part. After laying out gears of this kind, it is often advisable to make brass templets in order to ascertain by actual experiment if the gears are properly formed and give the required motion.

Intermittent Gearing for High Speeds. —The design of gearing illustrated by diagram B, Fig. 2, is considered pref-

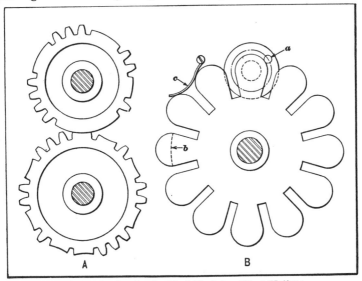

Fig. 2. (A) Gearing for Variable Intermittent Motion;
(B) High-speed Intermittent Gearing

erable to the forms previously described, where the driving member revolves at a comparatively high speed. With this gearing, the driven member is stationary during one-half revolution of the driver. The latter has a stud a or roller which engages radial slots in the driven gear while passing through the inner half of its circle of travel. The flat spring illustrated at c is used to hold the driven wheel in position so that the driving roller will enter the next successive slot without interference. The projections or teeth on the driven gear may have semi-circular ends as shown, or all of the ends may be concentric as indicated by the dotted line at *b*. If the semi-circular ends are not provided, there should be some form of positive locking device to insure alignment between the radial slots and the driving pin or roller. The corners should also be rounded to facilitate engagement of the roller.

Another form of intermittent gearing designed to eliminate shocks when operating at relatively high speeds is illustrated in Fig. 3. The speed ratio between the driving and

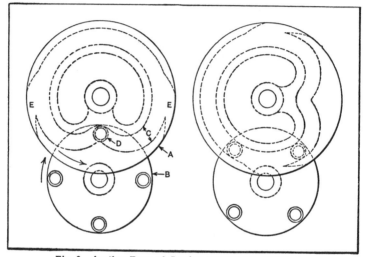

Fig. 3. Another Form of Gearing Designed to Eliminate
Shocks at High Speeds

driven members is 4 to 1, each revolution of the driver turning the driven wheel one-quarter revolution, or 90 degrees. The driver *A* has a cam groove *C* which is so shaped that the motion of the driven wheel *B* is gradually accelerated and retarded at the beginning and end of its movement. This groove is engaged by rollers *D* on the driven wheel. The rollers enter and leave the cam groove through the open spaces at *E,* and when the driven wheel is stationary, two of the rollers are in engagement with this groove, thus effectually locking the driven member. The illustration at the left shows the driven wheel at the center of its movement, and the view to the right shows the relative positions of the two parts after the movement is completed. As the roller at D is revolved 45 degrees from the position shown, the following roller enters the cam groove through the left-hand space E.

Rapid Intermittent Motion for Moving Picture Projector. — A very rapid intermittent motion is required on moving picture projectors. The film is not moved continuously, but each view or positive on it is drawn down to the projecting position while the shutter is closed, and the film remains stationary for a fractional part of a second while the picture is exposed on the screen; then, while the shutter is again closed, the next successive view is moved to the projecting position. It is apparent, therefore, that moving pictures are, in reality, a series of stationary pictures thrown upon the screen in such rapid succession that they are, in effect, blended together and any action or movement appears continuous. It is important to give the film a very rapid intermittent motion, because it is necessary to have the shutter closed when this movement occurs; and the length of time that the shutter is closed should be reduced to a minimum. This shutter is in the form of a wheel or disk, and it has a fan-shaped section which passes the projector lens while the film is being shifted. In order to avoid flicker on the screen, the shutter has two additional fan-shaped sections. With these three equally spaced sections, the light is not only shut off from the screen during each successive film movement, but twice between each movement at uniform intervals. By closing the shutter twice while the picture is on the screen, the flicker that would be visible and annoying if the shutter were only closed while moving the film is multiplied to such an extent that it becomes almost continuous and is practically eliminated as far as the observer is concerned, assuming that the projector is operated at the proper speed. The width and area of that section of the shutter which is passing the lens when the film is being moved is governed by the time required for the film movement. Theoretically, the area of each section or segment of the shutter should be equal, although, in practice, the two extra sections are made of somewhat smaller area than the main one, in order to increase the open space and the percentage of area left for the passage of light.

There is an important relation between this shutter wheel and the intermittent motion or gearing of the projector. This is due to the fact that the shutter must be closed while the film is being shifted. With the mechanism to be described, the film movement is very rapid so that the shutter blades may be proportionately reduced in area, thus leaving more open space for the light. The intermittent motion referred to is

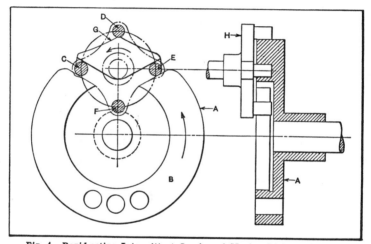

Fig. 4. Rapid-acting Intermittent Gearing of Moving Picture Projector

shown in Fig. 4. This mechanism is composed of a disk or wheel A, having an annular flange or ring B, which has two diagonal slots across it as shown; this wheel, which is the driver, imparts an intermittent motion to the follower H, which carries four equally spaced pins or rollers that engage the ring B on wheel A. Each time this wheel makes one revolution, the follower H is turned one-quarter revolution and in the same direction, as indicated by the arrows. The follower is stationary except when it is engaged by the slots or cam surfaces formed on one side of ring B. During this stationary period, the ring B simply passes between the four pins on the follower, two of these pins being on the outside and two on the inside of the ring.

The quarter-turn movement is obtained in the following manner: When the projection or cam surface G on the revolving wheel A strikes one of the pins, the rotation of the follower begins, and the pins are so spaced that one on the outside moves through a diagonal slot in ring B while a pin on the inside moves outward through the other slot. For instance, if the pins C and D are on the outside and E and F on the inside, pin D will first be engaged by cam surface G and, as the follower revolves, pin C will pass in through one diagonal slot while pin E is moving to the outside of the ring through the other slot. At the completion of the quarter-turn movement, pins C and F will be on the inside and D and E on the outside. As wheel A continues to revolve, ring B simply passes between these closely fitting pins which lock the follower against movement until projection G again comes around and strikes the next successive pin on the follower.

The follower operates a toothed wheel or sprocket which connects with the film and moves it downward each time the shutter is closed. Above and below the intermittent gearing there are other sprockets which rotate continuously, and these are so timed that a loop of film is formed above the intermittent gearing that is just large enough to provide for one film movement, which is equivalent to the length of one view or positive. As the film is drawn down rapidly by the intermittent mechanism, a loop is formed below it which is taken up by the lower sprocket as the film is wound upon the lower receiving reel. The normal speed of wheel A is sixteen revolutions per second, and it has been operated at two or three times the normal speed. The time required for turning the follower one-quarter revolution is approximately one-sixth of the time for a complete revolution, or 1/96 second, when running at normal speed. With the Geneva motion, which has been applied to many projectors, approximately one-quarter of the time is required for the intermittent action; therefore, the shutter blades must be of larger area than when the film movement occurs in one-sixth of the time. The mechanism shown in Fig. 4 is claimed to be superior to the Geneva motion in that there is less tendency to subject the film to injurious stresses. The locking of the follower during the stationary period is also more secure, especially at the critical time when near the operating point. The three holes drilled in the ring *B* are to compensate for the slots on the opposite side and to balance the wheel *A*.

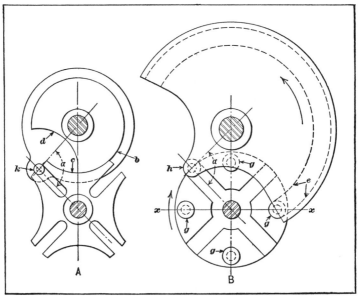

Fig. 5. Geneva Wheels which Vary in Regard to Method of Locking Driven Member during Idle Period

Geneva Wheel for Intermittent Motion. — The general type of intermittent gearing illustrated in Fig. 5 is commonly known as the "Geneva wheel," because of the similarity to the well-known Geneva stop used to prevent the over-winding of springs in watches, music boxes, etc. Geneva wheels are frequently used on machine tools for indexing or rotating some part of the machine through a fractional part of a revolution. The driven wheel shown at *A* in the illustration has four radial slots located 90 degrees apart, and the driver carries a roller *k* which engages one of these slots each time

.t makes a revolution, thus turning the driven wheel one-quarter revolution. The concentric surface *b* engages the concave surface *c* between each pair of slots before the driving roller is disengaged from the driven wheel, which prevents the latter from rotating while the roller is moving around to engage the next successive slot. The circular boss *b* on the driver is cut away at *d* to provide a clearance space for the projecting arms of the driven wheel.

The Geneva wheel illustrated by diagram *B* differs from the one just described principally in regard to the method of locking the driven wheel during the idle period. The driven wheel has four rollers *g* located 90 degrees apart and midway between the radial grooves which are engaged by the roller of the driver. There is a large circular groove *e* on the driver having a radius equal to the center-to-center distance between two of the rollers *g*, as measured on the center line *xx*. This circular groove engages one of the rollers as soon as the driving roller *h* has passed out of one of the grooves or radial slots. Each time the driver makes one revolution, the two rollers on the center line *xx* are engaged by the locking groove. The illustration shows the driving roller about to enter a slot and the locking roller at the point of disengagement. When the driven wheel has been moved 90 degrees from the position shown, the roller which is now at the lowest position will have moved around to the left-hand side so that it enters the locking groove as the driving roller leaves the radial slot.

When designing gearing of this general type, it is advisable to so proportion the driving and driven members that the angle *x* will be approximately 90 degrees. The radial slots in the driven part will then be tangent to the circular path of the driving roller at the time the roller enters and leaves the slot. When the gearing is designed in this way, the driven wheel is started gradually from a state of rest and the motion is also gradually checked.

Geneva Wheel Designed for Slight Over-Travel. —An ingenious special Geneva wheel was developed to operate the conveyor of an automatic weighing machine. This conveyor, by means of brackets attached to it, pushes the packages to be

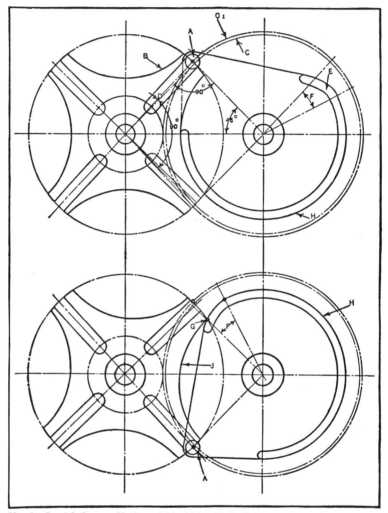

Fig. 6. Special Geneva Motion which Gives Slight Over-travel and Equal Return Movement to Provide Clearance

weighed on the scale platforms of the machine and then stops and remains stationary while the weighing takes place.

Originally, this conveyor was operated by an ordinary Geneva wheel, but the packages sometimes rubbed against the conveyor brackets, which seriously affected the accuracy of the weighing. To prevent this, it was necessary to have the conveyor moved back about 1/8 inch after delivering the packages to the scale platforms, thus providing clearance between the conveyor brackets and the packages. This clearance has been obtained by a slight change in the Geneva wheel, the result being that the conveyor is first given a small amount of over-travel and is then withdrawn a distance equal to this over-travel, thus providing the clearance desired.

The over-travel has been obtained by enlarging the diameter of the path traversed by the driving roller A (see Fig. 6). Circle C represents the normal path described by the center of roller A, or the radial position of the roller when the movement of driven wheel B is 90 degrees during each roller engagement. By increasing the diameter of this path, as represented by circle C_1, driven wheel B and the conveyor are given the required amount of over-travel. The center lines of the slots in wheel B are approximately tangent to circle C_1, so that engagement takes place without shock and the mechanism operates smoothly. Angle D is equivalent to one-half of the angular over-travel imparted to wheel B .

When roller A reaches the position shown by the lower view, it has moved wheel B 90 degrees plus twice the angle D. Now as the roller leaves its slot, cam surface E, which extends through angle F, comes into contact with corner G of the slotted wheel and pushes the wheel back an amount equal to twice the angle D, thus withdrawing the conveyor brackets from the packages during the time required for weighing. The driven wheel is locked during this dwell by the engagement of concentric surface H with arc J on the driven wheel, the wheel being released as soon as roller A again moves around into engagement with a slot, or in the position indicated by the upper view.

This simple method of preventing frictional resistance between the conveyor brackets and packages during the weighing operation did not introduce any difficulties in manufacturing the intermittent motion described.

Intermittent Motion for Dial Feed. —An intermitttent motion which is incorporated in an automatic station-dial machine for buffing brass shells is so designed that shaft A (see Fig. 7) revolves intermittently and has eight dwelling periods per revolution, each dwell being equivalent to 3/5 revolution of driving shaft B. The disk C is fastened to shaft A, and in it there are eight equally spaced hardened steel pins D and an equal number of larger pins E. As the sectional view shows, pins D are located on a higher level than pins E. As shaft

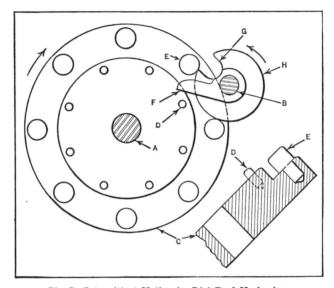

Fig. 7. Intermittent Motion for Dial Feed Mechanism

B revolves in the direction indicated by the arrow, a finger F first engages pin D and turns disk C until the larger pin E is engaged by notch G; when F leaves pin D, the positive drive between E and G continues until one-eighth revolution of disk C is completed. Then the concentric surface H is in contact with and tangent to two of the pins E, thus locking the driven disk in the dwelling position until the driver is again in position for an indexing movement.

One-twelfth Turn of Driven Shaft to One and One-quarter Turns of Driver. — The solution of an interesting problem in design is indicated in Fig. 8. The requirements were that for every one and one-quarter revolution of a continuously rotating shaft A, a second shaft L in alignment with the driving shaft must rotate intermittently, with equal velocity and in the same direction as shaft A, one-twelfth revolution or through an angle of 30 degrees. An eccentric bushing C is keyed to the driving shaft A. A 96-tooth gear D is loosely mounted on eccentric bushing C, but is prevented from rotating by lever E; the pitch-line of gear D, however, is always tangent to the pitch-line of the 120-tooth gear F and to that of the planetary pinion G. This pinion is carried by a double

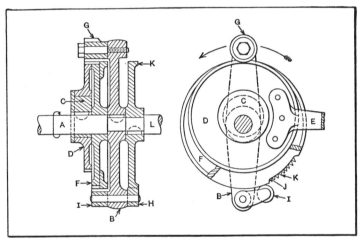

Fig. 8. Mechanism for Rotating Driven Shaft Intermittently and at Same Velocity as Driver

arm B which is also keyed to the driving shaft A. As arm B traverses the pinion around gear D, gear F is revolved on shaft A in the ratio of 120 to 96 or 1.25 to 1. The end of arm B opposite the pinion carries a link and roller I which runs on a flange of gear F until a depression in the periphery allows the roller to drop and permits pawl H to engage ratchet wheel J which is keyed to shaft L; each time the pawl engages the ratchet wheel, the latter is turned forward until roll I runs up on top of the flange again. As gear F advances one-fifth revolution for each revolution of the arm and pawl, and since 30 degrees equals one-twelfth revolution, the opening or depression for the roll must be shortened 1/5 × 1/12 of 30 degrees, or to 29½ degrees.

Two-speed Intermittent Rotary Motion. — The fast and slow motion of the pattern cylinder of a certain type of loom is derived from the reversible intermittent gearing shown in Fig. 9. The large gear A is mounted on the pattern cylinder shaft, and receives its motion either through the segment gear and crank combination B or through a similar combina-

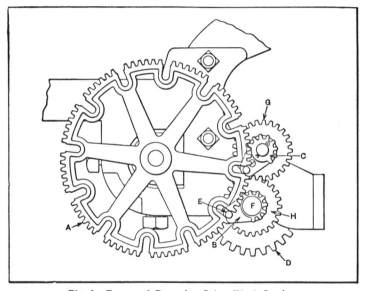

Fig. 9. Two-speed Reversing Intermittent Gearing

tion C, these two combinations being used to reverse the direction of rotation. Gear D is the driver for this train of mechanism. Whether the motion is transmitted from gear D to the pattern gear A through the crank and segment gear combination B or through combination C depends upon the position of a sliding key F. An intermittent fast and slow motion is obtained with either combination. When key F locks the crank and gear B to the shaft, the pattern wheel is rotated at a relatively slow speed when the segment pinion is acting as the driver, and at a faster speed when the crankpin E comes around into engagement with one of the radial slots in the pattern gear. When this direct drive is employed, the gears G and H revolve idly with the upper crank and gear combination C. When a reversal of motion is required, sliding key F is pushed in to engage gear H, which then drives gear G and the combination at C.

Adjustable Intermittent Motion. — The intermittent feed mechanism shown in Fig. 10 is so arranged that the intermittent action may be varied according to requirements by means of a simple form of "skipping" device. A pitman con-

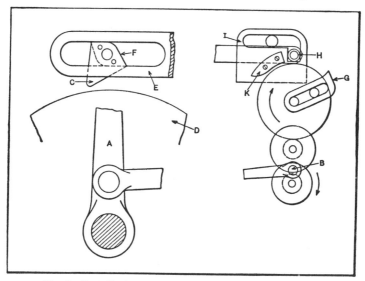

Fig. 10. Feed Mechanism with Skipping Device for Varying the Intermittent Motion

necting with crank B transmits an oscillating movement to lever A. This lever carries a stud on its free or upper end upon which is pivoted a fiber pawl C. This pawl engages the smooth periphery of disk D and turns the latter a fractional part of a revolution when lever A is moving to the left, unless the engagement of the pawl is prevented by the mechanism to be described. The pawl is formed of two pieces attached to opposite sides of a diamond-shaped block F. This block is within the slot and, being slightly thinner than the bar, causes the projecting sides C to frictionally engage the lower side of the bar. Any motion of lever A towards the right causes the pawl to turn to the position shown so that it clears the disk D for the return stroke. The reverse motion of lever A changes the position of block F so that the ends C grip the disk D, which is given the required feeding movement. The skipping of the feed is accomplished by a train of change-gears and a cam G. This cam serves to lift the pin H clear of its seat, so that the bar carrying pawl C is free to slide horizontally as lever A moves to the left;

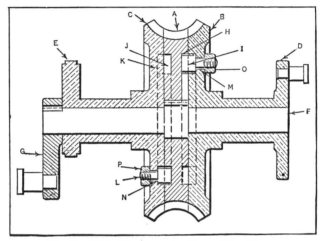

Fig. 11. Triple Worm-gear for Transmitting Three
Intermittent Movements

the result is that the pawl is not turned by frictional resistance to the gripping position, and it simply makes an idle stroke. The cam G is pushed in when it strikes dog K and is suddenly thrown outward by a spring after passing the dog; this sudden release disengages pin H from its seat, into which it drops again upon the return of bar E. The number of feeding strokes before an idle stroke are governed by the ratio of the change-gears.

Triple Intermittent Worm-Gear. — The mechanism here described was part of a barbed wire fence-making machine. While the accompanying drawing of the device (Fig. 11) is not made to any scale, the description will explain the principle of the design. This device transmits three distinct movements, all of which can have a dwell of different length, and this dwell can be varied in time to at least 180 degrees of the cycle.

The worm-gear is made up on three sections, A, B, and C. Section A is keyed to shaft F; section B revolves on shaft F and carries a disk crank D; section C also revolves on F and has an eccentric E. Shaft F has a crank G, and D, E, and G are the work levers. Worm section A has two concentric slots J and H on each side. In each of these slots there is fastened a stop (not shown). In sections B and C there are two concentric slots M and N cut long enough to meet the required adjustment of timing. Dog bolts L and I are held in the desired position by nuts O and P.

A section of the teeth long enough to make complete disengagement from the worm is cut from A, B, and C. We will assume that section C where the teeth are out is set central relative to the center of the worm; then section C would stay in this position until section A rotated enough to bring the stop in slot J into engagement with dog L. This action, of course, must be timed so that when engagement occurs, the worm teeth will register. By the varied settings of these two dog bolts, the timing of any section can be changed to meet the requirements of whatever machine is being designed.

Intermittent Rotary Motion Varied by Changing Cams. — Some novel features are embodied in the design of the intermittent motion mechanism shown in Fig. 12. Referring to the upper view, the shaft B revolves continuously, receiving its motion from one of the constantly rotating shafts of the machine on which it is employed. Another shaft C, imparts the intermittent movements obtained by the mechanism. With the cam pieces N and O, the shaft C makes ½ revolution, and then dwells while shaft B makes 7½ revolutions. Seven additional sets of cam pieces are provided, which can be used in place of those shown at N and O for obtaining different intermittent movements.

The driving shaft B runs in the bearings L and W. A spur gear V, pinned to the shaft B, meshes with the gear cut

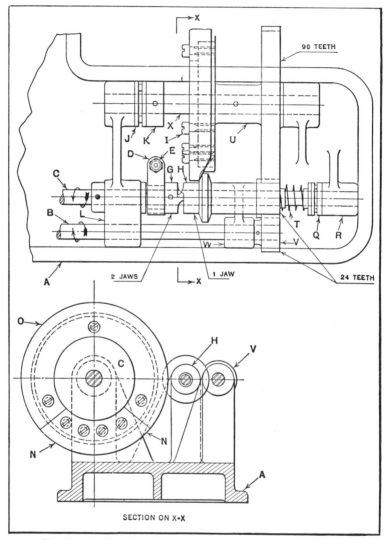

Fig. 12. Variable Intermittent Motion Obtained by Changing Cams.

integral with the sleeve H. The gear on sleeve H also meshes with a gear on the hub U. Hub U is fastened to shaft X by a pin, and has, on the opposite end from the large gear, a turned disk on which the removable cam pieces N and O are mounted. These pieces are fastened to the disk by screws I. The collar K is pinned to shaft X, and the end thrust resulting from the cam action is taken by the ball thrust bearing J.

Sleeve H, which slides and also turns on shaft C, has a shoulder or collar which is beveled on both sides and which acts against the cam pieces N and O. Sleeve H has a single jaw machined on one end which engages one of the two jaws on member G. The member G, which is fastened to shaft C, is equipped with a leather-lined band brake D. The friction drag of the brake is controlled by adjusting nut E, which closes or spreads the band as required. The spring T acts against sleeve H, and forces the beveled shoulder against the cam, the resulting thrust being taken up by the ball thrust bearing Q.

As there are eight different intermittent motions to be transmitted, the eight sets of cams must all be of different lengths. When a shorter cam than the one shown at O is used, the two cam pieces N are slid back, so that they make contact with each end of the shorter cam, and are then fastened securely in place by the screws I. Additional tapped holes are provided in member U for fastening the new cam pieces in place. It will be noted that the pitch circles of the gears on hubs H and U and of the cam and the beveled shoulder are the same, so that nearly all sliding action between the cam surfaces is eliminated.

The operation of the mechanism may be explained as follows: Sleeve H is driven by shaft B, and makes the same number of revolutions per minute. Shaft X revolves one-fourth times as fast as shaft B and sleeve H. With the particular cam O shown in the illustration, shaft C will dwell or remain stationary while shaft B makes $3\frac{1}{2}$ revolutions, after which shaft C will make $\frac{1}{2}$ revolution, thus completing one cycle. Cam N simply slides sleeve H back, thus disengaging the clutch until the cam again moves around to the clutch-engaging position. The friction brake serves to stop the shaft C as soon as the clutch teeth are disengaged, and also takes up any lost motion that may occur.

Mechanism for Controlling Length of Rotating and of Idle Periods. — An intermittent motion was required to drive a shaft at the rate of one revolution per second and have provision for adjusting the length of the periods at which the shaft remained stationary. The first work for which the mechanism was employed required the shaft to make one revolution in one second and remain stationary for fifty-nine seconds. The second class of work required the shaft to make three revolutions in three seconds and then remain stationary for fifty-seven seconds, while a third operation required the shaft to make five revolutions in five seconds and remain stationary for fifty-five seconds.

The mechanism is shown in Fig. 13 as arranged to meet the first requirement. Upon the machine base A is mounted the bearing stand R which is bored out to a running fit for the small end of the worm U. Shaft N, in turn, is made a running fit inside the worm U and transmits the intermittent motion to the machine proper, which is not shown in the illustration. The clutch member K slides on shaft N and is prevented from turning by a key L. On the face of the large end of worm U is a clutch tooth which engages the tooth on member K. The worm meshes with the worm-gear E, imparting a continual and uniform motion to the cam-plate F. The cam-plate pushes the lever C outward and thus disengages the clutch at the required intervals. The spring P provides for the return of lever C and the engagement of clutch K at the end of the period in which the shaft N is required to remain stationary. The brake-drum Y is keyed to shaft N, and is equipped with a steel brake-band M having a leather-lined contact surface. The pressure of the brake-band on the drum is maintained by a spring V.

When the mechanism is in operation, the grooved pulley S is driven by a belt from the lineshaft at a speed of 60 revolutions per minute. This pulley rotates the worm U, causing the worm-gear E and the cam-plate F to make one revolution per minute. As the worm has a single thread, a point on the circumference of the cam-plate will revolve a distance Z, **equal**

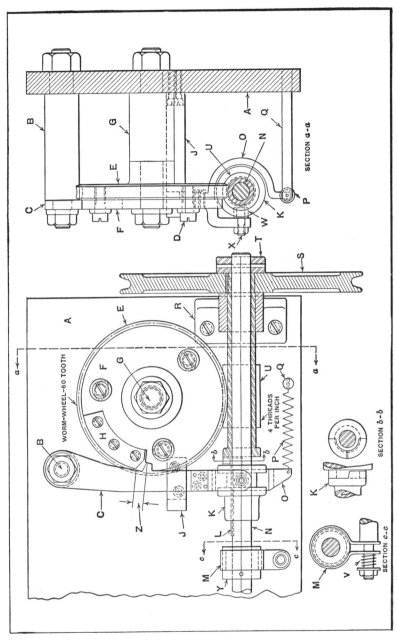

Fig. 13. Intermittent Mechanism which may be Arranged to Vary Length of Rotating and of Idle Periods

to the lead of the thread, or about ¼ inch, per revolution of the worm and of shaft N.

After the cam-plate has been properly adjusted, it will not require changing except to compensate for wear. At the instant the cam-plate has reached the end of the travel Z, the movement of lever C disengages the clutch K, so that the rotation of shaft N is stopped exactly at the end of one revolution. The friction brake consisting of drum Y and band M prevents the momentum of the clutch and shaft from causing a further movement of the shaft.

In order to obtain the necessary dwells required for the second and third operations, the distance Z must be increased. This necessitates the use of interchangeable cam-plates F. For the second job, the dimension Z is three times as great as for the first job, and for the third job five times as great. Any other desired length of dwell may be obtained by providing a suitable cam-plate.

Intermittent Motion which Automatically Increases and Decreases. — The mechanism described in the following is designed to turn a driven shaft through a small arc for every other revolution of the driving shaft and according to the following requirements: The feeding movement of the driven shaft is to increase by small amounts until a maximum feed is obtained; the feed then decreases to a minimum, again increases to a maximum, and at this point instantly begins at the minimum again. If a line is drawn representing these movements graphically, it will readily be seen that there are two periods of increasing feed and one period of decreasing feed for every cycle of movements. It was necessary to derive the feeding motion from a shaft running at twice the desired speed.

The principle upon which this mechanism operates is shown partly in diagrammatical form, in Fig. 14. The arm C carries a sliding block D which is connected to lever L by a pitman K. Block D is fed to or from the center of B by screw E, working in a divided nut. On the upper end of C are two intermittent motion star-wheels F and G, of six teeth each, with their planes at right angles. Wheel F is pivoted on the side of C, and G is fastened onto E. Wheel F has three projecting pins F_1, F_2, and F_3, placed on alternate teeth and denoted in the illustration by black dots. Suppose star-wheel F is rotated one tooth (denoted by the straight lines). If that tooth is one of the three with the projecting pins on the side, wheel G and screw E will be rotated one-sixth revolution. Conversely, if the tooth on F has no projecting pin, G will not rotate. This

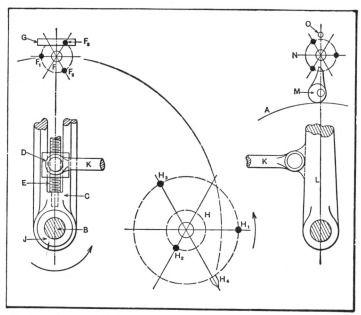

Fig. 14. Diagram of Mechanism for Automatically Increasing and Decreasing
Intermittent Motion of Driven Shaft

method gives the alternating feed through K and L, as the arm C revolves about the center of B.

The center of F revolves about B in a path as shown by the broken line, which passes midway between the two circles at H. These circles represent the controlling star with six teeth or points (denoted by straight lines). A pin (not shown) upon the block D is caused to engage two teeth of H, on two successive revolutions before it is retracted. This is done only when the slide is at the upper end of its travel.

Star-wheel H merely operates star F. Three projecting points (denoted by black dots) engage point after point of F as it comes around. Suppose it is projections H_1 or H_3 that engage with F. Then, by reason of their being on the outer side of the center of F, star-wheel F must revolve in an opposite direction to that in which it would revolve if H_2 were the projection engaging F, because H_2 is on the inner side of the center path of F.

It is now apparent that block D moves up (or down) through the action of H upon F and F upon G; and the motion takes place only for every other revolution of B. When block D reaches the outer limit, the pin upon the block is released as mentioned, which, in turn, revolves H two teeth before it is retracted, thereby engaging the opposite side of F, reversing the direction of rotation of F and G, and returning D toward the center. Pin H_3 is now brought into action and D goes outward again to the extreme position. The controlling star is rotated as before, returning H_1 to position and bringing H_4 into action to open the split nut at screw E, which allows a spiral spring in J to return the block D and pitman K instantly to the center, thus completing one cycle. Through link K, lever L, and pawl M, ratchet A is rotated. A star N is carried by L, which is operated by the hinge tappet O and which is provided with three projecting pins to lift the eccentric pawl from the ratchet every alternate stroke of L.

Constant Intermittent Rotary Motion from Variable Rotary Motion. — The feeding movement of a planer tool, which occurs at the end of each return stroke, is derived from a shaft which revolves in first one direction and then the other, the number of revolutions depending upon the length of the stroke which is adjusted to suit the work. The simple mechanism to be described makes it possible to obtain the same rotary movement for operating the feed-screw of the tool-slide, regardless of the number of revolutions made by the shaft which drives the feeding mechanism. A crank at the end of the driving shaft turns part of a revolution and then remains stationary while the shaft continues to revolve.

One method of securing this fractional part of a turn and then stopping the motion of the feed disk is illustrated at *A* in Fig. 15. The link *f* connects the crankpin of the feed disk with a rack which, through suitable gearing, transmits motion to the feed-screw. The main pinion shaft of the gear train for driving the planer table has attached to its end the cup-shaped casting *a*, which forms one part of a friction clutch. The crank disk *b* has a hub *c*, which fits into the tapering seat in part *a* and forms the other member of the clutch. If this

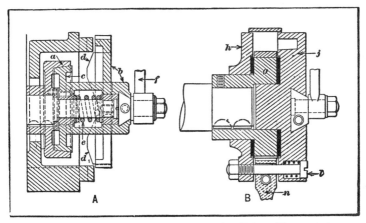

Fig. 15. Mechanisms for Deriving an Unvarying Rotary Movement from a Driving Shaft regardless of the Number of Revolutions Made by the Shaft

friction clutch is engaged when the planer is started, the crank disk *b* revolves until one of the tapered projections or cam surfaces *d* strikes a stationary taper lug, thus forcing part *c* out of engagement with *a* against the tension of spring *e*. The crank disk then remains stationary until part *a* and the driving shaft reverse their direction of rotation at the end of the stroke. This reversal of motion disengages the tapering surface *d* or *d_1*, as the case may be, and allows the friction clutch to reëngage; the crank disk is then turned in the opposite direction, until the other tapering projection strikes a second lug which again stops the motion of the feed disk. This intermittent action in first one direction and then the other is continued as long as the planer is in operation, and the feed disk oscillates through the same arc regardless of the length of stroke or the number of revolutions made by the driving shaft; consequently the feeding movement of the tool will not be varied by a change in the length of the stroke.

Another planer feed mechanism which operates on the same general principle as the one just described is illustrated at *B*, Fig. 15. In this case, the hub *g* is keyed to the shaft and the flange formed on this hub is between plates *h* and *j*. This flange does not come directly into contact with the plates, as there are leather washers on each side as indicated by the

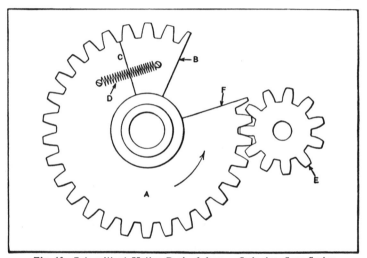

Fig. 16. Intermittent Motion Derived from a Swinging Gear Sector

heavy black lines. The plates *h* and *j* are held in contact with these washers by three bolts *l* having springs under the heads. The hub *g* is surrounded by a band which is split on the lower side and has lugs *n* into which is fitted a pawl of such shape that, when it strikes a fixed stop, the band is opened and released from the hub. This releasing of the band occurs after the crank disk has turned far enough to give the necessary feeding movement. The crank is held in position while the driving shaft continues to revolve, by the friction between plates *h* and *j* and the leather washers previously referred to. When reversal occurs at the end of the stroke, the hub *g* revolves in the opposite direction and the band again grips it until the pawl of lug *n* strikes the opposite stop.

Intermittent Gear with Swinging Sector. — The gearing illustrated in Fig. 16 has one period of rest for each revolution of the driver which has a sector *B* that is free to swing in the space provided for it, but is normally held in the position shown by a spiral spring *D*. The driver revolves at a uniform speed in the direction shown by the arrow and, when the sector *B* comes into engagement with the driven gear, the latter stops revolving while the sector is swinging across the open space or until side *B* strikes side *F*, when the driven gear is again set in motion. As soon as the sector is released by

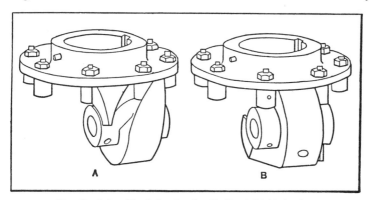

Fig. 17. Intermittent Gearing for Shafts at Right Angles

the driven gear, the spring draws it back to the position shown in the illustration, preparatory to again arresting the movement *E*. The resistance to motion offered by gear *E* should be great enough to overcome the tension of spring *D*, as otherwise the sector would not swing away from the position shown. In order to avoid shocks, this gearing would have to be revolved quite slowly; while the design is not to be recommended, the principle may be of some practical value.

Intermittent Gearing for Shafts at Right Angles. — When driving and driven shafts are at right angles to each other, intermittent gears which are similar to bevel gears in form, but constructed on the same general principle as the spur gearing illustrated in Fig. 1, may be employed. The smooth or blank space on the driving gear for arresting the motion of the driven member corresponds to the pitch cone and engages concave locking surfaces formed on the driven gear. Owing to the conical shape, such gearing is more difficult to construct than the spur-gear type.

A form of intermittent gearing for shafts at right angles to each other but not lying in the same plane is illustrated in Fig. 17. The driving member is in the form of a cylindrical cam and has a groove which engages, successively, the rollers on the driven wheel. Diagram *A* shows the cam in the driving or operating position, and at *B* the driven wheel is shown

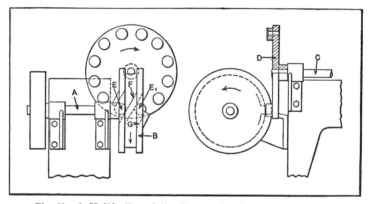

Fig. 18. A Modification of the Type of Gearing Shown in Fig. 17

locked against rotation during the period of rest. The locking action is obtained by parallel faces on the cam which fit closely between the rollers and are located in planes at right angles to the axis of rotation. This mechanism was designed for a high-speed automatic machine requiring an accurate indexing movement and a positive locking of the driven member during the stationary period. The gearing operated successfully at a speed of 350 revolutions per minute, and it was because of the speed that this design was used in preference to the Geneva-wheel type of gearing previously described. The curvature of the operating groove on the driving cam is such that the driven wheel is started slowly and, after the speed is accelerated, there is a gradual reduction of velocity. The driven wheel has no lost motion for any position and the mechanism operates without appreciable shock or vibration, and is practically noiseless.

Another form of intermittent drive for shafts located at right angles but not lying in the same plane is illustrated in Fig. 18. This mechanism operates on the same general principle as the one just described, but differs in regard to the form of the driving member or cam. This cam B is attached to the end of the driving shaft A and has an annular groove corresponding in width to the diameter of the rollers on the driven wheel D carried by shaft C. This annular groove is not continuous as there are inclined openings on both sides. When the cam revolves in the direction indicated by the arrow, the inclined surface F pushes roller E over to the left, thus causing disk D to turn; at the same time, roller E_1 enters the opening on the opposite side and is pushed over to the central position by cam surface G. This roller E_1 remains in the groove until the cam has made one revolution, thus locking the driven wheel against rotation. This locking roller then passes out at the opposite side and another roller is engaged by the groove. The ratio of this gearing, which was used to provide a feeding movement on an automatic machine, depends upon the number of rollers on the driven wheel.

Locking Plates for Intermittent Gears. — In the design of intermittent gearing, it is essential that the driven gear be provided with some form of locking device to prevent it from moving during a period of rest. If this provision were not made, the driven gear might not be in the correct position to mesh with the first tooth of the driver when the two gears come into contact. Various devices are used for this purpose. The simplest form of locking circle is formed by milling the blank space of the driving gear down to the pitch line as shown at a (Diagram A, Fig. 19). During each period of rest of the driven gear the locking circle on the driver rotates in one of the stops b on the driven gear and, of course, does not transmit any motion to that gear. In applications where the driven gear acts as an idler, driving a third gear and consequently containing a complete set of teeth, it is impractical to make use of a locking circle of the type just mentioned. In cases of this kind, a locking circle may be employed which is in a different plane from that of the gear teeth and which may be in the form either of a plate riveted to the gear or a flange cast integral with it. In these designs, the locking circle should not interfere with the action of the teeth.

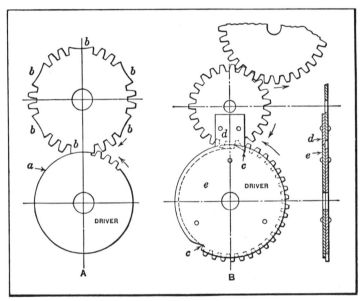

Fig. 19. Application of Locking Plates to Intermittent Gearing

The form of locking device consisting of plates riveted to the gears is used on light gearing, principally where the gears themselves are stampings. This construction is shown at *B*. When such plates are made use of, notches, as shown at *c*, must be provided in the locking circle in order to permit the corners of the plates which form the stops *d* to pass the plate *e* on the driving gear at the beginning and ending of each period of rest. The shape of this notch may be readily obtained by laying out the gears when in the respective positions of entering and leaving the periods of rest. Such notches are not necessary on gears of the type shown at *A,* the spaces between the teeth being such as to take care of this condition.

The form of locking circle, or ring, which is cast on the gears is the type which is used on heavier construction. The ring is cast integral with the driving gear. The stops are cast integral with the driven gear, and should be designed to allow sufficient cutter clearance, at the ends of the gear teeth for cutting the gear teeth. Notches are also provided in this type of ring to permit the stops to enter and leave just as described for the riveted-plate type of locking device.

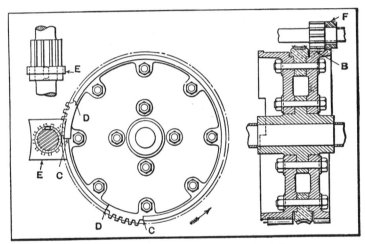

Fig. 20. Intermittent Gears with Locking Circles

Locking Circles Cast Integral with Intermittent Gears. — In Fig. 20 is shown a design for intermittent gears which is suitable for either light or heavy construction. An intermittent gear is bolted to each side of the worm-wheel and hub, one gear providing motion for indexing while the other operates a clutch mechanism. The portions of the rims which have no teeth are cast to a smaller diameter than the root diameter of the teeth, so as to allow sufficient clearance for the pinion teeth while the intermittent spaces on the gear rotate past the pinion. This clearance may be plainly seen at *B.* The locking circle of each gear is cast integral with the gear, but is broken between the points *C* and *D* where the pinion starts and stops rotating. It is not necessary for the diameter of the locking circle to be equal to the pitch diameter of the gear; it may be made to any convenient dimension, but it is well to make this diameter as large as possible, thereby reducing any possible binding on the locking plate. The cutting of the gear teeth presents no difficulties, but it is sometimes necessary to reduce the addendum of the teeth adjacent to points

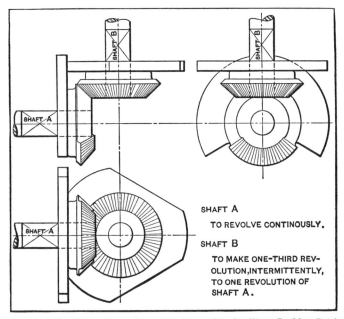

Fig. 21. Intermittent Bevel Gears Provided with Auxiliary Locking Device

C and *D* so that they will enter and leave the pinion teeth without interference. The locking plates *E* and *F* are secured to the pinion shafts at the sides of the pinions by means of feather keys. Gears of this nature run smoothly, and as their manufacture offers no difficulties to the machine shop they might well be used more extensively in the design of automatic machinery.

Auxiliary Locking Device for Intermittent Bevel Gears. — The intermittent bevel gearing illustrated in Fig. 21 is provided with auxiliary locking plates which regulate the motion of the driven gear and hold it stationary while disengaged from the driver. The driving gear is on shaft *A* and revolves continuously. It is only provided with enough teeth to rotate the driven gear and shaft *B* one-third revolution to one complete revolution of the driver. This mechanism is used to actuate feeding rolls requiring an intermittent motion. Formerly the gearing was used without the locking device to be described, but there were slight variations in the movements of the driven shaft so that the gears did not always mesh

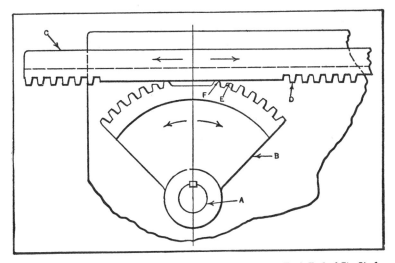

Fig. 22. Reciprocating Rack which Engages Segment Gear at Each End of Its Stroke

correctly, which caused them to break, and also interfered with the timing of the feeding movement. These defects were eliminated by applying locking plates to the shafts *A* and *B*, one plate being located just back of each gear. The plate on shaft *B* has three equally spaced flat sides or edges and the plate on shaft *A* is cut away to provide a clearance space for the protruding sections of the plate on shaft *B* when this shaft is in motion. As the plan view shows, the flat side of the plate on shaft *B*, during the idle period, is intercepted by the plate on *A* so that the driven shaft is not only locked but its motion is limited to one-third revolution for each complete turn of the driving shaft.

Intermittent Motion from Reciprocating Rack. — The intermittent motion to be described is part of a device for feeding brass shells to a dial press. In the operation of this feed mechanism, it was necessary to turn shaft A (Fig. 22) and segment gear B through part of a revolution at each end of the stroke of reciprocating rack C. The illustration shows the segment gear in the dwelling position. As the rack moves, say, to the left, the rack teeth beginning at D engage the segment gear and turn it. When the rack reverses, the segment gear is turned in the reverse direction until the rack teeth at the right leave it, and then dwell occurs until the teeth on the left-hand side engage the segment gear. Tooth E (and the corresponding tooth on the opposite side) is cut away to avoid interference as D comes into contact with F.

The "Beaver-tail" Stop Mechanism. — The "beaver-tail" stop mechanism is used in conjunction with a geared drive to prevent or minimize inertia shock or impact at some point in a repeated cycle where a clutch is thrown or tools are brought into contact with each other or with the work as in power press operation. The name "beaver-tail" is applied to this mechanism because of the shape of the cam which forms an important part of it. The driving pinion A (See diagram, Fig. 23) revolves continuously, and drives gear B through ordinary gear teeth except when the "beaver-tail" mechanism comes into action, at which time the motion of gear B is controlled by the two rollers R and R_1 and the "beaver-tail" cam located between the rollers. If driven gear B is to be stopped once during each revolution, only one cam is attached to it. If two stops per revolution are required, two cams located 180 degrees apart, are used. The teeth of the driven gear are cut away at each stopping position, and the large developed tooth or cam takes their place.

The rollers on the driving pinion are diametrically opposite each other, and their centers are on the pitch circle of the pinion. When the beginning of the blank space on gear

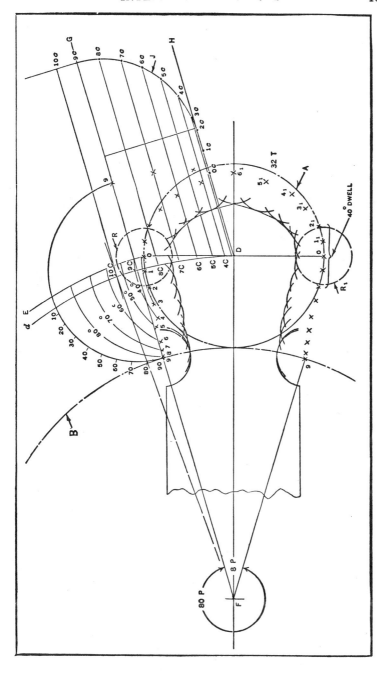

Fig. 23.   Development of a "Beaver-tail" Cam for a 3 to 1 Gear Ratio and a Dwell Equal to About 40 Degrees of Pinion Rotation

B reaches the pinion and during the next quarter revolution of the pinion, one roller moves along the "beaver-tail" cam and brings the gear to rest with a harmonic motion. The center of the roller at the point of engagement coincides with the point of tangency of the two pitch circles, so that engagement takes place without shock. The driven gear is locked during the brief dwell which occurs while the rollers are revolving about a concentric part of the cam. This stationary or neutral position is shown by the diagram. After the dwell, the other roller, during a quarter revolution of the pinion, engages the cam and accelerates the gear until it has the same speed as the pinion, when the gear teeth mesh and the ordinary gear drive is resumed.

There may be one, two, or more stops per revolution, depending upon requirements and the number of stopping positions provided. Both stopping and starting are accomplished with harmonic deceleration and acceleration, so that there is no shock to the mechanism (except from possible backlash) due to the reversal of strains. While the deceleration is taking place, the driven member acts temporarily as the driver, returning energy to the drive as the revolving parts are brought to rest. During a brief dwell period of twenty or thirty degrees of driver rotation, while the cam is in the neutral position, a clutch on the driving shaft may be thrown in or out without inertia or load effect from the driven parts. As the load picks up, the driver has temporarily a greater mechanical advantage, due to the position of the roll on the cam, than it normally would have because of the ratio of the gears.

Application of "Beaver-tail" Stop to Power Presses. — The "beaver-tail" stop is an invention of Charles R. Gabriel. It was developed about twenty-five years ago for use in connection with the feed mechanism of gear-cutting machines built by the Brown & Sharpe Mfg. Co. During the last five or six years this stop mechanism has been utilized very successfully by the E. W. Bliss Co. in conjunction with the drives of power presses designed for special purposes.

For example, a small press used for curing celluloid is arranged with a double "beaver-tail" stop to permit disengaging and engaging the clutch at the bottom of the stroke while the press is under full load. This stop mechanism has also been used on larger machines having shafts up to 5- or 6-inch sizes. Machines equipped with this stop are used either for curing celluloid or for other products which must be held under pressure for a period of time. The rolling key clutch is used on the back-shaft or driving shaft, and since the clutch is disengaged and engaged as the rollers are passing across the dwell or neutral position on the cam, it is protected from an otherwise unduly severe strain. Presses are sometimes built for this service without using the "beaver-tail" cam, but if the load is heavy, the clutch life is likely to be short.

A rather different use of the motion is its application for removing the impact load from the tools as on a press used for the extrusion of collapsible tin tubes. The "beaver-tail" is so placed that it brings the slide and punch to rest just as contact is made with the slug or blank to be extruded. The extrusion then takes place with an easy accelerated motion, the whole action resulting in various manufacturing advantages and an enormously greater tool life.

With the clutch on the back-shaft, the driven gear must be fitted with two "beaver-tails," one at the point of contact of the tools, and the other opposite it, to favor the clutch in stopping the press. A variation of this practice is to place the clutch in the gear on the main shaft and use only one cam at the tool contact position. This cam need not have any allowance for dwell.

Size of Driven Gear Used with "Beaver-tail" Stop. — The pitch circle velocity and travel of the gear and pinion are the same except during the slowing down and picking up periods, when the pinion moves half its circumference and the gear moves a much shorter distance. This difference in travel reduces the number of teeth or pitches on the gear (including those teeth that are actually cut away) and hence also reduces the gear diameter.

As an example to illustrate the procedure, assume that a pinion makes four revolutions to one of the gear. In this case, if the pinion has 32 teeth, there would normally be 128 gear teeth (4 × 32 = 128). If in the case selected the gear moves only the space of ten teeth during the slowing down and picking up periods, while the pinion moves a half revolution or sixteen teeth, then the difference (16 — 10 = 6) must be deducted, leaving 122 teeth or pitches as the gear circumference. The pitch diameter of the gear relative to the pinion is in the ratio of 122 :32.

Profile of "Beaver-tail" Cam.— The construction or development of this motion is worked out in Fig. 23. The method is based upon keeping the gear stationary and revolving the pinion about it as in planetary gearing, determining the relative progress of the center of the pinion (along the arc Dd) and then plotting the corresponding positions of the rolls which thus outline the proper shape of the cam.

The pitch circles of the pinion and gear are indicated at A and B. The rollers R and R_1 are shown in the neutral or locking position 0-0 which is their position when the center of the pinion coincides with the center line of the cam at D. This is one of the limiting positions. The other is either of the points 9, when the two rolls and the center of the pinion are on a radial line from the center of the gear. Here one roll center coincides with the point of tangency between the pitch circles of the gear and pinion, and the driving action shifts from the rolls to the gear teeth (or vice versa).

The spacing of points 0-0 depends upon the pinion, which we have assumed in this case to have 32 teeth (the number must be even) and, say, a 4-inch pitch diameter. The pinion moves through just half a revolution from the point where one roll engages the cam to the point where the other roll leaves it on the other side, the movement being equivalent to 16 teeth. The gear obviously moves a shorter distance, which we have assumed in this case to be 8 pitches. (A greater distance reduces the dwell but eases the accelerating or decelerating action.) This is 8 pitches less than the pinion movement. Accordingly, the pitch circumference or number of teeth of the gear for a 3:1 ratio is $32 \times 3 = 96 - 8 = 88$. The gear diameter equals $(4 \times 88) \div 32 = 11$ inches.

The order of construction, after determining the pitch diameters of the gear and pinion is as follows: Lay off the center line FD, the pinion pitch circle A about the center D, and the gear pitch circle B tangential to it, about the center F. Locate the roll centers 0-0 in their neutral position on the intersections of the pinion pitch circle and the vertical center line through D.

Locate the starting and finishing positions of the roll centers 9-9 on the pitch circle B of the gear—in this case, 8 pitches apart or 4 pitches each side of the center line FD. Draw the radial line FG from the center of the gear through one of the points 9. The centers of the pinion and both rolls must lie on this line when the roll R is at 9. Draw the arc Dd and the arc OE from the center F and the points D and 0.

Following the scheme of revolving the pinion about the gear to develop the shape of the cam, the center of the pinion for various positions of the roll R must move along the arc Dd. When the rolls are at the locking or stop position 0-0, the pinion center is at D. When the roll R is at the point 9, the pinion center is at $9C$, the intersection of the arc Dd and the line FG. Here the gearing is in normal mesh and the gear and pinion are moving relatively at full speed. Since the pinion moves through a quarter revolution (90 degrees) in accomplishing this change of speed, we have taken nine positions for the roll, representing a change of 10 degrees each on the pitch circle of the pinion. The corresponding positions of the center of the pinion must be on the arc Dd between $9C$ and D, and should be spaced to give harmonic deceleration. To get a starting point for this harmonic change, lay off from 9 on the gear circle B a distance equal to 10 degrees on the pinion pitch circle. (Note the small arc at the division marked 80). A radial line from the gear center F through this point, intercepts the arc Dd at the point $10C$. Then the distance $10C$-$9C$ represents the travel of the center of the pinion for 10 degrees of rotation (direct-geared, full speed).

To locate the pinion center positions below $9C$ for harmonic deceleration, the following approximate construction has been used: Draw the line DH from D parallel to FG. Swing the arc J from a center on FG, tangent to DH. From the point $9c$ at the intersection of the arc J and the line FG, lay off the distance $10c$-$9c$ about the arc J, continuing along the line DH. Transfer the points $8c$, $7c$, etc., on arc J, by means of lines parallel to FG, to the arc Dd. This gives the pinion center positions $8C$, $7C$, etc. Since the points $0c$, $1c$, and $2c$ all fall on the point D, there is a relative dwell in the motion during the roll positions 0, 1, and 2.

To locate the roll center positions, swing an arc equal to the pitch radius from the proper center and measure off the corresponding number of degrees from the arc $0E$ as a starting point. Thus the points 1 and 2 are on the radius from the center D, and are 10 and 20 degrees, respectively, from 0. The point 5, for example, is on a pinion radius swung from the center $5C$, and is measured off 50 degrees from the intersection of that radius with the arc $0E$. This gives the positions of the center of the roll R between the points 0 and 9. Corresponding positions of the roll R_1 may be obtained at the same time, as they are at the other extreme of the pinion diameter in each case. Thus the point 5_1 is at the intersection of the diameter line through the points 5 and $5C$ and the pinion circle arc swung from the center $5C$. The remaining positions may be obtained by repeating the construction on the other side of the center line FD, or since the cam is symmetrical, by transferring the points 0 to 9 and 0 to 6_1 across the center line FD on perpendicular lines to the corresponding points on the other side of it. The diameter of the rolls R and R_1 is determined to suit the case, and the profile of the cam is outlined, as shown, by drawing in a portion of the roll circumference, swung from each of the roll center positions.

Note that the dwell is about 40 degrees; that is, in laying out the uniform distance $10c$-$9c$ around the arc $9c$-D, the

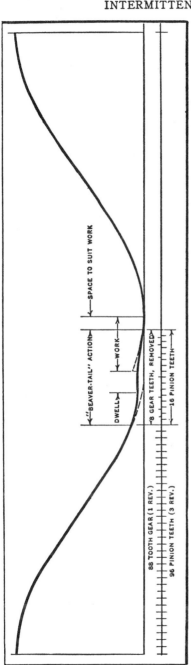

Fig. 24. Modified Crank Motion of an Extrusion Press Using the "Beaver-tail" Shown in Fig. 23

points $2c$, $1c$, and $0c$ all fall on the neutral center D, so that rotation of the pinion through the corresponding positions does not result in any motion relative to the cam. If it is desired to reduce this dwell, the number of pitch spaces between the points 9-9, assumed in this case as eight pitches, should be increased. For convenience in gear-cutting, this distance should always be a whole number of pitches. The action resulting from the use of the "beaver-

tail" cam in the drive of a crank press and the modification of the crank motion is shown by the curve in Fig. 24. This is based upon the design shown in Fig. 23. The motion of the pinion, which has uniform speed, is used as the time basis for the curve. The spacing of the crank is either fixed or adjustable to suit the work, which in this case may be assumed to be extrusion. Note that a close approximation to uniform motion during the working period results from the combination of harmonic acceleration of the "beaver-tail" and harmonic deceleration of the crank.

Automatic Indexing Mechanism. — The indexing or dividing of circular work requiring equally spaced grooves milled

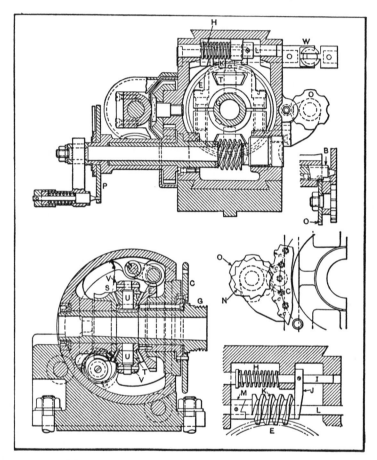

Fig. 25. Automatic Indexing or Dividing Mechanism

across the periphery may be controlled automatically by the dividing-head illustrated in Fig. 25. In addition to the transverse and longitudinal sections shown, there are three detail views which illustrate important features. The mechanism for controlling the indexing automatically derives its motion from a spindle L driven through coupling W from a special pulley carried on a bracket attached to the bed of the milling machine. The clutch M (see detail view) on spindle L locks worm K to the spindle when the worm is pressed against the clutch M by a spring H, acting through rod I and finger J. This engagement with clutch M occurs when lock-bolt B is withdrawn from plate C, so that worm-wheel E is free to revolve. The movement of plate C at each indexing is controlled by a counter mechanism consisting of a dividing-plate C having teeth on the periphery which engage the teeth of disk N, thus rotating stop-plate O which controls the engagement of lock-bolt B and the extent of the indexing movement.

The table of the milling machine on which this mechanism is used should be arranged to return automatically. When one groove has been milled across the work, the table returns and, when near the end of the return stroke, lock-bolt B is withdrawn by a suitable mechanism (not shown in the illustration). When this bolt is disengaged from dividing-plate C, the worm-wheel E is free to revolve. The pressure of spring H forces rod I, finger J, and worm K to the left, the worm engaging clutch M on spindle L which is constantly revolving. As worm K and worm-wheel E revolve, rotary motion is transmitted to dividing-plate C, and also to spindle G through epicyclic gearing consisting of bevel pinions T mounted on pins U attached to part V which is keyed to spindle G. The indexing movement continues until bolt B enters one of the succeeding holes in plate C. The movement of worm-wheel E is then arrested and the worm, as it continues to revolve, disengages itself from clutch M and stops rotating. The dividing-plate C has a number of teeth c, the number corresponding to the number of its holes. Whenever this plate is set in motion, these teeth engage disk N and turn the counter O. A solid portion of this counter is thus placed in front of lock-bolt B, which prevents the bolt from reëngaging with plate C until a rotation equal to the required number of holes has been completed. One of the concave notches in counter O then releases bolt B which engages plate C. The number of teeth in plates C and N and the notches in counter O depend upon the number of divisions required. This dividing-head may be used the same as the hand-operated design. The hand-operated indexing movements, as well as the automatic movements, are transmitted to spindle G through the train of epicyclic gearing previously referred to.

Combined Indexing and Locking Mechanism. — The automatic indexing and locking mechanism illustrated in Fig. 26 was designed for a multiple-spindle automatic screw machine. The motions of this machine are all controlled by cams on a

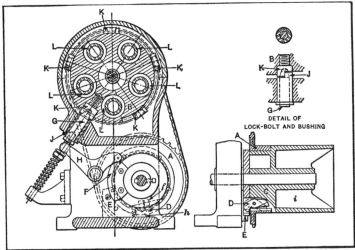

Fig. 26. Indexing and Locking Mechanism for Spindle Head of Multiple-spindle Automatic Screw Machine

camshaft which transmits motion for indexing by means of a chain and sprocket gearing. The sprocket wheel A on the camshaft is directly connected by a chain with a sprocket wheel fast to the spindle head B. Sprocket wheel A is normally loose on its seat on cam C but it is engaged with the cam for indexing by means of a dog D contained in a slot in the cam. One end of this dog is arranged to engage a recess inside the hub of sprocket wheel A and the outer or projecting end is in position to be acted upon by stationary cam E. Normally the dog D is out of engagement with the sprocket wheel, but for the indexing movement, cam E throws the dog into

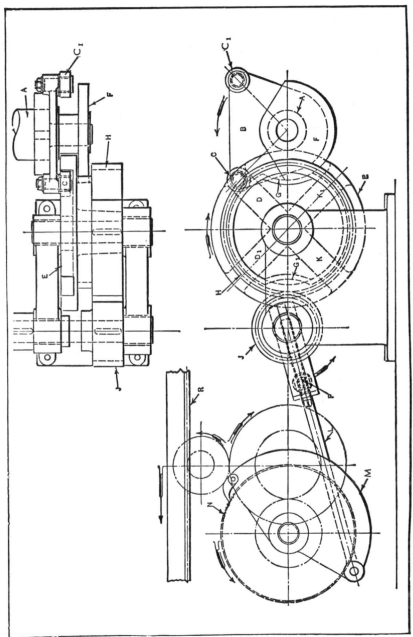

Fig. 27.   Intermittent Spacing Mechanism of a Large Perforating Press

engagement, thus revolving the sprocket and the spindle head to a new position. As there are five spindles in the head, in this case, the spindle head must be revolved one-fifth revolution at a time. After the indexing movement is completed, cam E disconnects the dog from the sprocket automatically.

As it is necessary on machines of this class to locate the spindle head very accurately each time it is indexed, some form of auxiliary locating and locking mechanism is employed. In this case, the locking bolt is at G, and is forced into its seat by the spiral spring shown. The action of the bolt is controlled by a lever H and cam F. This cam allows the bolt to drop in place as soon as the indexing motion is completed. The conical point of bolt G engages a seat of corresponding form in whichever plug K is in position. These plugs are spaced equidistantly about the periphery of the spindle head. The tapered seat for the end of bolt G is formed partly in the plug K and partly in the bushing J through which the plug passes, as indicated by the detailed view. With this arrangement, the location of the spindle head does not depend upon the closeness of the fit of the cylindrical part of the locking bolt in bushing J.

Intermittent Spacing Mechanism of Geneva Type. — The feeding or spacing mechanism to be described is used on a large perforating press. This mechanism serves to advance the plate automatically after each series of holes has been punched. It is designed to advance the plate as quickly as possible, so that the idle stroke of the punch will be reduced to a minimum. With this mechanism, the advancing or feeding movement occurs while the crank that operates the press slide makes one-fourth revolution. To accomplish this result, a modification of the Geneva movement is utilized.

The feeding mechanism is driven from shaft A (see Fig. 27) to which is attached plate B carrying rollers C and C_1. Roller C comes into engagement with groove D of the Geneva wheel when the main crank of the press is within 45 degrees of its upper position. After roller C has turned wheel E one-quarter of a revolution and as it is leaving groove D, roller

C_1 comes into engagement with groove D_1, so that the turning movement of wheel E is continued for another quarter revotion. It will be seen that while plate B makes one-half revolution, wheel E also turns the same amount; then as plate B turns another half revolution, wheel E remains stationary.

Attached to shaft A beyond plate B (see plan view) there is a circular cam or segment F which engages an arc G of corresponding radius, thus locking wheel E while rollers C and C_1 are out of engagement. During the second revolution of plate B, roller C engages groove K and roller C_1, in turn, engages K_1. During the following half turn of plate B, cam F is in engagement with arc G_1, so that Geneva wheel E is again locked. This wheel is integral with a 40-tooth spur gear H which drives a 20-tooth pinion J; consequently, for every half revolution of wheel E pinion J makes one complete turn.

On the shaft carrying pinion J there is a crank having an adjustable crankpin. This crank imparts an oscillating movement through link L to the segment-shaped part M to which is attached a pawl engaging ratchet wheel N. The intermittent motion thus imparted to the ratchet wheel is transmitted through a train of gearing to rack R, which is attached to the plate feeding table. By varying the radial position of crank P, and consequently the movement of the feed pawl, the advancement of the work can be varied as desired.

Since a quarter turn of wheel E causes one-half turn of pinion J, it will be seen that the engagement of roller C with wheel E results in the forward or feeding stroke of the ratchet wheel, whereas engagement of roller C_1 returns the feeding pawl to its starting position. This mechanism operates smoothly, because wheel E is gradually accelerated during one-eighth revolution and then gradually decelerated during one-eighth revolution as each roller successively comes into engagement with it. This is an important feature, since the primary object of this mechanism is to provide the required feeding movement in a minimum length of time.

Indexing Mechanism of Screw-slotting Machine. — The automatic machine to which the indexing mechanism shown in Fig. 28 is applied mills the screw-driver slots across the heads of screws. The screws to be slotted are placed in a slowly revolving hopper from which they are conveyed by a chute to the work-holder or turret *M*, which, in turn, locates each successive screw beneath the narrow cutter or saw, which mills the screw-driver slot. The work-holding and cutter-feeding movements are derived from a camshaft at the back

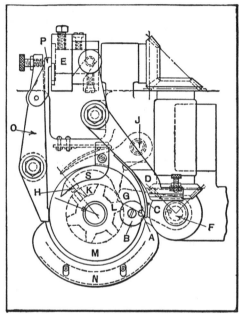

Fig. 28. Automatic Indexing Mechanism
of a Screw-slotting Machine

of the machine which is connected with the main driving pulley through change-gearing so that the rate of operation may be varied to suit the size of the work.

Fig. 28 is a plan view showing the turret operating mechanism. After a screw is released from the chute, it falls into the position shown at *A* where it is held between a seat in bushing *B* and spring *C*, which is attached to escapement lever *D*. This escapement permits the blank to fall into position in the work-holding turret and also holds the screw blank in place in bushing *B*. The lever *D* receives its motion from cam *E* on a camshaft at the rear of the machine, which is driven through change-gears. At *F* is a vertical shaft extending down through the bed of the machine, which is driven through bevel gearing from the shaft on which cam *E* is mounted. This vertical shaft carries a revolving arm *G* that strikes locking lever *H* pivoted as *J*, and raises it from the slot in disk *K* in which it is seated. All of these parts are shown by dotted lines, as they are located beneath the turret or work-holder of the machine. As arm G continues its movement, it strikes against one of the teeth of star-wheel L and revolves it one-sixth revolution. When this indexing movement has been completed, locking lever H again drops into a slot in K, thus locking the turret in position.

The turret carries six equally spaced bushings B, although only one is shown in the illustration. The slotting saw (not shown) is located on the side opposite bushing B and the screw blanks, after being placed in the work-holder at A, are indexed around to the saw by the intermittent action of the indexing mechanism, which movement occurs after each screwhead is slotted. As the screw blanks leave position A they are held loosely in place in the bushings by guard N which may be adjusted in or out to agree with the body diameter of the screw blank. As each screw arrives at the operating position beneath the slotting saw, it is held firmly against its seat in the bushings by the inner end of lever O. This lever receives its motion from the left-hand face of cam E. This cam does not bear directly against the end of the lever O, but acts through the intermediate lever P which is adjustable by means of the thumb-screw shown. It is thus possible to regulate the pressure with which O bears against the work, the adjustment being varied according to the size of the screw blank.

The slide carrying the slotting saw moves vertically and is fed downward by a cam as each successive blank is located beneath it. After the slot is milled, the saw is moved upward rapidly by a spring action, and lever O releases the slotted screw which drops through a chute and into a receptacle. If the screw blank does not release readily, the continued rotation of the turret brings it into contact with the curved edge of the ejector S. Incidentally, the bushings B are provided with a number of seats in their periphery so that, by simply turning these seats outward, the bushings are adapted for holding screws of a number of different sizes. The indexing mechanism is so arranged that any inaccuracy which may occur is in a direction lengthwise of the screw slots and not at right angles to the face of the saw, so that the centering of the slot in the head of the screw is not affected.

Semi-Automatic Indexing and Locking Mechanism. — The fixture illustrated in Fig. 29 was designed for a two-spindle

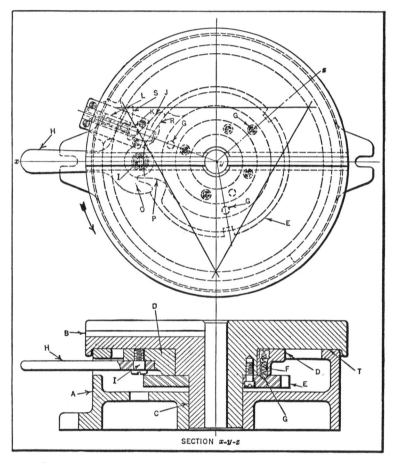

SECTION x-y-z

Fig. 29. Sectional and Plan Views of Semi-automatic Indexing Fixture

drilling machine. There is one drilling and one reaming position, and a third station directly in front of the operator for loading and unloading the work. The triangular center lines on the plan view represent these three positions. The operator has to give the indexing lever only two strokes, one forward and one return, to unlock the fixture and index the work. After indexing, the lock-bolt falls into the next notch and the fixture is locked until released by another movement of the operating lever.

The illustration does not show the work or fixtures, which are mounted on top of the rotary table. The base of the fixture is fastened to the machine table, lugs being provided at each side, as shown by the plan view. The table B is free to revolve on the base A around the central bearing C.

On the table hub, and free to rotate, is ring D to which is pivoted the indexing lever H. The index-plate E is attached to the table by screws and dowel-pins, and it has three index notches cut in its periphery. These two members D and E act as one unit during the forward stroke of the lever, owing to the fact that plunger F engages drill point spots G in the index-plate. Lever H, which is pivoted at I, projects through a slot in the base A. As this lever is pulled toward the operator to turn the table in the direction of the arrow, finger J (see plan view) comes into contact with pin K fastened in lock-bolt L, and releases this bolt from the index-plate E. Bolt L has one taper side to permit easy engagement and eliminate backlash. It is supported by the base and is backed up by a spring.

After bolt L is withdrawn to clear index-plate E, finger O contacts with surface P, which is part of ring D; then lever H and ring D act as a unit, and further movement rotates the index-plate. After the lock-bolt is released by finger J acting against pin K, this bolt rides on the periphery of the index-plate until it falls into the next notch. The motion of lever H is then reversed or pushed backward, so that finger O leaves surface P and finger J comes into contact with surface R on ring D. As this backward movement occurs, plunger F which is backed up by a spring, is lifted from spot G and slides over the surface of plate E until it reaches the next indentation. At the same time, finger S engages pin K and locks the bolt. Provision should be made for lubricating the movable parts.

CHAPTER IV

TRIPPING OR STOP MECHANISMS

THE different devices used for controlling motion may either be manually or automatically operated, and be adjustable for varying the time of disengagement, or non-adjustable so that the tripping action occurs at the same point in the cycle of operations. Tripping and disengaging devices also vary in that some operate periodically or at regular intervals, whereas others act once and then must be re-set by hand preparatory to another disengagement. The application of disengaging mechanisms varies greatly. With some classes of machinery, an automatic trip of some form is used to stop the machine completely after it has performed a certain operation or cycle of movements. On many machine tools, trips are used to disconnect a feeding movement at a predetermined point, not only to prevent the tool from feeding too far, but to make it unnecessary for the operator to watch the machine constantly, in order to avoid spoiling work. (When a feeding motion must be disconnected at a certain point within close limits, it is common practice to use some form of positive stop for locating a slide or carriage after the feeding movement has been discontinued by a trip acting through suitable mechanism.) The function of some tripping devices is to safeguard the mechanism by stopping either the entire machine or a part of it, in case there is an unusual resistance to motion, which might subject the machine to injurious strains.

The three mechanical methods of arresting motion which are most commonly employed are by means of clutches, by shifting belts, and by the disengagement of gearing. When the tripping' action is automatic, some design of clutch is generally used to disconnect the driven member from the driver or source of power. The action of the clutch is controlled in various ways. Shifting belts are not ordinarily applied to machines as a part of the regular mechanism, but are very generally used to control the starting or stopping of an entire machine; clutches are also used extensively for this purpose. Gearing which is engaged or disengaged to start or stop a driven member is used in some cases. Feeding mechanisms of some types have a worm-wheel driven by a worm which is dropped out of mesh when the feeding action is discontinued. The method of controlling motion may depend upon the speed of the driving and driven members, and the necessity of eliminating shocks in starting, or upon some other factor, such as the inertia of the driven part or the frequency with which starting and stopping is required.

How Tripping Action is Controlled. — Automatic tripping devices may be adjustable and be set beforehand to act after a certain part has moved a given distance, or they may only act when a machine begins to operate under abnormal conditions. The adjustable form of trip, if for a part having a rectilinear motion, may consist simply of a stop which is placed in such a position that it will disengage a clutch after the part under the control of the trip has moved the required distance. If a rotary motion is involved, the same principle may be applied with whatever modification of the mechanism is necessary. When the trip is designed to act automatically only when the machine is operating under adverse conditions, the action may be governed by variations of pressure or resistance to motion, or the product on which the machine is working may cause the trip to act in case the operation is not as it should be. The following examples will illustrate a few of the applications and the possibilities of tripping mechanisms of different types.

As various tripping devices are used in conjunction with reversing mechanisms to change the direction of motion, instead of stopping it entirely, Chapter VI on "Reversing Mechanisms" shows additional applications of automatic tripping appliances. Additional examples will be found in Chapter

VII on "Overload Relief Mechanisms" and "Automatic Safeguards."

Trip which Disengages a Clutch. — One of the simplest forms of automatic tripping mechanisms is illustrated diagrammatically at *A*, Fig. 1. This general type is applied to some classes of machine tools for disengaging the feeding movements of a tool-slide at a predetermined point. The tool-slide, which may be the carriage of an engine lathe, is moved along the bed by a feed-screw *a* or a splined rod which is rotated through a clutch *b*. The shifting member of this clutch is operated by a lever *c* the lower end of which connects with rod *d*. This rod extends along the bed a distance equivalent

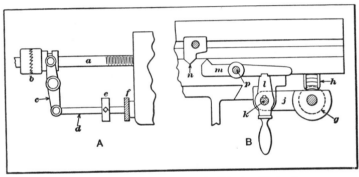

Fig. 1. Simple Forms of Automatic Tripping Mechanisms

to the carriage movement and carries an adjustable stop collar *e*, which is engaged by some projecting part *f* on the carriage; when this engagement occurs, the rod is shifted in a lengthwise direction, thus throwing the clutch out of mesh and stopping the feeding movement. Obviously, the point at which disengagement occurs depends upon the position of stop collar *e* which is set in accordance with the length of the part to be turned. There are other trip mechanisms of the clutch-shifting type which differ from the kind described in regard to the details of the mechanism for shifting the clutch.

Trip which Disengages Gearing. — Diagram *B*, Fig. 1, illustrates a form of automatic trip which serves to disengage worm gearing instead of a clutch. The worm *g* revolves worm-wheel h and the table feed-screw. This worm is carried by an arm j pivoted at k and held in position by the engagement of lever l with a notch in lever m. When the adjustable trip dog n, attached to the work table, strikes lever m and swings it about pivot p, the worm g drops out of engagement with worm-wheel h and the feeding motion stops. The point of this engagement may be varied at will by simply changing the position of the trip dog n. Some of the trip mechanisms on vertical drilling machines operate on this same general principle.

Many different designs of automatic tripping mechanisms, especially of the type used on machine tools for controlling feeding movements, are of the same principle as those described, in that trip dogs are attached either directly to the driven member or to some auxiliary mechanism such as a revolving disk geared to the driven part, and these dogs stop the feeding movement either by disengaging a clutch or gearing. If the feeding movement is intermittent and is obtained through ratchet gearing, the pawl may be prevented from engaging the gear teeth of the ratchet wheel after the latter has turned a predetermined amount. An example of this type of tripping device is described in Chapter II (see Fig. 9).

Automatic Stop for Drilling Machine. — Fig. 2 shows a side elevation and plan of an automatic stop or trip for a vertical-spindle drilling machine, which operates by disengaging a friction clutch. The feeding movement is transmitted to the spindle from the friction gear c to the disk d and through worm gearing at k to a pinion meshing with rack l attached to the spindle sleeve. The position of friction gear c is controlled by hand lever g which, through link e, lever f, shaft a, and collar b, moves the friction gear in or out of engagement with disk d. Lever g is held in the engaged position by the latch or trigger n. An adjustable stop collar h is set by means of graduations to automatically disengage the feed after a hole has been drilled to whatever depth is required. This collar acts by simply striking the end of latch n, thus releasing lever g and the friction gear c. Any wear in the friction clutch is compensated for by adjusting set-screw j in the end of connecting link e.

Duplex Automatic Tripping Mechanism. —Another form of tripping device for a vertical-spindle drilling machine is illustrated at B, Fig. 2. This stop may be set to disengage the worm e from the worm-wheel on the pinion shaft, or it may be utilized to disengage miter gear g which drives the worm-shaft. The tripping dog is attached to a bracket or

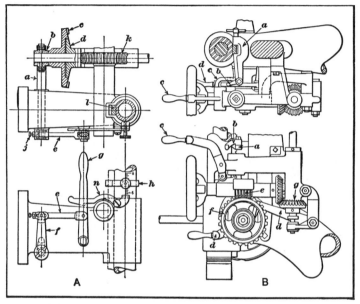

Fig. 2. Automatic Stop or Tripping Mechanisms of Vertical-spindle Drilling Machines

arm a clamped to the feed rack on the sleeve. The dog b may be swung so as to engage either levers c or d; as shown in the plan view, it is in the latter position. Lever c controls the engagement of worm e with wheel f, whereas d serves to disengage the bevel gear g. When the worm is out of mesh, the spindle may be moved vertically by the hand-feed lever, for facing or similar operations, after a hole has been drilled. Ordinarily, gear g is disengaged, but this does not leave the spindle free for rapid adjustment.

Adjustable Dial Type of Tripping Mechanism. — The automatic tripping mechanism shown in Fig. 3 was designed for drilling machines and may be adjusted to disengage the downward feeding motion of the drill at any depth up to 14 inches. The feeding movement is transmitted through the drill spindle from shaft A, through worm gearing, to shaft B which has a pinion engaging the rack on the spindle quill. The automatic disengagement of the feed is controlled by the engagement of pawl H with lever N. The distance that the spindle feeds downward before the feed is tripped is regulated by the gradu-

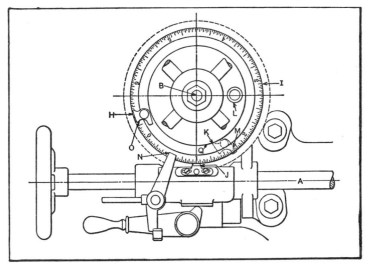

Fig. 3. Automatic Feed-tripping Mechanism having Graduated Adjusting Dial for Controlling Time of Disengagement

ated adjustable dial I. The graduations on this dial indicate 1/32 inch of the spindle travel, and one complete revolution represents 7 inches of spindle travel. The pawl H is so designed that it can be set to allow two revolutions of the dial before engaging lever N.

The operation of the mechanism is as follows: If the feed is to be tripped automatically in 7 inches or less, pawl H is set as indicated by the dotted lines at K; if it is desired to trip the feed at a distance greater than 7 inches, pawl H is turned to the position shown by the full lines. For example, if it should be required to automatically trip the feed at a depth of 3 inches, the knurled nut *L* would first be loosened and the graduated dial *I* turned until the figure 3 on it was opposite the mark on pointer *J*, after which nut *L* would be tightened. The pawl *H* would then be set in the position shown by the dotted lines, with the result that, when the drill had traveled 3 inches, the surface *M* would come into contact with the side *N* of the trip arm and disengage the feed. On the other hand, if it were required to drill to a depth of 9 inches before the feed was automatically tripped, the dial *I*

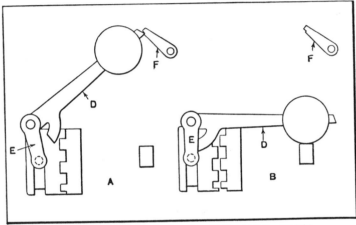

Fig. 4.   Releasing Mechanism in which a Falling Weight
Disengages a Toothed Clutch

would be set with figure 2 opposite the mark on pointer *J*, and pawl *H* would be turned to the position shown by the full lines. With the pawl in this position, the contact of surface *O* with lever *N* would not throw out the feed, as the pawl, being loose on its stud, would simply turn and pass the tripping arm without moving it. After the pawl had passed the arm, it would then be in the position shown by the dotted lines, that is, with the end in contact with a projecting sleeve, as at *K*, thus preventing further rotary movement, so that, when it again came around to the tripping lever, the feed would be disengaged. If the knurled nut *L* is loose, the feed cannot be automatically tripped at any point.

Clutch Release of Gravity Type. — The disengagement of a toothed clutch can be accurately accomplished at a predetermined point in the traverse of a sliding machine member by an arrangement similar to that shown at A and B in Fig. 4. This design enables disengagement to be accomplished so quickly that wear on the clutch teeth is reduced to a minimum. The mechanism consists of a weighted lever D connected at

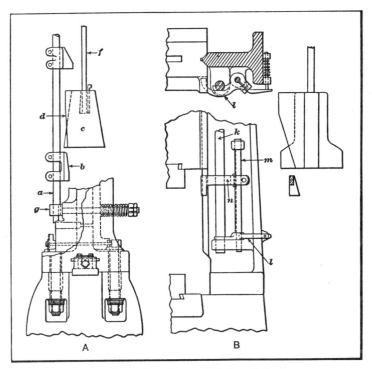

Fig. 5. Board Drop-hammer Tripping Mechanisms

one end to two links E which control the axial movement of the driven clutch member. When the clutch is engaged as shown at A, the opposite end of the weighted lever is supported by trip-lever F, but when lever F is tripped during the operation of the machine, the weighted lever falls, and thus forces the driven clutch member away from the driving member. The relative positions of the various details when the clutch is disengaged are illustrated at *B*. A similar mechanism may be used for releasing a worm from a worm-wheel.

Tripping Mechanism of Drop-hammer. — When a board type of drop-hammer has fallen and is rebounding, the friction rolls grip the board and elevate the hammer preparatory to the delivery of another blow. The eccentrically mounted gripping roll is moved inward against the board for elevating the hammer, when a "friction bar" is released by a tripping mechanism and allowed to fall. Most of these tripping devices operate on the same general principle as the design illustrated at *A* in Fig. 5. The friction bar *a* is attached at its upper end to a lever that controls the position of the eccentrically mounted friction roll; when the bar falls, the lifting board is gripped between this front roll and one at the rear that revolves in one position. Before the friction bar is released, the lower end rests upon a seat which prevents it from falling. When the hammer *c* descends, an incline surface *d* on it engages bracket *b* and pushes bar *a* off of its seat. The weight of this bar is sufficient to give the roll referred to the required gripping pressure on the board *f*, so that the hammer is lifted to the top of its stroke. As the hammer rises, it engages a lever and raises the friction bar which, in this particular case, is returned to its seat by a spring-operated guide *g*. In order to operate the hammer properly, it is necessary to release the friction bar at exactly the right time, which must be varied according to the thickness of the hammer dies.

The tripping mechanism must be so set that, as the hammer rebounds, its upward movement is continued by the action of the friction rolls. If the release of the friction bar occurs too soon, the rolls will grip the board either before the hammer strikes its blow or before it has had time to rebound. On the other hand, if the release occurs too late, the hammer will fall back after rebounding and the roll will have to pick up a "dead" or stationary load. The point of release depends upon the vertical position of bracket *b*.

A trip mechanism of the swinging latch type is shown at *B*, Fig. 5. In this case, the friction bar *k* is held in the upper

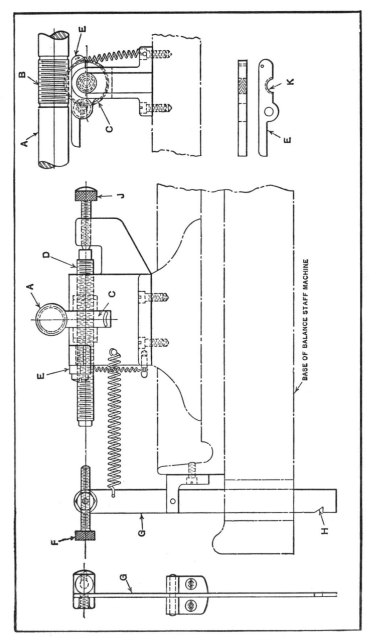

Fig. 6. Device for Automatically Stopping the Operation of a Balance-staff Machine when the End of the Stock is Reached

position by a catch l which engages a slot in the bar and is attached to the short vertical shaft m. This shaft also carries a lever n that extends out far enough to engage an inclined surface on the hammer. As the hammer descends, lever n, bar m, and catch l are turned, thus releasing the friction bar k and allowing the rolls to grip the board for elevating the hammer. The point at which release occurs may be varied by changing the vertical position of lever n.

Stop for an Automatic Machine. — An automatic stop designed to prevent the jamming of stock in the feeding chuck of a balance-staff machine when the end of the stock has been reached is shown in Fig. 6. Several automatic stops were tried out on these machines without success previous to the development of the stop shown. This stop is a modification of the type used on bench screw machines. Before proceeding with the description of the stop it may be well to mention that a balance-staff machine is used for turning the balance staffs for watches. The balance staff is the shaft on which the balance wheel of a watch is secured. At each end of this shaft is a fine pivot or tapering point. When assembled in the watch these pivots are supported by jewel bearings which are cupped out to receive the pivots.

The balance-staff machine is similar in design to a small bench screw machine. It consists of a spindle containing a chuck and a mechanism for feeding the stock forward, a camshaft with a set of cams for operating the chucking mechanism, for reversing and controlling the feed, and for operating the front and rear cross-slides. The front cross-slide carries the cutting-off milling cutter, and the rear cross-slide the milling cutters that form the pivots. It might be added that the balance staff is only rough-milled in this machine; it is afterward put into a pivot-turning machine and the pivots turned down to the required diameter and then polished. During the milling of the pivots and the cutting-off operation, the work is firmly held in a support. The spindle camshaft and the cutters in the cross-slides are all driven separately from the main driving shaft.

In applying the stop it was necessary to alter the regular camshaft A somewhat. A left-hand quadruple-thread worm having five threads per inch with a lead of 0.20 inch and a pitch of 0.050 inch was cut on the shaft at B. This worm is employed to drive the worm-gear C. When the machine is in operation the camshaft turns the worm-gear C which, in turn, revolves shaft D. A keyway is cut the entire length of the threaded part of shaft D. A key held in worm-gear C is made a sliding fit in this keyway. Now as shaft D has a left-hand thread which fits the threaded part K of piece E (see view in lower right-hand corner), shaft D will move toward the set-screw F when the machine is in operation. When lever G is pushed back by the end of shaft D coming in contact with screw F, the stopping device of the machine, which is controlled by a latch held in V-slot H, is released and the machine stopped. The time at which this releasing movement occurs may readily be varied by adjusting screw F.

The stock used in the manufacture of the balance staffs comes in three-foot lengths, and the stopping device is set to stop the machine automatically when about 2 inches of stock is left in the feeding chuck. This prevents the jamming of the stock between the cutters and enables the operator to easily take out the piece left in the chuck and to feed in a new piece of stock by lifting part E sufficiently to disengage the thread of shaft D. The latter member may be pushed back into contact with the stop-screw J, when lever G will spring back to its former position as shown in the illustration. The machine is then started by lifting up the starting latch so that it will be engaged by the V-slot H.

Stop Mechanism which Operates After Predetermined Number of Revolutions. — This mechanism was applied to a coil spring winding machine for turning a shaft a predetermined and exact number of revolutions and then stopping it instantly. To obtain the exact number of turns needed, the motion is transmitted through a multiple-disk drive arranged as shown in Fig. 7. The drive is from pulley A through friction disks B to shaft C, which rotates part D, having a hole at E in

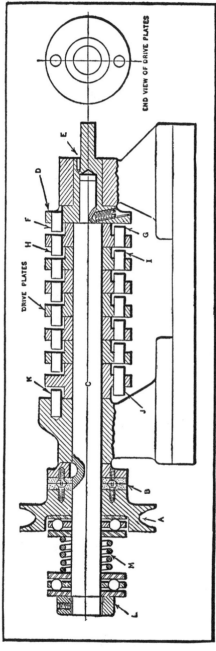

Fig. 7. Accurate Stop Mechanism Used on a Spring Coiling Machine

which one end of the wire to be coiled is inserted. In the flange of part D, there is a pin F which strikes pin G, causing pin H to turn and strike pin I. In this manner, the pin in each successive disk strikes a pin on the disk following, after whatever part of a revolution is needed to bring the successive pins into contact. Finally, pin J in the last disk engages the fixed pin K, which locks the entire combination of disks and stops shaft C immediately as the friction disks B slip, permitting the pulley to continue revolving. The coil spring is then removed, and the drive, which must be equipped with a reverse countershaft, is reversed, the disks being unwound or turned backward until they are again locked in the reverse position. It will be noted that the frictional resistance between disks *B* can be varied by means of nut *L,* which is used to regulate the pressure from spring *M* located between two thrust bearings.

Quick Stop Mechanism for Spring-winding Machine. — A quick stop mechanism which has been used successfully

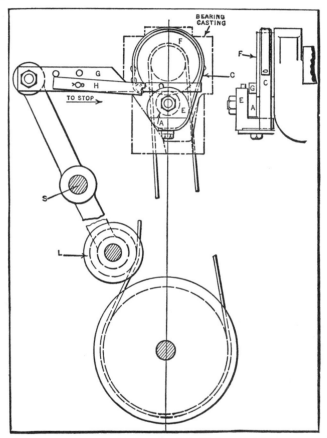

Fig. 8. Stop Mechanism which Slackens Driving Belt and Applies Brake

on a machine for winding springs has an operating shaft *S* (see Fig. 8) which withdraws idler pulley *L* and advances arm *G* on which pawl *H* is attached. This pawl pushes against a pin in eccentric *E* and turns it, thus tightening brake-band *C* on brake-drum *F*. It will be seen, therefore, that as soon as the driving belt is slackened by withdrawing pulley L, the brake is applied. Arm G is supported by an extension which rests on collar A.

When the brake-band is adjusted to the proper tension, the pawl will tighten it and then slip over the pin, allowing the brake to release and leaving the spindle free after stopping it; or if desired, the adjustment can be such that the pawl will not slide over the pin, in which case the spindle is stopped and held stationary. The spindle speed is 7,000 revolutions per minute and there is very little wear on the parts. The operating shaft S has a handle on one end, conveniently located for the operator, but this motion could, of course, be applied automatically if desired.

Device for Stopping a Machine Momentarily. — It is an easy matter to provide a means of stopping a machine with its own power by the use of a cam or some similar mechanical device, but it is a more difficult problem to design a device that will automatically stop a machine and start it again so that it is only momentarily brought to a full stop. With the usual type of device designed simply to shut off the power, there is no stored up energy that can be utilized to start the machine again without attention on the part of the operator. Fig. 9 shows a simple and automatic device for shutting off the power momentarily.

The shaft A is a part of the machine that rotates while the machine is in operation. When it is desired to stop the motive power, latch B is moved to the left, pivoting about pin P. The machine is started again by moving latch B back to the right-hand position. The problem is to provide a means of throwing latch B to the left or stopping position and then back again to the running position without attention on the part of the operator. The latch can be readily moved to the left by means of a cam, such as shown at C, but the instant that the cam has so acted the machine will stop, and the cam will be left in such a position that the lever cannot be moved back to the starting position, either through the action of a spring, such as shown at D, or by other means.

To overcome this difficulty, shaft A was provided with a flange member E secured by the pin F. The cam member C is free to turn on shaft A, but its movement relative to flange E is limited by two pins G and H, which are driven into flange E and allowed to project through the slots J and K in the cam member. On the cam member is a stud L to which is secured one end of a spring M, the other end being secured to some part of the machine.

When shaft A travels in the direction indicated by the arrow, the flange E will drive the cam member C in the same

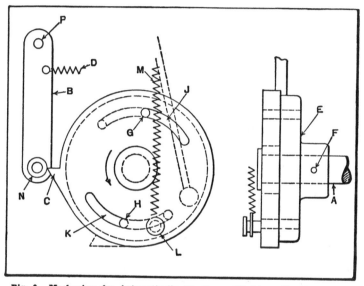

Fig. 9. Mechanism for Automatically Starting a Machine after Stopping it Momentarily

direction. As the pin L reaches its highest point and begins its descent, it will act against the pressure exerted by spring M until the instant that pin L passes a vertical line through the center of the shaft. At this point spring M will accelerate the movement of the cam member and not only push lever B to the left, thus bringing the machine to a stop, but will also carry the point of the cam past the roller N, so that lever B will be returned to its original position by the action of spring

D. The machine will thus be started automatically after having been stopped for an instant.

Mechanism which Stops Rapid-moving Slide at Top of Stroke. — A stencil-cutting machine used for cutting odd-shaped openings in stencils of various fibrous materials has a fixed knife *A* (see Fig. 10) and an upper knife *B* which is given a rapid reciprocating motion. The stop mechanism to be described is so arranged that when a foot-lever is depressed, the cutter-slide is disconnected from the driving member, which allows the cutter-slide to stop instantly and always at the top of its stroke, thus permitting the work to be moved either away from the knife or from one opening to another.

The machine is driven by a constant-speed motor through a belt connecting with a pulley on the rear end of shaft *C*. The foot-lever which controls the action of the cutter-slide is pivoted in the machine base and connects with the rear end of lever *D*. The drive to the cutter-slide is through two hinged fingers *E* and *F* (see end view which shows cover plate removed). These clutch fingers are pivoted to a cross-head which is given a rapid reciprocating movement by crank-pin *G*. The foot-lever normally is held in its upper position by a spring (not shown). When the foot-lever is depressed, clutch fingers *E* and *F* are forced out of engagement with angular notches in the cutter-slide, and as soon as the latter is released, it is pulled to the top of its stroke by spring *H*. (The spring *H* also takes up any lost motion that there might be in the reciprocating cross-head or connections.) The cutter-slide remains in this upper position until the clutch fingers on the reciprocating cross-head are released so that they are again free to engage the angular notches.

The mechanism for controlling the disengagement and the engagement of the clutch fingers operates as follows: Lever *D* is pivoted at *J* so that its forward end moves upward as the foot-lever is depressed. The round end of lever *D* connects with and transmits this upward movement to bar *K* and to cross-arm *L* (see end view). Cross-arm *L* is secured to bar *K* by a set-screw, which allows the cross-arm to be ad-

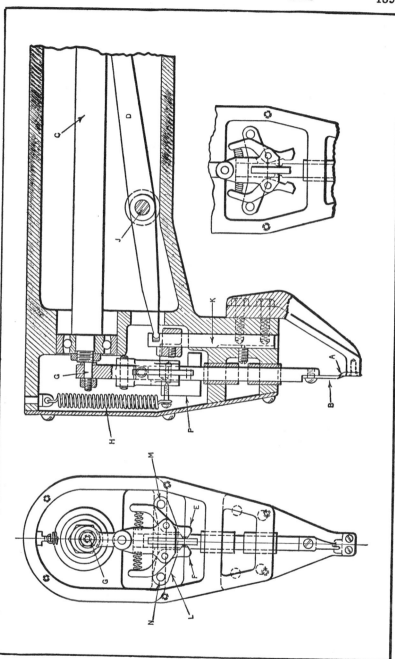

Fig. 10. Mechanism for Quickly Stopping Cutter-slide of Stencil-cutting Machine, at the Top of its Stroke

justed up or down to accommodate any variation in any other part of the machine. This cross-arm has pins N and M, which act against the upper ends of the clutch fingers, compressing the springs shown and releasing the fingers, as indicated by the detail view in the lower right-hand corner of the illustration.

When the foot-lever is released, it is pulled upward by its spring, which, through the mechanism described, moves cross-arm L downward, and engages a tongue on the cutter-bar, forcing it down and thus allowing the springs to close the clutch fingers upon the cutter-bar so that they can engage the angular notches. A little clearance is provided in the bottom of each notch to allow for any wear on the tooth or the notch. The upper end of the cutter-bar fits into a hole in the cross-head. The cutter-bar also has a tongue P which serves as a guide for the upper end, the cross-head being slotted to receive this tongue, which also acts as a stop when the clutch fingers are released.

With this mechanism, starting and stopping is almost instantaneous, with the main drive shaft of the machine running at high speed. The knives may be formed to various curved shapes to facilitate cutting any fancy scroll work or small round holes. This machine, incidentally, is used in cutting leather gaskets, design stencils, and various other articles not required in large enough quantities to warrant the cost of a punch and die. The machine may also be used to advantage in trimming and cutting odd-shaped patterns and uneven edges of leather goods after the stitching operation. The upper knife has a sharp point or pilot on the lower end for piercing the work when starting the cut.

Clutches that Automatically Disengage. — The clutches used on power presses are designed to automatically disengage after making one or more revolutions. The clutch connects the flywheel or driving gear of the press with the driven shaft, whenever it is tripped, by pressing down a foot-treadle. As long as this treadle is held down, the clutch remains in engagement and the press continues to run; if the treadle is released, the clutch is disengaged when the ram or slide of the press is approximately at the top of its stroke. The downward movement of the treadle releases a pin, key, or some other form of locking device which quickly engages the driving member; when the treadle is released, the locking device encounters some form of trip or cam surface which withdraws it and stops the press. There are many designs of clutches of this general type.

Automatic Clutches of the Key Type. — Fig. 11 shows a clutch of the type having a key which is engaged or disengaged with the hub A of the flywheel. This flywheel revolves freely on the shaft until the dog D is pulled down by the action of the foot-treadle; then the key C is forced downward into engagement with the flywheel by a strong steel spring E. When

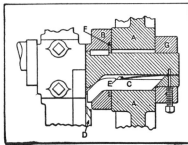

Fig. 11. Automatic Clutch of Shifting-key Type

The foot-treadle is released, the dog D is forced up, and when key C comes into contact with the dog, it is pushed back into the shaft, thus allowing the flywheel to again run freely. If the treadle is depressed and then released, the press will make one revolution before stopping, but if the treadle is held downward, the press will continue to run. This clutch is equipped with a safety device to prevent the ram or slide of the press from descending unexpectedly while setting dies or making adjustments. This safety device consists of a steel ring F having a keyway or slot in it for receiving the key C. When the press slide is at the top of its stroke and dog D is up, the key is entirely within the shaft and may be held in this position by turning ring F, thus preventing accidental engagement of the clutch. Ring F has an extension arm that enables it to be turned readily.

A clutch is shown in Fig. 12 that has a rocking key instead of one that moves radially. This key A extends across the shaft and, when the press is not in motion, the key rests in a semi-circular seat and occupies the position shown in the end view. When in this position, the lever B at one end of the key is in engagement with the latch C, which is connected with the foot-treadle. As soon as latch C is swung out of the way by depressing the treadle, lever B and the key tend to turn as they are acted upon by the compressed spring E. When the flywheel has turned far enough to bring one of the recesses F opposite the key, the latter, by making a quarter turn in its seat, engages the recess and locks the flywheel and

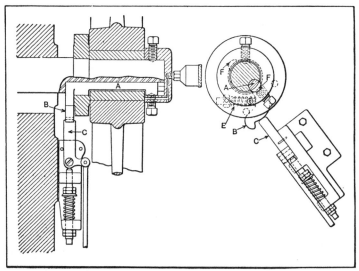

Fig. 12.   Automatic Clutch of Turning-key Type

shaft together. If the treadle is immediately released, thus allowing latch C to swing back to the vertical position, it will engage lever B when it comes around and force this lever and the key back out of engagement with the flywheel.

Clutches Engaged by a Wedging Action. — Some designs of automatic clutches are engaged by a wedging action of some locking member between cam or eccentric surfaces, instead of employing pins or keys. An example of the cam type of clutch is shown at A. Fig. 13. A cam a, having a series of eccentric or cam surfaces, is keyed to the crankshaft, and surrounding this cam there is a slotted ring c containing rollers *b,* which, in turn, are surrounded by a hardened tool steel ring *d.* These parts are inserted in a recess formed in the hub of the flywheel. On the slotted ring *c,* there is a lug *f* which is in engagement with the pivoted stop lever *e* when the press is not in operation. As soon as the stop levei is drawn downward by means of the foot pedal, the rollers are carried around by the action of the flywheel until they are wedged tightly between the cam surfaces and the outer ring *d;* the crankshaft is then driven with the flywheel and continues to revolve until lever *e* is released and, by striking

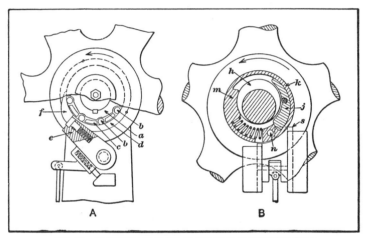

Fig. 13. Automatic Clutches of the Cam or Wedging Type

stop *f,* throws the rollers out of engagement. The slotted ring *c* has a spring attached to it (not shown) which turns the ring and rollers toward the high points of the cam when the ring is released by the lowering of lever *e.*

The design of clutch illustrated at *B* is equipped with an eccentric *h* which is solid with the crankshaft and a wedge-shaped member *j* which serves to lock the flywheel and crank-shaft together. This wedge *j* is located between the eccentric and a ring *k* inserted in a recess in the hub of the flywheel. The ring is split and compressed somewhat so as to exert a pressure against the wall of the recess. When the stop *s* is in engagement with pin *n,* the flywheel simply revolves about the expansion ring *k*. When stop *s* is withdrawn, the ring *k* expands, and, as it begins to revolve with the flywheel, the wedge *j* is forced between the eccentric *h* and the inside of the ring; consequently, the flywheel, expansion ring, and the shaft are firmly locked together. When the foot pedal is released and stop *s* engages pin *n*, the ring contracts and remains stationary while the flywheel continues to revolve. The surface at *m* serves as a brake, so that the crank-shaft is stopped when the slide is approximately at the top of its stroke.

Clutch of One-revolution Coiled-spring Type. — A spring clutch designed to disengage automatically at the end of one

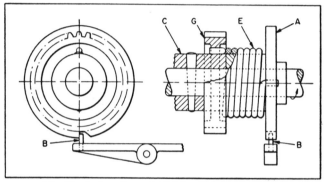

Fig. 14. Spring Type of Clutch which Automatically
Disengages After Making One Revolution

revolution of the driving shaft is shown in Fig. 14. Wheel *A* is free to rotate on the driving shaft; gear *G* is free to rotate on bushing *C* which is fastened to the drive shaft; gear *G* and wheel *A* are attached to opposite ends of spring *E*; and spring *E* is a snug fit on bushing *C* but is not fastened to it.

With the drive shaft rotating, gear *G* will not move as long as wheel *A* is held from rotating by dog *B*; but when the dog is withdrawn momentarily, wheel *A* is dragged around by the spring which now grips bushing *C* firmly and hence rotates with it. Since gear *G* is attached to the spring, it too will rotate at the same speed as the bushing and drive shaft. When the rotation of wheel *A* is stopped by the re-engagement of dog *B*, the spring will open slightly, slip on the bushing, and no longer drive gear *G*.

By making wheel A large and heavy, like a flywheel, the clutch will take hold gradually and with a smooth action. When the clutch is disengaged, the driving spring will be opened slightly, due to the rotating action against the stationary wheel. When wheel A is released, it rotates forward a slight amount and causes the spring to grip the drum. As the driven members of the clutch start to rotate, their motion will be resisted by the inertia of the heavy wheel, which, in turn, prevents the clutch from engaging quickly.

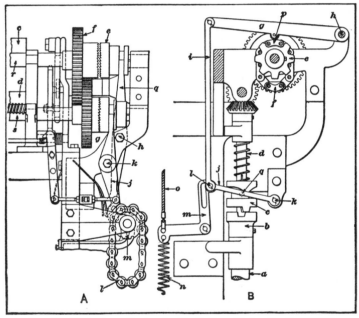

Fig. 15. Mechanisms Equipped with Endless Chains for Controlling Engagement and Disengagement of Clutches

Variable Clutch Control by "Pattern Chain." — The ingenious method of controlling clutches illustrated at A in Fig. 15 is applied to a textile machine known as a "twister." The variations in the yarn are obtained by controlling the action of two sets of delivery rolls. The lower rolls r and s of each set support the upper rolls c and d. Splined to the end of roll r is a shifting clutch member e which revolves the roll when engaged with the clutch teeth on the hub of gear *f*. A similar clutch and gear combination is located at *g* for driving the lower set of delivery rolls. The upper clutch is connected with lever *q* pivoted at *h*, and the lower clutch, with lever *j* pivoted at *k*. The action of these clutch levers is governed by a pattern chain *l* suspended on a drum *m*. As this drum revolves, the rollers of the pattern chain come into engagement with the lower ends of the clutch levers, thus shifting the clutches in and out of engagement. By changing the position of the rolls or risers of the pattern chain, the pattern of the yarn may be varied and different fancy effects be obtained. The chain drum is revolved by means of change gearing for varying the speed according to requirements. The clutch gears are rotated continuously, and the delivery rolls are only stopped when a knob or knot is being formed, both sets of rolls being rotated while the yarns are being twisted together between the knots.

Another application of an endless chain for controlling the engagement and disengagement of a clutch at predetermined intervals is illustrated at *B*, Fig. 15. This mechanism is applied to a loom. The vertical shaft *a* is driven through bevel gearing (not shown) at the lower end, from the driving shaft of the loom. The upper end of shaft *a* carries a clutch member *b*, which is engaged by the shifting clutch member *c* splined to shaft *d*. Shaft *d*, through the bevel and spur gearing shown, is connected with the chain drum or cylinder *e* carrying the clutch controlling chain *f*. Above this chain, there is a lever *g* pivoted at *h* and connected by link *i* with another lever *j* pivoted at *k*. The pin *l* connecting the link and lever engages a slot in bellcrank *m*, the movements of which are controlled by a spring *n* and a connector *o* which extends to another part of the machine. The vertical slot in lever *m* has a short horizontal section at the upper end.

The action of the mechanism is as follows: When the clutch members are engaged, the chain drum and chain revolve, and when one of the links *p* engages lever *g*, the lower lever *j* is raised, thus locating pin *q* in the upper part of the annular groove of the shifting clutch member. As soon as pin *l* at the end of lever *j* reaches the upper end of the vertical slot, the bellcrank lever *m* swings over under the action of spring *n*, thus engaging pin *l* with the horizontal part of the slot and locking the lever *j* in the upper position. As soon as the lever *j* is raised, a projection engages pin *q* and disconnects the clutch, thus stopping the rotation of shaft *d*. The link *p* on the pattern chain is no longer under the roller of lever *g*, but this lever is still held in the upper position, by the engagement of pin *l* with the horizontal slot in bellcrank lever *m*. The clutch remains disengaged until the connector *o* swings the vertical part of lever *m* to the right, thus allowing

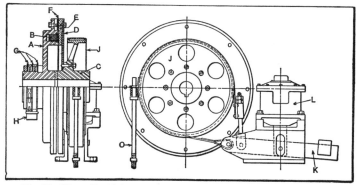

Fig. 16. Magnetic Clutch Equipped with Automatic Band Brake which
Operates when Clutch Releases

the upper clutch member *c* to re-engage the lower part. The movements of the connector are controlled by another chain which operates on the same general principle as the one referred to.

Magnetic Clutch with Automatic Band Brake. — The magnetic clutch illustrated in Fig. 16 is equipped with an electrically operated brake which acts automatically when the clutch is released, in order to stop the driven part as quickly as possible. The driving shaft carries the field *A* which is provided with a magnetizing coil *B*. The hub *C* on the driven shaft has attached to it a flexible spring-steel disk or plate *D*. This plate carries the armature *E* which is prevented from coming directly into contact with the magnetizing coil by a ring of frictional material at F. This friction ring, which is made of woven asbestos and brass wire, provides a frictional surface for driving. The ends of the winding of the magnetizing coil are attached to the rings G which are in contact with a pair of brushes H connected with the electrical circuit. The automatic brake, which is of the band type, engages drum J, and the ends of the band are pivoted to lever K at two points as shown. The plunger of a solenoid enclosed in cylinder L is attached to lever K, and at the outer end of this lever there is a weight which serves to apply the brake when the clutch is disengaged.

In the operation of this clutch, the current is gradually admitted to the magnetizing coil by means of a rheostat. The magnetic attraction between this coil and the armature causes the friction ring F to be held firmly against the driving member, so that motion is transmitted between the driving and driven shafts. The solenoid is also energized so that lever K is pulled upward and the band brake about drum J released. This brake is held in circular form and out of contact with the drum by a spring and rod O. As soon as the circuit is broken, the clutch is released, and the solenoid allows the weighted lever K to fall, thus supplying the brake automatically to the driven part. This feature is of particular advantage when the driven side of the clutch is connected to some part which tends to revolve quite a long time after disengagement.

Multiple-disk Clutch Equipped with Brake. — It is sometimes necessary to start and stop machines or certain parts of machines smoothly, with great rapidity, and in synchronism with other moving parts. With light or slow moving apparatus, the problem is relatively simple, but the difficulties multiply as weight and speed are increased. The clutch mechanism shown in Fig. 17 has proved very efficient for the class of service mentioned. This design of clutch is used on a machine transmitting a load of 20 horsepower and operating about 3600 times per day, under unusually trying conditions

This machine picks up its load from dead rest, makes three revolutions and comes to rest again in three-fifths second, or at an average rate of 300 revolutions per minute, without the slightest shock or effort. When it is considered that the clutch drum is driven at only 340 revolutions per minute, and the engagement is only a fraction of a second, it will be seen that the slip is very slight indeed. The absence of shock may be attributed to the perfect cushioning of the pressure applied to the clutch and to the liberal friction area provided,

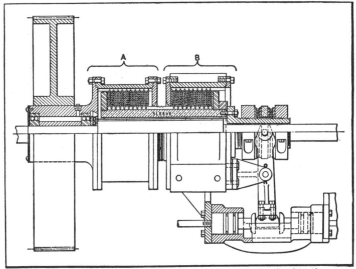

Fig. 17. Quick-acting Multiple-disk Clutch and Brake Combination

there being nearly a square inch for each pound of pull at the average radius of the disks.

The device consists essentially of two multiple-disk friction clutches of the dry type mounted tandem on a single sleeve which is fitted to slide, but not to turn, on a shaft that is directly coupled to the intermittent load. The driving clutch is shown at *A* and the brake clutch, at *B*. The two clutches are built up in the usual form for disk clutches, that is, with two alternate series of disks, one keyed to the driving member and the other to the driven member; one set is preferably faced with friction fabric. One series of disks in a set is provided with internal projections to engage longitudinal slots on the sleeve, while the other disks have external projections loosely fitting the internal slots of the driving and braking clutch drums. The projecting lugs on the disks are reinforced to provide greater bearing surface on the sides of the slots in which they travel. As both clutches are mounted on the same sleeve, and the outer part of the driving clutch is continuously driven, the sleeve becomes the driven member of the driving clutch and the driving member of the brake clutch. The driven member of the brake clutch is solidly bolted to the frame of the machine of which the clutch constitutes a part, so that, in reality, it is not driven, but acts as a brake to bring the sleeve to rest when this clutch is engaged.

Both the clutch drums are built in skeleton form to facilitate the egress of material wearing off the friction facings, and to permit of the easy application of castor oil to the facings. If this treatment is not neglected, a set of facings may last two years or more in constant service, but, if the facings are allowed to become entirely dry, they will be less durable. The sleeve is provided with a flange on each end so that, when it is moved endwise, the disks of one of the clutches will be clamped between one of the sleeve flanges and the head of one of the clutch drums, while the pressure on the disks of the other clutch will be released. Movement of the sleeve in the opposite direction will release the disks of the first clutch and clamp those of the second. In the illustration, the parts are shown in the position of rest, or with the driving clutch disengaged and the brake clutch set.

The controlling mechanism is operated pneumatically and may be made automatic by connecting with other moving parts to actuate the valves. The actual movement of the sleeve which engages and disengages the clutches is derived from two opposed pneumatic cylinders and the connections shown. It will be apparent that the cylinders must work alternately, that is, when one is under pressure the other must be open or free to exhaust. The distribution of air is controlled by two valves, together with a series of interconnecting pipes. With the valves in the "up" position, compressed air is free to pass through the pipe to one of the cylinders and to the top of the other valve for forcing it down, cutting off the air supply of the cylinder it serves, and opening it to exhaust.

A small hole near the live-air inlet leads to the annular space below the valve proper, around the stem, and is open continuously, admitting air to hold the valve in the "up" position when so placed. As the only connections between the controlling valves and the cylinders are pipes, the control may be somewhat remote and placed in any convenient position. Experiments have been made to determine the practicability of operating the valves magnetically, and also of moving the clutch sleeve by means of magnets, but both have been found far less efficient and much slower than air, the slowness of the electrical operation being due to the time required for the magnets to "build up." The drift of the shaft after the operation of the stopping valve has been found to be very small and practically constant, the shaft stopping within a few degrees of the same position every time. Any wear of the friction disks or their facings is automatically compensated for by additional travel of the pneumatic pistons, so that mechanical adjustments are rarely required.

CHAPTER V

ELECTRICAL TRIPPING MECHANISMS

WHEN an automatic machine must be safeguarded in some way, electrical tripping mechanisms are sometimes preferable to purely mechanical devices. For some purposes a mechanical tripping mechanism would be so complicated as compared to an electrical device as to be inferior to the former, if not entirely impracticable. Another advantage of the electrical devices lies in the fact that they may in some cases, be used as a check on the accuracy of preceding operations and thus avoid finishing pieces of work that are defective. The application of electricity to automatic mechines may be regarded as a complication in itself, but this is far from being true if the tripping devices are properly applied.

The following examples are typical applications of electromagnetic tripping devices to automatic machines, and by studying these designs, one may readily understand how similar tripping mechanisms could be applied to other classes of machinery. In most of these examples, the tripping devices constitute part of attachments for standard machines that were converted into "automatics."

Methods of Closing Electromagnetic Circuit. — A metallic cartridge shell is shown in Fig. 1 (view to right) in place on a machine which pierces the primer hole, and a shell is illustrated at H on which the piercing operation has been performed. After the hole has been pierced, the primer is inserted in the primer cavity J. These operations are performed on a standard Waterbury-Farrel cartridge primer. The shells were formerly placed on dial pins by hand and indexed under the cross-head for piercing and inserting the primer; they were then removed from the dial pins automatically. An improvement was made in the method of operation by applying an

148

automatic feed mechanism to place the shells on the dial pins, but this did not dispense with the necessity of an operator for each machine, as there are three possible conditions that may result in the production of imperfect work: 1. The feed mechanism might fail to deliver the shell to the dial pin, or the supply of shells might become exhausted, while primers would continue to feed and thus be wasted. 2. The piercing punch might break and the machine would then continue to place primers in the cavities of shells which had not been pierced, and such shells would obviously be useless. 3. The primer

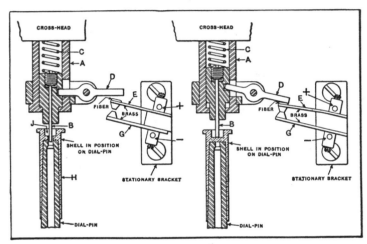

Fig. 1. Mechanism for Closing Circuit and Stopping Machine in Case Punch Fails to Pierce the Shell

feed might fail to work properly, or the supply of primers might become exhausted.

The application of a suitable electromagnetic tripping mechanism to this machine takes care of all of these contingencies. First, consider the possibility of the feed mechanism failing to deliver a shell to the dial pin. Referring to diagram *A*, Fig. 2, it will be seen that the shells are carried on pins on the dial and are indexed under the punch *a*. If a shell is in its place on the dial pin, it contracts the spring *b* when the ram descends, but should the mechanism fail to deliver a shell to the pin, the sleeve *g* passes down over the dial pin and pushes the upper contact d of the tripping mechanism down upon the lower contact e. This closes the electrical circuit and stops the cross-head on the upstroke. The contacts are fastened to the frame of the machine and the method by which the tripping mechanism operates will be described in detail later.

The way in which the piercing operation is safeguarded by the electromagnetic tripping mechanism is illustrated in Fig. 1. The punch-holder A is located at the index point immediately after the completion of the piercing operation. If a shell is pierced, the pin B descends through the hole in the shell, as shown at the left of the illustration; but if the pierc-

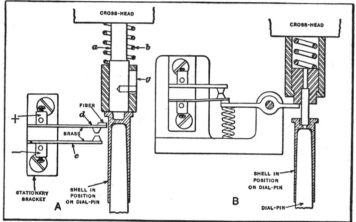

Fig. 2. Circuit-closing Device which Acts when Shell has not been Placed on Dial Pin; (B) Trip which Prevents Passing a Shell without a Primer

ing operation does not take place, the punch is held in the position indicated in the right-hand illustration, thus contracting the light spring C and throwing the lever D against the contact E. This closes the electrical circuit and causes the machine to be stopped so that shells cannot have primers inserted in them when the primer hole has not been properly pierced.

The failure of the machine to feed a primer into the primer cavity of the shell is guarded against by the mechanism illustrated at B in Fig. 2. The design of this tripping mechanism is practically the same as that used to control the piercing operation, and will be readily understood without further description.

Fig. 3 illustrates the mechanism used on a press for assembling the brass cups A and B, the cup A being inserted in the cup B. These cups are held in hoppers on each side of the machine from which they are taken by notched dials. The

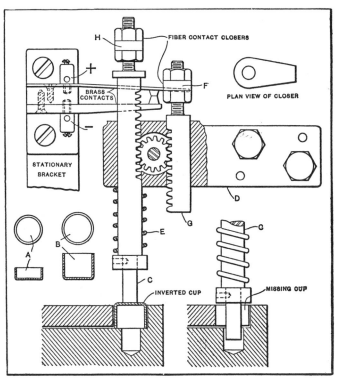

Fig. 3. Circuit-closing Device Used on Machine for Assembling Shells A and B

cups A are dropped into holes in the machine dial which passes over the dial carrying the cups B. The operation of the machine will not be described; it should be mentioned, however, that a plunger descends in such a manner that the cup A is forced into place in cup B. Several conditions may occur that will result in loss or damage. The feed mechanism could fail to deliver either one or both cups to their respective dials, or it could deliver them to the dials in an inverted position. Either the absence or inversion of either or both cups is detected by an electromagnetic tripping device which automatically stops the machine until the error has been corrected. The punch C is located at an index point preceding the assembling punch, and is carried by a bracket which is fastened to the cross-head. In the case of an inverted cup, the punch C is held on the bottom of the cup and pulls the rod G down through the action of the pinion, which engages with rack

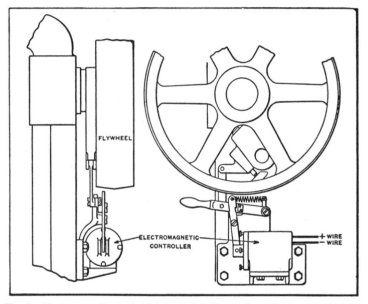

Fig. 4. Electromagnetic Controller Applied to Power Press for Operating Clutch

teeth cut in the rods C and G. The descent of the rod G causes the contact closer F to pull down the upper electrical contact until it closes the circuit and causes the machine to be stopped.

The detail view at the right shows the punch and die when the feed mechanism has failed to deliver a cup to the dial plate. In this case, the upper electrical contact is pulled down by the contact closer H and causes the machine to be stopped as previously described.

Electromagnetic Controller. — Fig 4 shows the electromagnetic tripping device used on the machines referred to in the foregoing, for stopping the machine. In this illustration, the tripping device is shown in place on a power press equipped with a Horton clutch. The arrangement of the tripping mechanism will be more readily understood by referring to Fig. 5, which shows an end and cross-sectional view. This tripping mechanism is self-contained and can be applied to any style of press or type of machine. The bracket *A* carries the magnet *B*, pole-piece *C*, and levers *D* and *E*. The brass pole *G* is wound with No. 14 double-covered wire and the con-

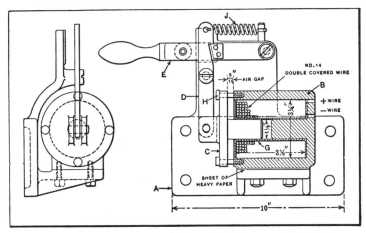

Fig. 5. End and Cross-sectional Views of Electromagnetic Controller

necting wires extend through the back of the spool. The brass pins *H* help to support the pole-piece *C* and provide adjustment for different widths of air gap, which should be as small as possible.

In order to start the press, the lever *E* is pulled down. This engages the flywheel clutch (see Fig. 4) and allows the spring *J* to pull the lever *D* over the hardened knife-edge, thus setting the pole-piece *C* at the proper working distance from the magnet. The inside dimensions of the device are given in Fig. 5. When the magnet is energized by two dry cells, it gives an initial pull of from twelve to fifteen pounds. As the dry cells are used on open-circuit—except for the fractional part of a second during which the contacts meet—they have a long life.

The initial pull provided by an electromagnet of this kind varies with the material used for the magnet and the pole-piece. Where cast iron is used, the pull of the magnet can be calculated by the formula:

$$NI = 3000 \, Z \sqrt{P \div D},$$

in which

$N =$ number of coils of wire on the spool (ampere-turns);

$I =$ current in amperes;

$P =$ pull in pounds;

$Z =$ air gap in inches;

$D =$ diameter of plunger in inches.

The electromagnet shown in Fig. 5 was designed to give a pull of 15 pounds, and it will be seen that $Z = 5/16$ inch and $D = 1.125$ inch.   Then:

$$NI = 3000 \times 5/16 \sqrt{15 \div 1.125} = 3423.19 \text{ ampere-turns.}$$

Assuming that there are 375 turns of wire on a spool, the amount of current required will be found to be $\dfrac{3423.19}{375} = 9.14$ or, say, 10 amperes.   Two good dry cells connected in series will average 15 amperes during their useful life and give a considerably higher current when new.   As 10 amperes is sufficient to enable the electromagnet to do the work required of it, it will be seen that an ample factor of safety is provided.   When designing devices of this kind, moving wires and moving contacts should be avoided and the mechanism should be made as simple as possible.   The dry cells should be used on open circuit, the contacts carefully insulated from the machine, and covers provided for contacts and terminals.

Electric Stop for Drawing Presses. — The automatic electric stop to be described was designed for use on roll-feed double-action presses employed in the production of screw shells such as are used on incandescent electric lamp bulbs

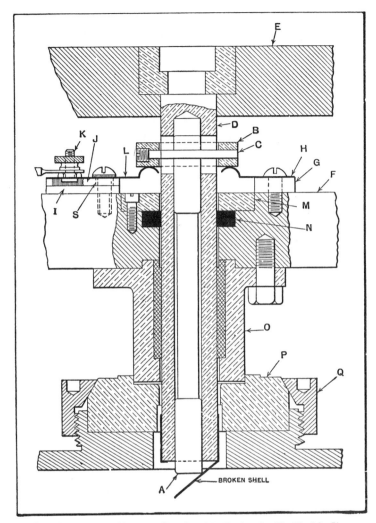

Fig. 6. Shell Blanking and Drawing Die Equipped with Electric Stop

and sockets. Each press was equipped with a gang or mul·
tiple die designed to blank and draw five shells at one stroke.
In drawing the shells from 0.010-, 0.008- and 0.006-inch
metal it occasionally happened that the bottom of a shell
would be pushed through, as shown in Fig. 6. If the press
was not stopped immediately when this occurred, the scrap

shells would pile up on the die, frequently stopping the press. In such cases it was necessary to employ a long lever to release the dies. The dies were often broken or otherwise made unfit for future work as a result of this treatment.

The electric stop eliminated this trouble by automatically stopping the press whenever the bottom of a shell pushed through, thus making it possible for one operator to attend four presses, where it had previously been necessary to have an operator for each press. An electric bell was also placed on the presses so that the operator's attention would be called to a machine as soon as it stopped. The construction of the automatic stop is as follows: A collar B, Fig. 6, is attached to each of the gang punches, and held to the push-pin A by a threaded dowel-pin C. This collar is designed to make contact between the brass springs H and L when the bottom of the shell is pushed through. Spring H is fastened to a brass block G, which is grounded on the press through the cutting punch-holder F. Spring L is prevented from grounding on the cutting punch-holder by the fiber blocks J and I. In order to prevent the screw that holds blocks J and I in place from making a short circuit between the punch-holder F and the contact L, an insulating bushing S was made of fiber and the screw inserted through this bushing. At the outer end of spring L is placed a terminal post K to which is fastened a wire. This wire, in turn, is connected in series with contacts L on the other four punches of the same press as indicated in Fig. 9.

Each press is also equipped with a make-and-break device, such as shown in Fig. 7. This device is provided with a plunger-holder A fastened to the blanking slide by the bracket C. When the slide is down within 1/16 inch of the bottom of the stroke, the spring-actuated plunger G makes contact with the brass plug D which is held in the fiber insulating block E. The fiber block is fastened to the frame of the press by the bracket F. By completing the circuit in this way, the electrically operated latch shown in Fig. 8 can be tripped only when the press is within 1/16 inch of the bottom of the stroke.

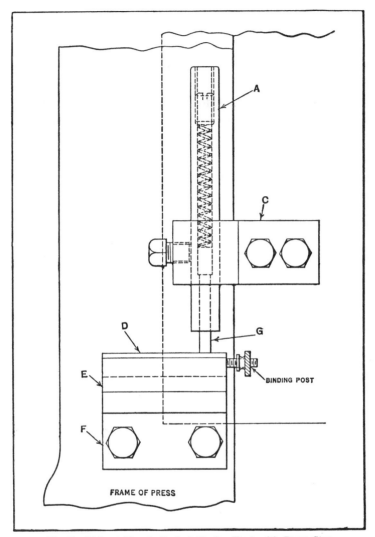

Fig. 7. Make-and-break Contact Device Used with Press Stop

The plunger-holder *A* is adjustable, however, so that this distance can be decreased or increased, as the nature of the work may require. The rod that connects the clutch key and treadle of the press was replaced by a tripping mechanism. The clutch trip lever *U* is connected to the lever *W* by rod *V*. The treadle is also connected to lever W by rod X. When the treadle is pressed down by the operator's foot in starting the press, lever W is caught and held under roll Y on the electric latch. The clutch trip lever U is thus drawn down so as to clear the clutch key. When the latch is released, the levers fly up into the position shown by the dotted lines. The clutch trip thus engages the key and stops the press. If it is desired to stop

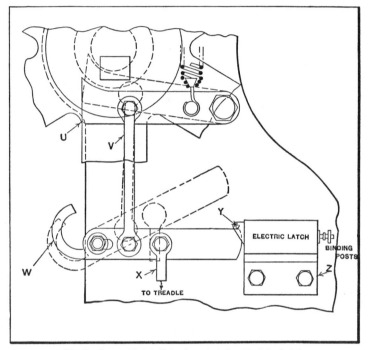

Fig. 8. Press Clutch Trip Operated by Electric Latch

the press without releasing the electric latch, all that is required is a slight pull on the hooked end of lever W which has an elongated hole in it that permits a lengthwise movement of about 3/8 inch.

The operation of the stop is as follows: Referring to Fig. 6, the metal is fed under punch O which cuts the blank and holds it in the drawing die P; drawing punch D then descends and draws the shell. If the bottom of the shell is not pushed out, the press continues to operate. The make-and-break device shown in Fig. 7 makes a contact when the blanking slide is down within 1/16 inch of the bottom of its stroke and continues to maintain this contact until the slide is moved upward about 1/16 inch on the return stroke. The press is so timed that although the drawing slide has a longer stroke, both slides start upward together and maintain the same speed for about 1/4 inch of the upward stroke.

It is clear, then, that if the bottom of the drawn shell is not pushed out, the push-pin A, Fig. 6, will be held flush with the

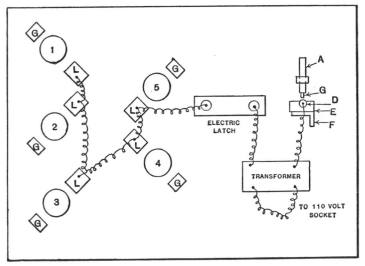

Fig. 9.   Wiring Diagram of Electric Stop System on Gang Punch

bottom of the drawing punch. Pin A must project from the punch end about 1/8 inch in order for collar B to make contact with springs L and H. As the drawing slide returns, the shell is stripped from the drawing punch by the bottom of the die P, and before the push-pin can drop so that it projects 1/8 inch below the punch, the contact made by the make-and-break device is broken. It requires a movement of only 1/16 inch to break the contact so that it is impossible to trip the press automatically until contact is made again at the end of the next downward stroke.

If the bottom of the shell is pushed out on the drawing stroke, push-pin A is permitted to drop and the collar B to make contact with the springs H and L, which results in releasing the electric latch Y shown in Fig. 8. When the electric latch is released, the tripping levers on the press take the position shown by the dotted lines, causing the clutch trip to throw out the clutch and stop the press before another revolution is made. Some difficulty was experienced at first from the lubricant used on the metal being forced into the contacts and causing short circuits. A rubber washer N, Fig. 6, placed in the punch-block F prevented the lubricant from reaching the contact points and thus eliminated this trouble. When the end of a roll of metal is reached, the contacts act in the same way as when the bottom of the shell is pushed out, and stop the press.

CHAPTER VI

REVERSING MECHANISMS FOR ROTATING PARTS

A REVERSAL of motion is essential to the operation of many different forms of mechanism. Machine parts having a rectilinear or straight-line motion must, of necessity, reverse their movement, and many rotating parts also revolve first in one direction and then the other. The reversal in some cases is applied to a single shaft or slide and, in other instances, an entire train of mechanism is given a reversal of motion. The types of reversing mechanisms vary considerably, both as to principle of operation and as to form or design. Some are so arranged that the reversal of motion occurs at a fixed point in the cycle of movements, whereas, with other designs, the point of reversal may be changed by means of adjustable dogs or tappets which are attached to the movable part and control the action of the reversing mechanism. The adjustable type is required on some machine tools for varying the length of the stroke made by a cutting tool or machine table so that the stroke will conform to the length of the work. Reversing mechanisms also differ in that some are hand-controlled and others are operated automatically.

Intermediate Spur Gears for Reversing Motion. — A simple method of obtaining a reversal of motion by means of spur gears is shown at A and B in Fig. 1, where the reversing gears used on some designs of lathe headstocks are illustrated diagrammatically. The two intermediate gears b and c are mounted on a swiveling arm which can be adjusted for engaging either one of the intermediate gears with the spindle gear. When the gears are in the position shown at A, the drive is from a through c to d. When the arm carrying the intermediate gears is shifted as indicated at B, the motion is transmitted through both intermediate gears or from a through

161

b and *c* to *d*, thus reversing the direction of rotation. This mechanism, as applied to a lathe, is used for reversing the rotation of the lead-screw when cutting left-hand threads.

Another method of obtaining a reversal of rotation by means of an intermediate gear is illustrated by diagram *C*. In this case, there are two sets of gearing between the driving and driven shafts. For the forward motion, the drive is from gear *e* to *f*. When the rotation of the driven shaft is to be reversed, gear *e* is shifted to the left and into mesh with the

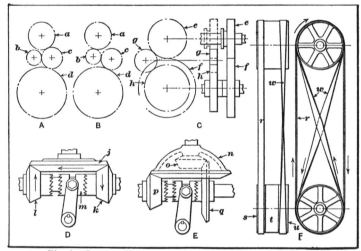

Fig. 1.   Common Methods of Obtaining a Reversal of Motion

intermediate gear *g*, as shown by the dotted lines, so that motion is transmitted through *e*, *g*, and *h*. This general arrangement for obtaining a reversal of rotation is applied extensively to the transmission gearing of automobiles.

Bevel-gear Type of Reversing Mechanism. — A combination of three bevel gears, as illustrated by diagram *D*, Fig. 1, is applied to many different classes of mechanisms for obtaining a reversal of motion, especially when the reversing action is automatically controlled. With the usual arrangement, gear *j* is the driver and it is constantly in mesh with the bevel pinions *l* and *k*. These bevel pinions are loose upon the driven shaft and have a clutch *m* interposed between them. This clutch is free to move endwise along the shaft, but it slides along a key or feather which compels it to revolve with the shaft. Each bevel pinion has teeth corresponding to clutch teeth, so that the engagement of the clutch with either pinion locks it to the shaft. Since these bevel pinions revolve in opposite directions, as indicated by the arrows, the rotation of the driven shaft is reversed as clutch m is shifted from one gear to the other. When the clutch is in the central or "neutral" position, it does not engage either gear, and no motion is transmitted to the driven shaft. Many of the reversing mechanisms which are equipped with this bevel gear combination differ in regard to the method of operating the clutch. For instance, clutch m might be shifted by the direct action of a slide or table having a rectilinear motion, or an auxiliary mechanism might be utilized to give the clutch a more rapid movement at the point of reversal. Some of these auxiliary features will be referred to later.

Two-speed Reversing Mechanism of Bevel-gear Type.— On some classes of machinery, it is desirable to have a relatively slow motion in one direction followed by a rapid return movement, in order to reduce the idle or non-productive period. One design of reversing mechanism of the bevel-gear type, by means of which a slow forward speed and a rapid return speed may be obtained, is illustrated at E in Fig. 1. In this case, there are two driving as well as two driven gears. The larger driver n is made cup-shaped so that a smaller driver o can be placed inside. When the clutch engages the smaller driven gear p, the fast speed is obtained, and, when the clutch engages gear q, the speed of the driven shaft is reduced an amount depending upon the ratio of the slow-speed gearing. Reversing mechanisms of this general type are not adapted for reversing the motion of heavy slides or work tables nor for fast-running machinery, because of the excessive shocks and stresses incident to a sudden reversal of movement in case of high velocities or heavy loads.

Reversal of Motion with Friction Disks. — When motion is transmitted between shafts located at right angles to each other by the type of frictional transmission shown in Fig. 11, Chapter XI, a reversal of rotation is easily obtained. As disk B is shifted inward along the face of disk A, the velocity ratio is gradually reduced, and when disk B passes the axis of disk A, the direction of rotation is reversed. This form of transmission has been applied to the feeding mechanisms of certain types of machine tools, and to other classes of machinery, especially where simplicity of design and ease of operation and control are essential factors. One method of arranging this form of drive, as applied to an automobile transmission, is to mount the driving member on a sliding shaft which enables the driving and driven disk to be readily disengaged, thus combining in one simple mechanism the clutching, speed-changing, and reversing functions.

Reversal from Open and Crossed Belts. —Shafts are often connected with open and crossed belts for permitting a reversal of rotation. The arrangement is illustrated by the diagram F in Fig. 1. There are three pulleys on the driven shaft. The central pulley t is keyed or attached to the shaft, whereas the outer pulleys s and u are loose and free to revolve upon the shaft. When the "open" belt r is shifted onto the tight pulley t, the driven shaft revolves in one direction and its rotation is reversed when the crossed belt w replaces the open belt on the tight pulley.

This form of drive is sometimes modified by having two pulleys on the driven shaft and a clutch interposed between the pulleys, so that either of them may be made the driven member. Thus, when the clutch is engaged with the pulley connecting with the open belt, the rotation is the reverse of that which is obtained when the clutch engages the pulley driven by the crossed belt. The countershafts for engine lathe and other machine tools which may require a reversal of movement are commonly arranged in this manner. Open and crossed belts are also applied to belt-driven planers for reversing the motion of the platen or work table. Many planer drives have pulleys which are so proportioned as to give a rapid return movement. A common arrangement is to place a central or tight pulley on the driven shaft which has two steps or diameters, the smaller one of which is for obtaining a fast return motion. Many modern planers are driven by motors which transmit power direct to the planer transmission. Incidentally, belt drives of the type referred to are often used in place of gearing, for reversing heavy or fast running parts, because the belts slip somewhat if the load becomes excessive, due to the stopping and starting at the points of reversal, and this slipping action automatically protects the mechanism from injurious shocks or stresses.

Operation of Reversing Clutches. — When a reversal of motion depends upon the action of a clutch which is shifted from one gear to another revolving in an opposite direction, it is essential to operate the clutch rapidly and to secure a full engagement of the clutch teeth. Provision should also be made against disengagement of the clutch as the result of vibrations incident to the operation of the machine. There are two common methods of controlling the clutches used in connection with the bevel-gear type of reversing mechanism illustrated at D in Fig. 1. One form of control may be defined as the swinging-latch type and the other as the beveled-plunger type. The general principle of operation is the same in each case, and is as follows: When the work table, or whatever part is to be reversed, approaches the end of its stroke, a spring is compressed, and then a latch or trip allows this compressed spring to suddenly and rapidly throw the reversing clutch from one gear to the other. Reversing mechanisms of this general design are often called the "load-and-fire" type, because the spring is first loaded or compressed and then tripped to secure a rapid movement of the clutch and a reversal of motion at a predetermined point within close limits. The action of the compressed spring also insures a full engagement of the clutch teeth and prevents the clutch from stopping in the central or neutral position, which might occur if a spring were not used and the momentum of the part to be reversed were insufficient to carry the clutch across the space intervening between the two reversing gears.

Reversing Mechanisms of the Face Gear Type. — An interesting application of the face gear is shown in Fig. 2. The double spur pinion *B*, mounted on and keyed to shaft *C*, may be moved along its axis, thus engaging the "face gear" with step *D* as shown, or step *E* with the face gear at *F*. This is an economical and substantial means of reversing the direction of rotation.

A face gear may be cut on a Fellows gear shaper. The teeth are cut on a face at right angles to the axis of the gear. They are produced by a cutter provided with involute teeth

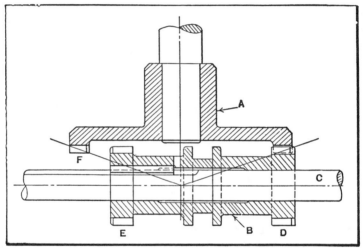

Fig. 2. Reversible Face Gear Drive with Sliding Double Pinion

corresponding in number and outline to the teeth in the pinion that is to mate with the face gear. The cutter and work are geared to rotate in the ratio of their respective teeth, and as the cutter reciprocates in contact with the work teeth are developed in the face gear by the molding-generating process.

In order to operate properly with the face gear, the axis of the mating pinion must be located at exactly the same distance from the face of the gear as the axis of the producing cutter. Proper tooth action is obtained, regardless of the axial position of the pinion, assuming, of course, that it is not moved out of engagement. This is an important point since cone adjustment may be disregarded. These gears are adapted to ratios of 3 to 1 and higher. Drives of 1 to 1 ratio are unsatisfactory, since the active face width is too limited. **Latch Type of Reversing Clutch Control.** — The reversing mechanism illustrated in Fig. 3 is a bevel-gear type equipped with the swinging latch form of clutch control. This mechanism is applied to a cylindrical grinding machine for reversing

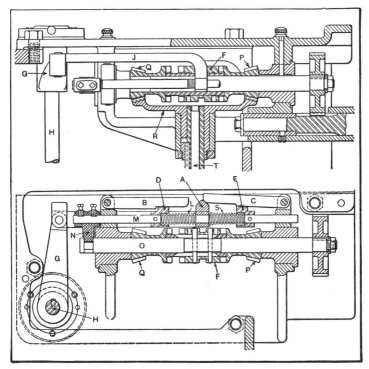

Fig. 3. Spring and Latch Type of Reversing Clutch Control

the motion of the work table, and is located at the rear of the machine. The rockshaft *H* extends through to the front of the machine and has attached to it a lever which is engaged by dogs on the work table, the distance between these dogs being varied according to the length of stroke required. At the rear end of rockshaft *H* there is a lever *G* which, by means of link *J*, transmits motion to the reversing mechanism. As the work table approaches the end of its stroke, lever G swings either to the right or left as the case may be. If the motion is to the left, tappet A, connected to link J, compresses spring L on the rod M and forces block D against a square shoulder on the lower side of latch B. Continued movement of tappet A to the left causes the beveled side of A to lift latch B, thus releasing block D, which, with rod M, is thrown rapidly to the left under the impulse of the compressed spring L.

After the movement of shaft M to the left, the shoulder on latch C drops in behind block E. The fork N on rod M also throws shaft O to the left and with it the reversing clutch F which is keyed to this shaft. The motion which prior to reversal was transmitted through bevel pinion P to the main gear R is now from pinion Q to R so that the movement of the work table is reversed. When the work table approaches the end of its stroke in the other direction, tappet A is moved to the right, thus compressing spring S. Then latch C is lifted by the beveled edge on A and the parts M, N, and O are quickly shifted to the right by spring S, thus again reversing the motion.

If the operator desires to stop the traversing movement at the end of the stroke, this may be done by the movement of a knob located in the center of the table-traversing handwheel at the front of the machine. This knob is connected with a plunger T which, by pressing the knob, may be held under pressure against the reversing clutch F. When this clutch is shifted at the end of the stroke either by springs L or S, plunger T drops into a groove in clutch F, thus holding it in central or neutral position. The knob previously referred to may be set at any part of the stroke to stop the traversing movement at the end of that stroke. The withdrawal of the knob again starts the traversing movement without requiring any further action on the part of the operator. The shaft connecting with bevel gear R extends to the front of the machine and, through suitable gearing, transmits a rectilinear motion to the work table of the grinding machine. This mechanism is an example of the "load-and-fire" type referred to.

Beveled Plunger Control for Reversing Clutch.— An example of the beveled plunger type of clutch control for a reversing mechanism is shown in Fig. 4. This design is also intended for a cylindrical grinding machine. The point of reversal is controlled by the tappets A which are adjusted along the work table to vary the length of the stroke. These tappets alternately engage lever B at the ends of the stroke and, by swinging this lever about its pivot, shift bar C which transmits motion to the reversing clutch. If the work table is moving toward the right, the tappet at the left engages

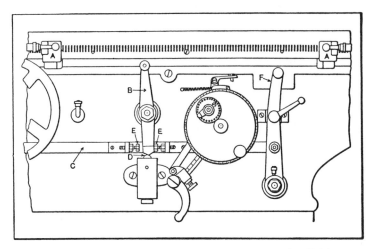

Fig. 4. Spring and Beveled Plunger Control for Reversing Clutch

lever B as the table approaches the end of its stroke. The movements of the lower end of reversing lever B towards the left forces the beveled plunger D downward, thus compressing a spring that is located beneath it. When the point of the V-shaped end of lever B has passed the point of plunger D, the latter is suddenly forced upward by the compressed spring and lever B, rod C, and the reversing clutch are shifted rapidly.

There is a certain amount of lost motion between the studs E on bar C and the reversing lever B. As the result of this lost motion, the clutch is not entirely disengaged until the V-shaped point of the reversing lever has passed the point of plunger D; the reversing clutch is withdrawn slowly from the bevel pinion which it engages until the sudden action of plunger D causes it to shift rapidly into engagement with the opposite bevel pinion. The clutch is held in engagement until the next reversal of motion by the upward pressure of the plunger against the beveled end of the reversing lever B. With the particular design illustrated, the point of reversal can also be controlled by hand lever F which is connected to rod C; by placing this lever in a central position, the clutch is shifted to neutral and the movement of the work table discontinued.

Controlling Point of Reversal by Special Mechanisms. — The points of reversal for a reciprocating slide are usually controlled by trip dogs mounted directly on the slide and adjusted to give the required length of travel or stroke. It is not always convenient, however, to control the reversal in this way. For instance, if the operating slide is at the rear of a machine where the trip dogs cannot be adjusted readily, some form of mechanism which operates in unison with the slide may be used to permit locating the trip dogs at the front of the machine. A simple method of controlling the points of reversal from the front of the machine is applied to a certain design of cylindrical grinding machine. The wheel slide travels along ways at the rear of the machine and the length of stroke is regulated in accordance with the length of the work by two trip dogs mounted on a wheel or circular rack at the front of the machine. The shaft carrying this wheel extends through the machine and is connected by gearing, so that it has an oscillating or turning movement in first one direction and then the other, which movements correspond to, and are in unison with those of the wheel carriage at the rear. Worm teeth are formed on the periphery of the trip-dog wheel and the dogs are held in position by worms which may be lifted out of engagement when the dogs are to be adjusted considerably. The dogs alternately strike a tappet or lever which controls the movements of the reversing clutch.

Another method of controlling the reversing points of a rear slide is by means of a shaft connected through gearing with the reciprocating slide and having at the front end a pinion meshing with a sliding rack carrying the trip dogs. As the rear slide operates, it turns the pinion shaft in first one direction and then the other, which imparts a reciprocating motion to the rack. The trip dogs attached to the rack, by engaging a lever, cause a reversal of motion by means of a clutch-and-gear type of reversing mechanism.

An indirect or independent method of controlling the points of reversal on an automatic bevel gear cutting machine is

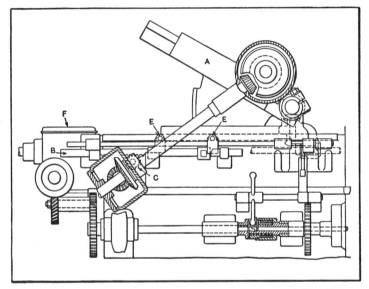

Fig. 5. Independent Method of Controlling Reversal of an Adjustable
Slide on a Bevel Gear Cutting Machine

illustrated in Fig. 5. The cutter-slide A must be set at an angle corresponding to the inclination of the gear teeth to be cut, so that it would be difficult to have the trip dogs attached directly to this slide. To avoid such an arrangement, a sliding rack B is employed. This rack meshes with a pinion C which rotates in unison with the feeding of the cutter-slide, since this pinion and the slide derive their motion from the same shaft. As pinion C rotates in first one direction and then the other, it traverses the rack B, which, by means of the adjustable dogs

E, controls the action of the reversing mechanism enclosed at
F. With this arrangement, the traversing movement of the
rack can be made less than the travel of the cutter-slide, if
this is desirable because of limited space.   On the other hand,
if the traversing movement of the slide is to be very short
and it is essential to reverse it at a given point within close
limits, the movements of the reverse controlling rack can be

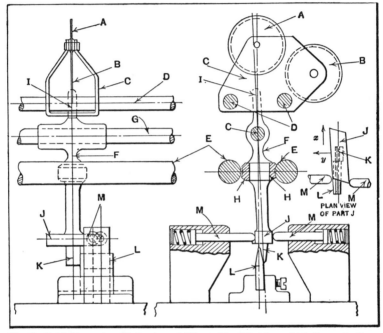

**Fig. 6. Quick Reversing Mechanism of Wire-coiling Machine, which Prevents Turns
of Wire from Piling Upon One Another at the Coil Ends**

increased considerably as compared with the motion of the
cutter-slide.

Quick-reversing Mechanism for Wire Coiling. — The rapid
reversing mechanism to be described is embodied in a machine
for winding coils having a number of layers one upon the
other, as used in the electrical trade.   As the device that guides
the wire on the coil arrives opposite one end of the coil, instantaneous reversal of direction is necessary in order to

avoid the turns of wire piling upon one another. The results obtained with this reversing mechanism are highly successful.

The wire is fed from the supply drum over pulley A (see Fig. 6) and under pulley B and then on the coil spool. As the latter is not driven in a special manner, it is not shown. The two pulleys are kept very thin, so as to get close up to the flanges of the spool. These V-pulleys are mounted in the carriage C, which is capable of sliding along the guide rods D. Beneath these rods are located two lead-screws E of a pitch equal to that of the wire when wound on the spool. These are rotated in opposite directions.

Suspended between the lead-screws is the lever F, which slides along the guide rod G. Embodied in this lever are two half-nuts H which, as the lever swings from one side to the other, engage alternately with the lead-screws E. Thus the lever will travel to the right or left according to which screw it is in mesh with.

The top of the lever carries an extension I which engages in a slot in the base of the carriage C, thus giving it the necessary motion. It will be obvious that the peg I must be free to swing as the lever is moved, but must have no side play. Lower down on the lever is fixed a hardened steel rhombus-shaped part J. This is located at an angle, as seen more clearly in the plan view. Beneath this is a hardened steel wedge K which is kept in contact with another inverted wedge L by means of the spring plungers M.

It will be seen, then, that when one half-nut is in contact with a lead-screw the steel wedges either slide or are ready to slide against each other. The method of securing the reversal will be more easily seen by examining the plan view of part J. Here part J, fixed to the lever, has moved along in the direction of arrow x. By so doing and due to its angular location, it has compressed the right-hand plunger. At the same time it has reached the end of its traverse and the two wedges K and L are about to separate.

Thus, at this moment, part J will be forced by the plunger, in the direction of arrow y, and the half-nut will mesh with the opposite lead-screw. The lever F will then immediately reverse its direction, and the cycle indicated will be repeated.

For winding coils of varying length, it is merely necessary to increase or decrease the length of the lower wedge L. As this device is automatic, except for the placing and removing of bobbins, and the starting of the wire, one operator can take charge of several wire coiling machines of this type.

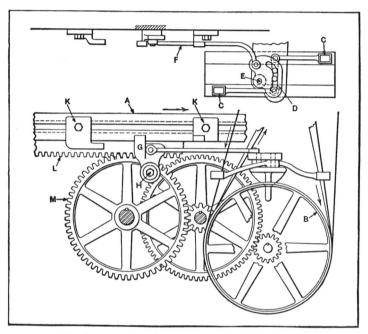

Fig. 7.   Reversing Mechanism of a Belt-driven Planer

Mechanism for Shifting Open and Crossed Belts. — The open and crossed belts illustrated by diagram F, Fig. 1, are shifted automatically for obtaining a reversal of motion, when used to drive such machines, as planers, broaching machines, or other classes of mechanisms which are designed for continuous operation and equipped with this form of drive. A side elevation and plan of the automatic belt-shifting device used on a planer is illustrated in Fig. 7. The shaft on which the belt pulleys B are mounted transmits motion to the planer table A through a train of gearing which gives a suitable speed reduction. In order to reverse the motion of the work table, this entire train of gearing is reversed by alternately shifting the open and crossed belts onto the central pulley, which is attached to the shaft. The length of the stroke is governed by the distance between the two dogs K which may be adjusted along a groove at the side of the table. The position of each belt is controlled by a guide C having an opening at the end through which the belt passes. These two guides or shifters, which are in the form of bellcranks, are pivoted and the inner ends carry small rollers that engage a groove in the cam-plate D. This cam-plate is pivoted at E and is connected by a link F with the arm or lever G which is pivoted at H.

When the planer is in operation, the table moves in a direction depending upon which belt is on the tight pulley. When this movement has continued far enough to bring one of the dogs K into contact with arm G, the latter is pushed over about its pivot, thus imparting a swinging movement to the cam-plate D. The groove in this cam-plate is so formed that the belt on the tight pulley is shifted to the loose pulley and the other belt is moved over to the driving position on the tight pulley. At the end of the return stroke, the other dog engages arm G, thus swinging the cam-plate in the opposite direction and again reversing the motion.

Reversal of Motion through Epicyclic Gearing.— A train of epicyclic or differential gearing may be designed to give a reversal of motion. This form of transmission has been applied to some automobiles of the smaller sizes. The principle governing the operation of one of the earlier designs is shown by the diagram, Fig. 8. Two sets of differential gears, indicated at A and B, are mounted inside of drums. These drums may be revolved independently for obtaining the slow forward speed and a reverse motion, or they may be locked together so as to revolve as a unit with the crank-shaft for obtaining the direct high-speed drive. The central gear a is the driver in each case, and is keyed to the crank-shaft. The slow forward speed is obtained with the combination illustrated at A. To obtain a reduction of speed, the internal gear b is held stationary by the application of a brake-band to its periphery; the pinions c carried by the driver member are then forced by the driving gear a to roll around inside of the internal gear, thus transmitting a slow rotary motion to the driven member attached to the pinions. In order to obtain a reversal of motion through the combination of gearing illustrated at B, the disk carrying the pinions is prevented from rotating by the gripping action of another brake-band, so that the pinions merely revolve on their studs and rotate the internal gear in a reverse direction. In this case, the in-

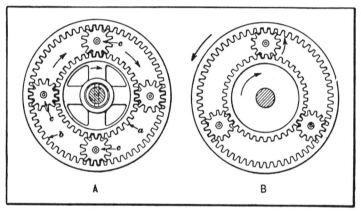

Fig. 8. Diagram Showing Arrangement of Epicyclic Gearing for Obtaining Forward and Reverse Motions

ternal gear is the driven member and transmits motion to the driving sprocket.

A reversal of motion may also be obtained with the train of epicyclic gearing shown in Fig. 9. In this case, there is no internal gear. Gear A is mounted on the sleeve of sprocket A_1, gear D is keyed to shaft K, and gear F is attached to the extended hub of drum H. The three gears, B, C, and E are locked together and revolve upon a pin carried by drum G. A duplicate set is also located on the opposite side of the drum, as the illustration shows. When this drum is held stationary by a brake-band, gear A and sprocket A_1 are driven at a slow forward speed through gears *D, C,* and *B,* gears *D* and *A* revolving in the same direction. The direct high-speed drive is obtained when clutch *J* is engaged, the whole mechanism then revolving as a unit with shaft *K*. When drum *H* is held stationary by a brake-band, gear *D* causes gear *E* to revolve about the stationary gear *F* in a direction opposite to the rotation of *D;* consequently, gear *A* is forced to follow in the same direction in which drum *G* and the planetary gears *B,*

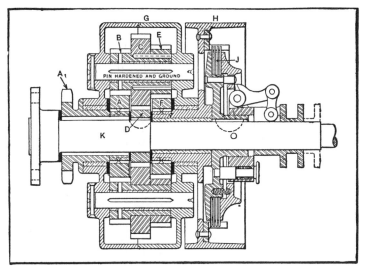

Fig. 9. Another Arrangement of Epicyclic Gearing which Gives Forward and Reverse Motions

C, and *E* are moving, thus reversing the motion of gear *A* and the sprocket.

Operation of Reversing-ratchet Pawl. — Fig. 10 shows a simple change in a ratchet movement which resulted in a more positive action. Originally, the arrangement was as follows: The ratchet wheel *A,* which is keyed to shaft *F,* served to impart motion to a sliding table (not shown). The motion of the shaft *F* was imparted to the table by means of gearing engaging with a rack on the under side of the table. The lever *B* was given an oscillating motion by a crank, to which it was connected by the connecting-rod *E*. The pawl

C, pivoting on the stud *I*, engaged the ratchet wheel *A*, the weight *H* serving to keep the pawl engaged. Originally pawl *C* was not slotted as shown, but carried pin *K*, the part *D* with its weight *G* not being used. As the table moved along, a dog on the edge struck pin *K*, and swung pawl *C* to the opposite

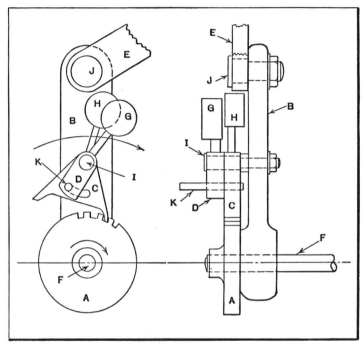

Fig. 10.   Reversing Ratchet, the Pawl of which is Operated
by a Swinging Weight

side, reversing the rotation of shaft *F* and the movement of the table.

In general, this arrangement operated satisfactorily, except on rare occasions, when the position of the dog on the table would be such as to swing pin *K* just far enough to balance pawl *C*, when the movement of the table would stop until reversed by the operator.

In order to overcome this condition, pawl *C* was slotted as shown, and the part *D* was added, pin *K* being a press fit in part *D* and sliding in the slot in pawl *C*. With this arrangement, pawl C, operating independently of part D, remains in contact with ratchet wheel A until the table has moved far enough to push pin K over so that weight G falls on the opposite side, the unbalanced effect causing pawl C to swing on stud I, and engage on the opposite side, producing a more positive effect than by swinging the pawl direct.

Automatic Ratchet Reversing Mechanism. — The simple design of ratchet reversing mechanism illustrated in Fig. 11 enables a ratchet wheel to be automatically reversed after making a predetermined number of revolutions, and the arrangement is such that the time of reversal or the number of revolutions made by the driven ratchet prior to reversal may

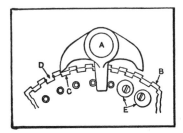

be varied at will throughout a wide range. The double pawl A is carried by an oscillating arm (not shown), and this pawl engages the driven ratchet B. Mounted concentrically with B there is a smaller controlling ratchet C which is normally restrained from rotating by suitable frictional resistance. The larger diameter of ratchet B

Fig. 11. Ratchet Mechanism which will Automatically Reverse after Making a Predetermined Number of Revolutions

prevents pawl A from engaging the smaller ratchet C, except when the deep notch D is reached by the pawl which then drops down into engagement with C.

The reversal of motion is effected by the engagement of the extension on pawl A with one of the trip dogs E. The number of revolutions made by ratchet B prior to reversal depends upon the number of deep notches D and the position of the trip dogs E. When this mechanism is in operation, ratchet B receives an intermittent motion from the oscillating pawl A and the controlling ratchet C remains stationary until one of the deep notches D is engaged by pawl A; then ratchets B and C rotate together an amount depending upon the motion of the pawl. Controlling ratchet C then remains stationary until another deep notch is engaged. The repeated movements of ratchet C each time the pawl drops into a deep notch, finally bring one of the trip dogs E into contact with the projection on the pawl; the latter is then swung around so that its opposite end engages ratchet B and, consequently, the direction of rotation is reversed. The time of reversal may be controlled by varying the distance between the trip dogs and by having one or more deep notches in the driven ratchet.

Combined Reversing and Feeding Movements.— Some reversing mechanisms are so designed that the longitudinal movement of a reversing rod is accompanied by a rotary

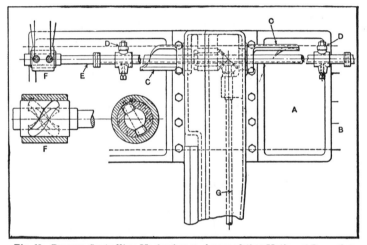

Fig. 12. Reverse Controlling Mechanism so Arranged that Motion of Reversing Rod is Accompanied by a Rotary Movement for Feeding Tool

motion for imparting a feeding movement at the time reversal occurs. A reversing device of this kind, as applied to a Richards' side-planing machine, is illustrated in Fig. 12. The saddle A is traversed along the bed B by means of a screw, the rotation of which is reversed by open and crossed belts that are alternately shifted from loose pulleys to a tight pulley attached to the screw. The two projecting arms C which are bolted to A strike dogs D mounted on rod E, which, by its longitudinal movement, actuates the belt-shifting mechanism. When rod E is shifted in a lengthwise direction, it is also given a rotary motion in the following manner: Within the bearing F there is a bushing having cam grooves cut into it, as shown by the enlarged detailed view. These grooves receive rollers carried by a pin that passes through the rod E. With this arrangement, any endwise movement of rod E is accompanied by a rotary motion resulting from the engagement of the rollers with the helical grooves in the fixed bushings of bearings F. This rotary movement is transmitted through bevel gears to a rod G which imparts a downward feeding movement to the feed-screw of the tool-slide, through the medium of ratchet gearing.

Automatic Control of Spindle Reversal. — Fig. 13 represents a sectional view through the bed of an automatic screw machine, beneath the headstock, and illustrates the mechanism for automatically controlling the reversal of the spindle rotation. This machine is driven by a single belt pulley rotating at constant speed. The various movements of the machine, other than revolving the spindle, are derived from a shaft at the rear which rotates at a constant speed. On this shaft H is mounted a series of automatically-controlled clutches which are similar in action to those used on punch-presses. These clutches control the feeding of the stock, the opening and closing of the chuck, the revolving of the turret, the reversing of the main spindle and the changing of the speed from fast to slow, or *vice versa*. This back-shaft H is connected by change-gearing through a worm drive, with a slow moving camshaft A at the front on which are mounted the cams for the turret and cross-slide movements and a series of dog carriers B carrying tappets or dogs which control the action of the different clutches on the back-shaft. The ratio of the change-gears previously referred to determines the duration of the cycle of operations and, consequently, the length of time it takes to make a given piece of work.

The main spindle is reversed by a clutch located between two clutch members revolving in opposite directions. The carrier B has an annular T-slot in which adjustable dogs like the one shown at C are mounted. These dogs engage a tappet D on lever E, the rear end of which carries a screw F, the cylindrical point of which enters a cam groove in clutch G. This clutch is mounted loosely on shaft H which revolves con-linuously. A plan view of the cam is shown in detail above the end view. The cam groove is exactly the same on the other side as on the side shown, the clutch being arranged to engage each half revolution and then automatically disengage. The normal position of the pin F is in the recess at a. When it is lowered entirely out of the groove by the action of dog

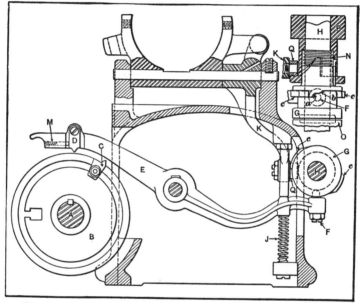

Fig. 13. Arrangement for Automatically Controlling Spindle Reversal

C on tappet D against the pressure of spring J, this releases clutch G, which is forced forward by a spring N coiled about the shaft, until it engages a mating member O, fastened to shaft H, and begins to revolve. Meanwhile dog C has passed tappet D, allowing pin F to drop into the cam groove again. The clutch G, as it revolves, brings inclined face b of the groove (or a similar incline on the opposite side) into con-tact with F, and the continued revolution of G, through the action of this inclination on the pin, forces the clutch teeth out of engagement, stopping G again with the pin in position *a* as at the start. A cam *P*, also loose on the shaft *H*, is keyed to *G*. This cam engages a roll *Q* on the end of lever *K*, which operates a clutch fork, controlling the position of the main spindle clutch. When it is time to again reverse the spindle, another dog *C* is set in the proper position, and the clutch is tripped, revolving for a second time a half revolution and stopping, thus operating lever *K* and the spindle clutch to change the direction of the spindle rotation. This represents the normal procedure in cases where the time taken to make one piece is short enough so that the rotation of dog carrier *B* is reasonably rapid. For many pieces, however, this movement is so slow that dog *C* does not come out from under tappet *D* in time to allow pin *F* to drop into the cam groove before the clutch has made the required half revolution. In such cases, incline *b* would pass without disengaging the clutch and pin *F* could not enter until the next recess came around and the next incline. *b;* hence the clutch would be stopped at the end of one revolution instead of a half revolution. This difficulty has been very simply overcome by the following means :

Tappet *D* is pivoted to lever *E* as shown, and is forced back against a shoulder to the position. indicated, by a spring *M* located in a drilled hole and pressing against a plunger bearing on *D*. This spring is of such strength as compared with spring *J* that the first effect of dog *C*, when it strikes *D,* is to move the latter backward without raising lever *E*. When *D* has been pressed so far back that it strikes the shoulder at the left, further movement being impossible, *E* is raised, pin *F* is withdrawn from the cam slot in the clutch *G,* and the latter is allowed to engage fixed member *O* on the shaft *H,* and starts to revolve. A cam surface *c* is provided on *G* which, immediately after the clutch begins to rotate, strikes pin *F* and depresses it still further, thus raising tappet *D* clear above the point of dog *C,* and allowing it to swing back to its normal position against the shoulder at the right under the influence of spring *M*. Lever *E* is then ready to drop instantly, as *D* and *C* are entirely clear of each other. As soon as the end of cam projection c passes, F drops into the groove and the rotation of the cam is arrested after a half revolution, as required. When it is known that shaft H revolves at 120 revolutions per minute, so that the half revolution of G occupies but one-fourth second, it will be seen that the device has a difficult duty to perform, but operates in a very satisfactory manner.

Automatic Variation in Point of Reversal. — One of the many interesting mechanisms found on textile machinery is the one employed on fly frames for controlling the winding of the roving on the bobbin. The bobbins are driven at a decreasing rate of speed as the diameter increases a n d they not only revolve but are given a vertical reciprocating motion, in order to wind the roving onto them in successive helical layers. This winding of the roving onto the bobbin involves, in addition to decreasing the speed as the diameter increases, a decrease in the traversing speed of the bobbin and a gradual

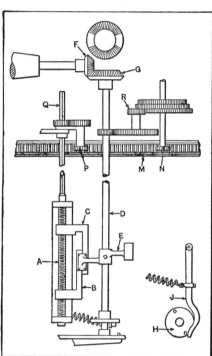

Fig. 14. Mechanism for Varying Point of Reversal and Speed of Rotation

shortening of the bobbin travel as one layer of roving is wound upon another. The bobbin should move a distance equal to the diameter of the roving while it rotates relative to the "flyer" a distance equal to one revolution; therefore, as the bobbin speed gradually diminishes, it is also necessary to decrease the rate of traverse, so that each layer of the roving will be coiled closely. The change in the point of reversal in order to shorten the stroke as the bobbin increases in diameter is required in order to form conical ends on the wound bobbin and a firm winding that will not unravel and cause trouble, such as would be the result of attempting to wind each layer the full length of the bobbin. These changes occur simultaneously, although they will be referred to separately in describing the "builder motion" illustrated diagrammatically in Fig. 14.

The plates B and C engage a screw A, which has a right-hand thread extending along one-half its length and a left-hand thread, along the remaining half. These plates and the screw are traversed vertically with the bobbin carriage. The vertical shaft D carries a dog E having two arms located 180 degrees apart. At each end of the stroke, shaft D makes a half turn which motion is utilized for reversing the motion, for shifting the cone belt slightly in order to decrease the speed, and also for shortening the stroke of the bobbin. As the plates B and C move vertically, one end of the tumbling dog E bears against them until it slides off at one end. Prior to the disengagement of the dog with one of the plates, gear F on the cone-pulley shaft revolves idly in a space on the rim of gear G where the teeth are omitted. There are two of these spaces located 180 degrees apart, as the illustration indicates. One of the projecting pins on the disk H at the lower end of shaft D is in engagement with a lever J which has attached to it a spring that holds the lever aganst a pin and tends to turn shaft D. When dog E slides off one end of a plate, shaft D is turned far enough by the action of the spring and lever J to bring gear G into mesh with pinion F; consequently, gear G is revolved one-half turn or until pinion F engages the space on the opposite side where there are no teeth. The partial rotation of shaft D shifts the reversing gears through a connection at the lower end and starts the bobbin carriage and plates B and C in the opposite direction. As the opposite end of the tumbling dog E swings around, it engages one of the plates and again causes a reversal of motion as it slides off of the opposite end.

The shifting of the belt on the cone-pulley at each reversal, for gradually decreasing the speed as the bobbin winding increases its diameter, is obtained by connecting rack M with shaft D through the pinion N and the train of gearing shown. This rack M has a fork attached to it that connects with the cone-belt, and it is traversed slightly each time dog E slides off a plate and allows shaft D to turn one-half revolution. The reduction in the length of the carriage traverse is obtained by revolving screw A at each reversal and thus shortening the distance between the plates B and C. This rotation of the screw is effected by pinion P which engages rack M and transmits motion through the other gears shown to the extension Q on the screw, which is made square and is free to slide through the gear hub as the carriage moves vertically. As the plates B and C are moved toward each other, the tumbling gear E has a shorter surface to traverse before it is disengaged. These two plates both move the same distance, so that the point of reversal decreases at each end and the bobbin is wound conical at both ends. The roving delivered by the front roll is either tightened or slackened by engaging pinion R with one of the three gears shown.

Oscillating Mechanism which Varies Point of Reversal. — The intermittent reversing mechanism shown in Fig. 15 is an important part of a power-driven valve-grinding machine designed for automatically grinding-in any desired number of valves at one time. The grinding movements imparted to the valve consist of rotating the valve approximately two-thirds of a revolution in one direction, then reversing the direction of rotation for one-third of a revolution and so on, the point of reversal being constantly changed, a feature which tends to prevent the cutting of grooves by the abrasive grit.

In addition to the reversing or oscillating movements, the valve is raised intermittently to allow the grinding or abrasive material to flow in between the seat and the valve. The various parts of the mechanism are designated by the same reference letters in the different illustrations.

The shaft C of the reversing mechanism imparts the oscillating movements to the valve-grinding spindles through bevel gears which are not shown as they do not affect this mechanism. A shaft driven from the reversing mechanism, is

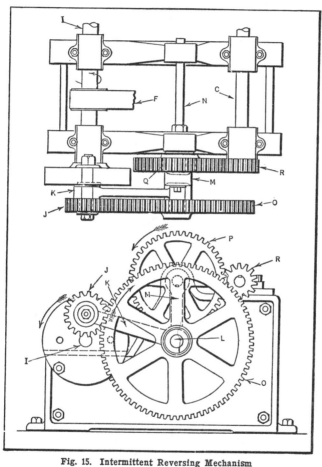

Fig. 15. Intermittent Reversing Mechanism

equipped with cams which transmit the intermittent valve-lifting movements to studs in contact with the valve stems.

Shaft I is connected with the lineshaft by a direct belt drive and is revolved continuously in the direction indicated by the arrow. Keyed to one end of shaft I is a disk having a small gear *J* secured to it in an off-center position with respect to the shaft. The connecting-rod or link *K* has a bearing on the solid stub shaft of gear *J* at one end, and a bearing on the shaft *L* at the other end. Shaft *L* is journalled on an oscillating arm or pendulum *M*, the upper end of which pivots on shaft *N*. The gear *O* keyed to shaft *L* is thus kept in

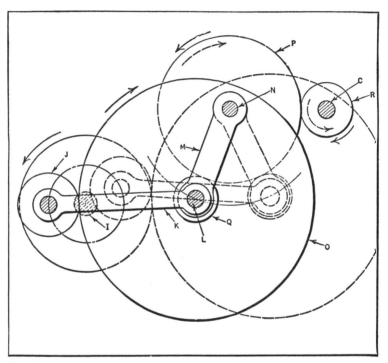

Fig. 16. Diagram of Mechanism Fig. 15, Showing Parts at Beginning of Forward Driving Movement in Full Lines and at Beginning of Reverse Movement in Dotted Lines

mesh with gear *J* at all times and has an oscillating movement similar to the pendulum of a clock, which results from the movement given shaft *L* by connecting-rod *K*.

In addition to the oscillating movement, gear *O*, with its shaft *L*, is rotated slowly by the driving gear or pinion *J*, one revolution of shaft *I* being equivalent to one revolution of gear *J* under the usual operating conditions. Keyed to the rear end of shaft L immediately back of the oscillating arm M, is a pinion Q (see Fig. 16) which meshes with a gear P journalled on shaft N on which arm M is pivoted. The slow drive transmitted to gear O by gear J will also be transmitter to gear P. In addition to this drive, gear P is also revolved by the oscillating movement of gear Q keyed to shaft L, being revolved first in one direction and then in the other. The design of the gearing is so calculated that when the connecting-rod K moves forward, gear P will also be revolved forward; that is, when the connecting-rod K moves from the position shown by the full lines in Fig. 16, to its extreme position to the right, there will be added to the slow driving movement of pinion J, through gears O and Q, the driving action of the pendulum or swinging movement also transmitted by the teeth of gear Q. Thus, gear P is driven forward in the direction indicated by the full line arrow until the connecting-rod starts its backward movement at approximately the position shown by the dotted lines in Fig. 16.

On the back stroke of rod K, gear P will be revolved in the opposite direction, as the driving speed in the reverse direction, resulting from the oscillating movement, is much greater than the slow forward driving movement of the pinion J. Thus, the driven gear R on shaft C is revolved first in one direction and then the other, the gear ratios being so calculated that the valve-driving spindles will revolve approximately two-thirds of a revolution in the forward direction, then in the reverse direction for one-third of a revolution, but never reversing in the same position due to the constant flow forward drive of gear J. These oscillating gear movements can be adapted for purposes other than that described. The complete valve-grinding machine is patented and was designed for a large concern engaged in the manufacture of tractors.

Reversing Shaft Rotation After Eight Revolutions. — The shaft reversing mechanism to be described is used in the construction of a washing machine. The principal driving members consist of the commonly employed gear and double pinions, the latter being located diametrically opposite each other.   The reversing parts are of simple construction and are designed to eliminate wear and friction as far as possible. They have been found to work faultlessly and are positive in their action.   The same reference letters are used on the different views of Fig. 17 to denote the same parts.

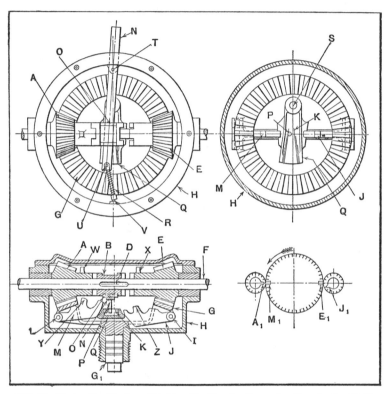

Fig. 17. Mechanism Used on Washing Machine for Reversing Barrel after Eight Revolutions in Each Direction

The cast-iron housing H carries the vertical shaft G_1, which connects the main gear G with the drum on the washing machine.   The two pinions A and E run on the horizontal shaft F which is connected with the motor or power source.   This shaft is provided with a key-slot and key D upon which the double-ended clutch jaw B slides.   The center of this clutch jaw is recessed to receive a semicircular bushing O.   This bushing has a square section on its outer side which is in contact with the starting and stopping lever N. The bushing O is provided with flanges on both sides which keep lever N in its proper position.

Another lever Q swivels on a stud S driven into gear G. Movement is imparted from lever N to lever Q by a pin P which is driven into lever N. This pin extends down into a recess in lever Q and works against two shock-absorbing springs K, one spring being located on each side of the recess. Gear G carries, besides stud S and reversing lever Q, the two cams J and M located in slots which extend below the gear and toward the center. These cams are arranged diametrically opposite each other and are pivoted on their respective studs I and L.

On the upper side of the cams are located lugs Y and Z which at intervals come in contact with lugs W and X, the latter being cast on the extensions of pinions A and E. The purpose of the cams is to push lever Q over the center, one cam coming into action when the main gear is revolving to the right and the other when it is revolving to the left. Thus when cam M is in the position shown with lever Q over to the left of the center, lug Y comes in line with lug W on pinion A at a certain moment. When this happens lug W pushes the cam down and this, in turn, forces lever Q over the center to a similar position on the right, carrying with it the other movable parts—pin P, lever N, bushing O, and clutch jaw B. The latter member is now disengaged from pinion A and engaged with pinion E, thus imparting the reciprocating or reversing motion to gear G.

When lever Q moves over to the right it also pushes cam J up into a new position, ready to act similarly to the other cam. To obtain a good working action, there must be a certain ratio between the gear and the mating pinions, and this ratio must be based on the number of revolutions which the washing machine barrel is required to make before reversing. In this case eight revolutions are chosen for half a period, or double that (sixteen revolutions) for one whole period. To obtain

this, the ratio must be 8 to 23.   The number of teeth in this
case is 16 for the pinions and 46 for the gear.   Another thing
to be taken into consideration is the location of the lugs on
the pinions.   The diagram in the lower right-hand corner of
the illustration shows the relative positions of these lugs in
regard to each other and also to the cams.   Assume that the
barrel of the washing machine has completed one period of
sixteen revolutions and is ready for the next period, and that
the gear will now revolve in the direction indicated by the
arrow for the next eight revolutions.   After completing the
eight revolutions, the direction of rotation is reversed by
means of lug J_1, which comes in contact with cam E_1.   To
obtain this action, the gear and pinions must be so assembled
that lug J_1 lacks just two tooth spaces of making contact with
cam E_1, when lug A_1 is in contact with cam M_1, as illustrated.
When the required eight revolutions are completed, lug A_1
again comes in contact with cam M_1 and the motion is reversed
again for the next eight revolutions.

A coil spring R, holds lever N in the required position on
each side of the center.   This spring is seated at point V on
the housing, and is guided on the lever by means of a pin U.
Lever N is mounted on a pin T located diametrically opposite
point V on the housing.   The lever is extended outside the
casing so that the operator can get a good grip on it for starting and stopping the mechanism.

**Reversal of Motion after Predetermined Number of Revo-
lutions up to 100,000.** — With the mechanism illustrated in
Fig. 18, a driven shaft may be reversed after making any
predetermined number of revolutions from 1 to 100,000 and
the motion may be discontinued entirely after the shaft has
made any given number of reversals up to 10,000.   This
mechanism was applied to a textile machine.   The reversing
shaft A is driven from the vertical shaft shown through either
one of the miter gears B which revolve in opposite directions
and are alternately engaged with the shaft by the sliding
clutch C.   The reversal of rotation after a predetermined
number of revolutions is controlled by a system of ratchets

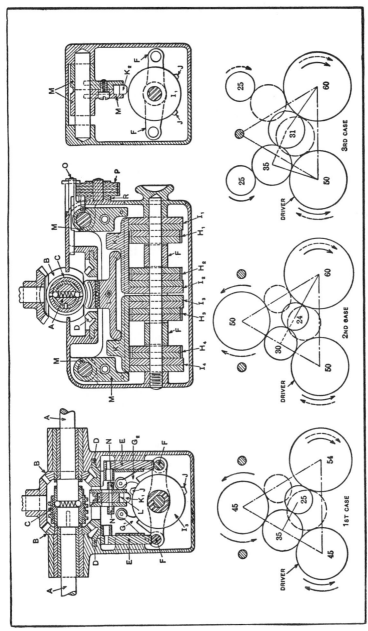

Fig. 18. Mechanism for Reversing Motion after any Predetermined Number of Revolutions between 1 and 100,000, and for Stopping Driven Shaft after any Number of Reversals up to 10,000

and pawls. Another ratchet-and-pawl mechanism is also utilized for stopping the rotation of A after a given number of reversals, by placing the shifting clutch C in the central or neutral position.

The lower miter gear D has a cam which engages the rollers on the upper ends of bars E. The lower ends of these bars oscillate the rockers F which carry two sets of pawls, G_1 and G_2. The set of four pawls G_1 at the left is in the operating or working position, as shown in the illustration. There are four pairs of ratchets H_1, H_2, H_3, and H_4. The teeth of each pair of ratchets are cut oppositely and the four pawls on one side of the ratchet shaft are for engaging the ratchets which control the number of revolutions made by shaft A in one direction, whereas the four pawls on the other side of the shaft are for operating the reverse motion ratchets. These ratchets operate progressively and transmit motion to disks I_1, I_2, I_3, and I_4. These disks have projections or cam surfaces J, which serve to shift the reversing clutch C after shaft A has made a predetermined number of revolutions, which number is regulated by adjusting the ratchets before the mechanism is put into operation. This system of cam disks and ratchets will be referred to as the "combination."

Each ratchet has 20 teeth with one deep cut or tooth. Each tooth of ratchet H_4 is equivalent to 8000 revolutions of shaft $A;$ each tooth of ratchet H_3 is equivalent to 400 revolutions; each tooth of ratchet H_2, 20 revolutions; and each tooth of ratchet H_1 is equivalent to one revolution of shaft A. The mechanism is set for a given number of revolutions by turning each ratchet so that the deep tooth is away from the operating pawl a certain number of teeth, the number depending, in each case, upon the number of revolutions of A represented by each tooth. For instance, to set the combination for a reversal of motion after shaft A makes 49,763 revolutions, ratchet H_4 is so located that there are six teeth between the operating pawl and the deep tooth; these six teeth are equivalent to 48,000 revolutions of A. Ratchet H_3 is then set at four teeth, representing 1600 revolutions of $A;$ ratchet H_2

is set at eight teeth, equivalent to 160 revolutions of A, and, finally, ratchet H_1 is set at three teeth, representing three revolutions of A. The mechanism is now set for a total of 48,000 + 1600 + 160 + 3 = 49,763 revolutions.

After the mechanism has been set in the manner described, its action is as follows: The pawl G_1 which is actuated once for every revolution of gear D, drops into the deep tooth or notch of ratchet H_1 after engaging three teeth on H_1, since, in this particular case, this ratchet was adjusted so that there were three teeth between the pawl and the deep tooth. As soon as this deep tooth is engaged by the pawl, the ratchet H_2 is turned a distance equivalent to one tooth; ratchet H_2 then remains stationary until H_1 has made a complete turn and its pawl again drops into the deep tooth, when H_2 is again moved one tooth. The pawls are so located that the first one must engage the deep notch before the next successive pawl can engage its ratchet at all, and the relation between the other pawls is the same. Ratchet H_2 continues to be moved a single tooth for each complete revolution of H_1 until it has moved eight teeth, in this particular instance. The pawl of H_2 then drops into a deep tooth and ratchet H_3 is moved one tooth. Ratchet H_3 now remains stationary until H_2, by the continued action of H_1, makes a complete revolution, when H_3 is moved another tooth. After H_3 has moved a distance equivalent to four teeth, its pawl, in turn, drops into the deep notch, and ratchet H_4 is turned one tooth. A complete revolution of H_3 turns H_4 another tooth and, when H_4 has moved six teeth, in this case, the shaft A will have made a total of 49,763 revolutions.

This result will be verified in order to more clearly show the action of the mechanism. As previously mentioned, each ratchet has 20 teeth. Each tooth of H_1 represents one revolution of shaft A and the movement of three teeth prior to engagement with the deep notch equals three revolutions. Since H_2 is set for a movement of eight teeth, H_1 will have to make eight complete turns, which will be equivalent to 160 additional turns of A. Now the four complete turns of H_2 necessary for moving H_3 four teeth require $4 \times 20 \times 20$ or 1600 additional turns of A, giving a total of 1763 revolutions. Finally, the movement of H_4 six teeth requires $6 \times 20 \times 20 \times 20 = 48,000$ additional turns of A, so that the total number of revolutions made by A prior to reversal equals 49,763, when the ratchet mechanism is set as previously described.

The progressive action of the ratchets gradually revolves the cam disks preparatory to shifting the reversing clutch. The cam J on disk I_4 first engages and lifts the floating lever K_1 at the left-hand end and the lever L one-half as much. When the other cam disks act upon K_1 and K_2, these levers, together with part L, are lifted the full amount and spring balls in L cause it to be thrown quickly into mesh with the clutch C. This clutch is threaded and two threads are also formed on the upper side of part L. As levers K_1, K_2, and part L are contained in a carriage M, all are constrained to move parallel to the axis of shaft A, because of the action of the screw threads on clutch C. This results in breaking the combination; that is, the floating levers are all removed from the cams, the four pawls G_1 are disengaged from their ratchets, and the idle set of pawls G_2 comes into action, thereby reversing the rotation of the controlling mechanism. As soon as the travel of carriage M is completed, which requires 1½ revolution of the clutch, the latter is constrained to act along the threads on L while making one revolution, until feathers attached to the clutch over-ride the spring balls in shaft A; the clutch is then instantly thrown out of mesh with one bevel gear and into mesh with the other, thereby reversing the rotation of shaft A. Just as the clutch starts this rapid shifting movement, the cam on gear D engages one of the rollers N on part L and throws the levers all down and the threads on L out of mesh with the clutch.

The rotation of shaft A is stopped after a predetermined number of reversals by means of a separate mechanism which arrests the movement of carriage M midway of its travel. The worm threads on clutch C then act upon part L and withdraw the clutch until it is out of engagement on one side and cannot engage on the other. Carriage M is stopped by a pin O which drops into a groove in M after the four ratchets P, having ten teeth, are properly aligned as regards a deep notch in each ratchet. These ratchets are operated consecutively by a stepped four-fingered pawl on R through the medium of a pin connecting with M.

The three diagrams in the lower part of Fig. 18 illustrate the systems of gearing controlled by the mechanism described in the foregoing. The requirements, as illustrated by these three diagrams, are as follows: 1. That a reversing gear shall drive two others continuously in the same direction but in opposite directions relative to each other. 2. That a reversing gear shall drive one of the two gears continuously in the same direction and the other in the same direction as that of the reversing gear. 3. That a reversing gear shall drive one gear continuously in the same direction and shall drive two others alternately in the same direction as itself. The full arrows and the full circles on these diagrams belong together, and likewise the broken or dotted lines and arrows. The full lines connecting the centers indicate that those gears are linked, whereas the broken lines denote spring connections. The movement of the reversing driver and the friction of the links swing the idler gears.

CHAPTER VII

OVERLOAD RELIEF MECHANISMS AND AUTOMATIC SAFEGUARDS

Some tripping or stop mechanisms are designed for machines requiring an automatic safeguard either against overloads and resulting excessive strains on the machine parts or as a means of protecting the machine against some other dangerous operating condition. A tripping mechanism of this general class may be so arranged that it will transmit power under normal conditions but disconnect the driving and driven parts if the resistance or strain becomes excessive, thus safeguarding the machine from injury. Automatic trip mechanisms are also utilized in some cases to prevent excessive speeds, or possibly for stopping the machine automatically if a part being operated upon is not in the proper working position. The following examples illustrate typical applications of different types of automatic safeguards.

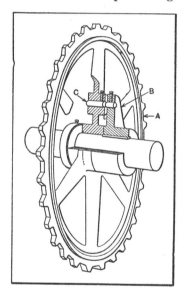

Fig. 1. Sprocket Driven Through Pin which Breaks in Case of Excessive Overload

Breakable Pins to Prevent Excessive Overload.— Some types of machines are so arranged that any unusual resistance to motion will automatically stop either the entire machine or whatever part is affected, in order to prevent damaging the mechanism or straining it excessively. A simple form of safety device consists of a pin which shears off or breaks in

198

case the overload becomes excessive. The sprocket A shown
in Fig. 1 is provided with a pin of this kind. This pin C connects the driving hub B with the hub of the sprocket. The
sprocket, instead of being keyed to the shaft, is loosely
mounted on it, and the hub B is keyed to the shaft instead.
The pin C is grooved or reduced in diameter an amount depending upon the maximum amount of power to be transmitted. If this pin is subjected to an unusual strain, it will
break, thus leaving the wheel free and protecting the driven
parts.

This same method of protection against overload has been
applied in various ways, and, while it is simple, there are certain disadvantages. In order to avoid replacing a broken
pin, the machine operator sometimes inserts a pin that is
stronger than it should be to afford adequate protection against
injurious strains. The ideal safety device is one which does
not break in case of overload, but simply disengages and is
so arranged that it can readily be re-engaged. In electrical
work, this principle has been applied by substituting circuit-
breakers for fuses which melt when the current becomes
excessive.

Automatic Clutch Control to Prevent Overload. — The principle governing the operation of an automatic device for disengaging a clutch when the overload becomes excessive is
illustrated by the diagram, Fig. 2. This mechanism was
applied to a metal-cutting machine, the object being to automatically disengage the feed in case the resistance to the rotation of the tool becomes abnormally high. The mechanism
is also arranged to reverse the feeding movement if, for any
reason, the excessive resistance should continue after the feed
has been disengaged. The spindle to which the cutting tool
is attached is represented at A. This spindle is driven through
worm-wheel M and worm L from the driving shaft B, which
receives its motion from a countershaft through a belt operating on pulley K. The driving shaft B is free to move in a
lengthwise direction within certain limits. The clutch C is
keyed to this shaft so that it will rotate and move axially with

the shaft. The gears D and F on each side of clutch C are free to revolve upon the shaft, but are prevented from moving in a lengthwise direction. The inner side of each gear is provided with clutch teeth corresponding to those on clutch C, which is used to lock either gear to shaft B. The shaft I, which transmits feeding movement to the cutting tool, is driven either through gears D and E or through gears F, P, and H. When clutch C engages gear D, the cutting tool is fed forward by shaft I, and a reversal of the feeding movement is obtained when clutch C is shifted into engagement with gear F.

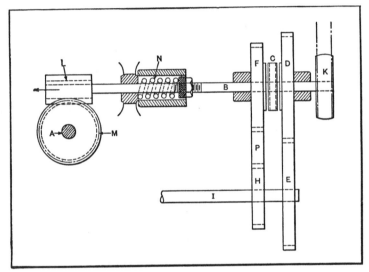

Fig. 2. Device for Automatically Stopping Feeding Motion when
Resistance to Rotation Becomes Excessive

When clutch C engages with gear D, excessive resistance to the motion of the cutting tool will cause the clutch to be shifted to the neutral position, thus stopping the feeding movement. This automatic action is obtained as follows: The shaft B is normally held by spring N in such a position that clutch C engages gear D, so that the feeding movement is forward. The tension on this spring is regulated by the nut shown. In case the resistance to the rotation of the cutting tool and spindle A should become excessive, the pressure between the teeth of the worm *L* and the worm-wheel *M* causes the worm to move in the direction indicated by the arrow, the worm-wheel acting somewhat like a nut. This lengthwise movement of worm-wheel *L* and shaft *B*, against the tension of spring *N*, disengages clutch *C* from gear *D* and stops the feeding movement. If the resistance to rotation again becomes normal, clutch *C* is automatically returned into engagement with gear *D*. On the other hand, if the resistance to rotation increases, clutch *C* may be drawn over into engagement with gear *F*, thus reversing the feeding movement.

Overload Relief for Worm Drive. — Other mechanical devices for automatically disengaging the driven member whenever the resistance to motion increases excessively are shown at *A* and *B* in Fig. 3. These devices operate on the same general principle as the one previously described, but differ somewhat in regard to the arrangement. The mechanism illustrated by diagram *A* is designed to allow a worm-wheel to make on revolution and then stop; the movement, however, may be discontinued before the revolution is completed, if the resistance to rotation becomes excessive. The sleeve *a* is revolved constantly by a pulley on its outer end. The inner end of this sleeve has clutch teeth intended to engage corresponding teeth on the end of sleeve *b*. The latter is attached to the shaft and both are free to move slightly in an endwise direction. The body of sleeve *b* is threaded to form a worm which engages worm-wheel *c*. The spring *e* tends to shift sleeve *b* to the left and into engagement with clutch teeth on sleeve *a*. The stop at *d* is utilized in this particular case to disengage the driving clutch after the worm-wheel has made a revolution. If stop *d* is withdrawn, the spring *e* revolves the worm-wheel slightly and moves the worm and clutch *b* to the left and into engagement with the constantly revolving clutch *a*. The worm-wheel then begins to revolve and continues until the lug *g* strikes the stop *d* or until some unusual resistance too great to be overcome by the spring is encountered; then, as the worm-wheel remains stationary, it forms a nut for the worm which screws itself out of engagement with clutch a. The strength of spring e is proportioned with reference to the safe or maximum load to be transmitted. One of the advantages of this type of mechanism is that the motion is positively transmitted until an excessive load causes the driving clutch to be disengaged. Provision may readily be made for the adjustment of spring e so that the tension can be varied according to conditions.

Diagram B, Fig. 3, illustrates a modification of the same general type of mechanism. The shaft m is free to move

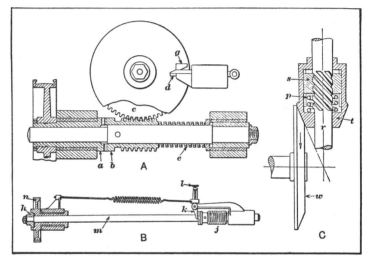

Fig. 3. (A and B) Devices for Automatically Disengaging the Driven Member whenever Resistance to Motion Increases Excessively; (C) Friction Gearing Designed to Vary Contact Pressure According to Load

slightly in an endwise direction and is keyed to the tapering disk h, which fits into a seat of corresponding taper in the hub of gear n, thus forming a friction clutch. Motion is applied to gear n and is transmitted by worm j to a worm-wheel (not shown), for any desired purpose. Shaft m turns freely in the hub of gear n, but is attached to worm j. The lever k, which has a spring fastened to it above the fulcrum or pivot, supplies the necessary amount of thrust to keep h in engagement with n under ordinary conditions. This thrust may be regulated by the thumb-screw l which changes the position of the block to which the spring is fastened. If the resistance to the motion of the worm-wheel becomes excessive, the worm moves bodily along the teeth of the wheel, as though it were a nut, and, by moving shaft m and disk h to the left, disengages the friction clutch. The endwise thrust from lever k might be obtained by means of a weight instead of a spring.

Pressure of Friction Gearing Varied According to Load. — A novel design of friction gearing, in which the pressure between the two friction wheels is automatically regulated by the amount of power transmitted, is shown at C in Fig. 3. The wheel w which is the driver revolves in the direction shown by the arrow. The driven pinion t is free to either rotate or slide in a lengthwise direction upon shaft r within certain limits. This shaft has a screw of coarse pitch which passes through nut s. This nut slides in grooves in the friction pinion t so that the pinion and nut revolve together. A spiral spring p inserted between nut s and the pinion forces the latter against the driver w with a pressure depending upon the position of the nut. If wheel w is revolving in the direction shown by the arrow and the driven shaft meets with an unusual degree of resistance to rotation, as soon as shaft r lags behind or stops revolving, nut s moves downward, owing to the action of the screw, and increases the compression on spring p and also the pressure between pinion t and wheel w. When the resistance to rotation again becomes normal, the spring moves the nut slightly upward and reduces the endwise thrust. While this device may not be entirely practicable, it embodies an interesting principle.

Cam and Spring Type of Overload Release Mechanism. — The valve driving mechanism of a certain rotary valve type of gasoline engine is provided with an automatic release mechanism as a safeguard in case the rotors should seize, due to a failure of the oil or water supply. The crankshaft transmits motion to hollow shaft A, Fig. 4, through a pair of bevel gears, and this motion is transmitted through pin B, sleeve C, and pin D to shaft E, which turns the rotors through another pair of bevel gears. Hollow shaft A fits over an extension on shaft E, and is held against endwise movement by a thrust collar.

Pin B engages a slot in sleeve C, this slot being similar in form to the well-known bayonet lock. Pin D in shaft E engages a cam surface formed on the end of sleeve C. The sleeve is held firmly against pin D by spring F, the tension of which may be adjusted by changing the position of collar G. A ball thrust bearing H is located between the spring and sleeve C.

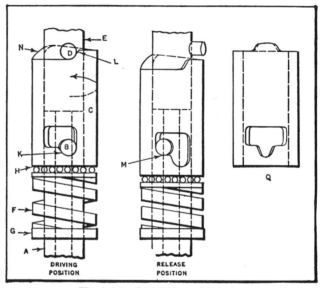

Fig. 4. Overload Release Mechanism

Normally, the entire assembly rotates in the direction indicated by the arrow, and it will be evident that sleeve C tends to slide downward, as pin B bears against an inclined surface K and pin D bears against an inclined surface L. Ordinarily, however, sleeve C is prevented from sliding downward by the upward pressure of spring F.

If the torque reaches the danger point, the spring thrust is overcome and sleeve C moves downward far enough to release pin D, which moves along a slight rise and finally disengages pin B from slope K, thus allowing the level surface M to engage pin B, which now rests against the end of the slot. Pin B holds sleeve C in this released position, and rotation of A and C continues while shaft E remains stationary.

To re-engage the drive, the crankshaft is given a few backward turns. As shaft A revolves sleeve C, due to the upward pressure on pin B and the frictional resistance between B and C, surface N comes into contact with pin D. At this point, any further turning movement would force sleeve C downward. However, the resistance between pin B and surface M is slight in comparison with the force required to rotate surface N beneath pin D. Therefore, shaft A and pin B turn until the pin engages the lower part of this slot and is again in the driving position. The clicking sound made by the dropping of pin B shows that it is unnecessary to continue turning the crankshaft backward, although more turns than required will do no harm, because each succeeding revolution simply results in forcing the sleeve downward as surface N engages pin D, the sleeve snapping upward into place again when pin D passes surface L.

This mechanism automatically retimes the rotary valves as well as the ignition, because the various parts always occupy the same relative positions when the forward drive begins. The special sleeve shown at Q is intended to provide the automatic release, regardless of the direction of rotation. While this design Q is not actually being used, it doubtless would be satisfactory, as the principle of operation is the same. The cam surfaces are altered to provide release in either direction of rotation and to permit re-engagement by turning backward relative to the direction for driving.

Sensitive Tripping Clutch for Delicate Machinery. — A sensitive tripping clutch designed for use on delicate machines of various classes, and whenever quick disengagement is required, is so arranged that the tripping action can be controlled automatically by whatever method is most suitable for the particular application.

The driving shaft D, Fig. 5, transmits motion to the driven sleeve A through engaging clutch teeth which give a positive drive. When the clutch is to be tripped to stop the rotation of the driven sleeve, a tripping pawl engages a notch in trip-ring H, which instantly disengages the driving and driven members, as described in detail later.

One arrangement of the tripping pawl T is shown in Fig. 7; this view includes ring H of the clutch in order to illustrate clearly the general arrangement. When pawl T engages one of the notches in ring H, as shown, the clutch is tripped. Just how this tripping action occurs will be apparent by referring to the details of Fig. 5. The driving shaft is connected to

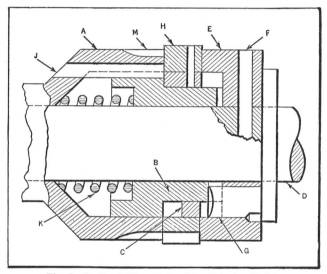

Fig. 5. Sectional View of Sensitive Tripping Clutch

sleeve E by pin F, and in sleeve E there are four driving pins G. These pins engage teeth on the small diameter of throw-out cam B. The cam (which is shown in detail at B, Fig. 6) is secured to the driven sleeve A by four pins J, Fig. 5, in such a manner as to allow cam B to slide axially. When in the running position, cam B is held in engagement with driving pins G by spring K.

Trip-ring H and ring C are connected by pins, so that they function as a single part, which can revolve but cannot move in any other direction. Clutch teeth on ring C are in alignment with corresponding teeth on the larger part of cam B (see detail of ring C in Fig. 6). When the tripping pawl, which normally is out of engagement with ring H, snaps into the tripping position, rings H and C stop revolving. Throwout cam B, however, and the driven sleeve continue to revolve until the teeth engaging ring C slide backward far enough to disengage the teeth engaging driving pins G, thus stopping

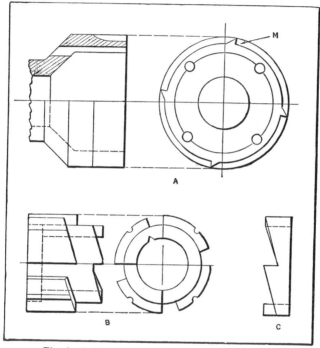

Fig. 6. Detail Views of Parts A, B and C, Fig. 5

the rotation of B and the driven sleeve. It will be seen, therefore, that the teeth on ring C act as cams to disengage the teeth on the smaller diameter of B which mesh with the driving pins.

To prevent continued rotation of the driven sleeve, as the result of friction, small teeth M are provided in the driven sleeve (see also detail view A, Fig. 6). These teeth M are also engaged by the tripping pawl, but they are so spaced in assembly that there is time enough after the stopping of ring *H* for the driving and driven teeth to disengage before a tooth *M* on the driven sleeve comes around into engagement with the tripping pawl. The end of the driven sleeve can be fitted with either a gear or sprocket, according to the driving requirements.

The method of tripping the clutch depends upon how the clutch is applied. The diagram Fig. 7 represents an arrange-

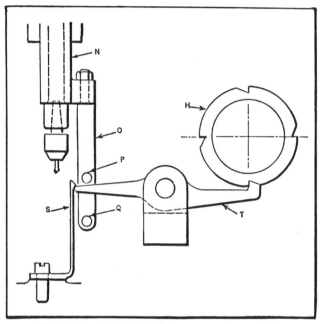

Fig. 7. One Method of Tripping Clutch Automatically

ment intended for a special drilling fixture. The rack sleeve *N* is free to move vertically, but it does not revolve. The attached bar *O* carries the trip-pins *P* and *Q*. These pins might be held in a T-slot to provide adjustment if necessary. It will be apparent that pawl *T* moves into engagement with ring *H* when pin *P* pushes the end of *T* down past the apex of the tripping spring *S*. The tripping pawl snaps quickly into engagement with ring *H* after its pointed end passes the apex of the tripping spring. The disengagement of the tripping pawl and the engagement of the clutch occurs when pin Q strikes the pawl as it ascends. By varying the angle of the engaging points on pawl T and spring S, the sensitiveness of the tripping action may be varied as required. This general method of tripping can be applied when tapping to certain depths and for various similar purposes.

An entirely different tripping arrangement is illustrated by the diagram Fig. 8. This particular arrangement might be utilized in connection with fine wire drawing or whenever the tripping of the clutch is to occur automatically in case of wire or thread breakage. The diagram represents wire drawing, although it is evident that the same principle might be applied to other processes. The winding drum is indicated at

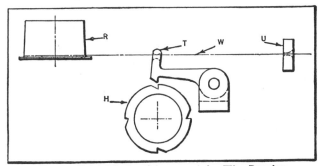

Fig. 8. Clutch-tripping Arrangement for Wire Drawing or Similar Process

R, the drawing die at U, and the wire by line W. The wire passes through a bell-mouthed hole in tripping pawl T. The pawl, in this instance, is supported by the wire under normal working conditions. If the wire should break accidentally, or if the end passes through the die, the pawl drops by its own weight and engages the trip-ring H. It will be evident from the foregoing examples that the tripping of the clutch might be controlled in many other ways.

Automatic Relief Mechanisms for Forging Machines. — Forging machines are equipped with a tripping or relief mechanism which prevents excessive straining or breakage of the parts controlling the motion of the movable die, in case the stock to be forged is not placed in the grooves of the dies, but is caught between the flat faces. These relieving mechanisms differ somewhat in design, but the object in each case is to temporarily and automatically release the movable die from the action of the driving mechanism, in case the operating parts are subjected to a strain or pressure that is abnormally high. The release may be obtained by inserting bolts or "breaker castings" in the mechanism, which will shear off or break if there is an excessive strain; another type of relief mechanism depends for its action upon a spring which is proportioned to resist compression for all ordinary strains but to compress sufficiently to release the pressure on the dies when that pressure increases beyond a safe maximum.

Spring and Toggle Relief Mechanism. — The plan view of a forging machine, shown in Fig. 9, illustrates one method of arranging a spring and toggle relief mechanism. When this machine is in operation, the stock is gripped between the stationary die A and the movable die B. The heading slide C, which carries a ram or plunger for performing the forging operation, is actuated by a crank on the crankshaft D. The gripping slide E to which die B is attached is moved inward for gripping the stock and outward for releasing it, by means of two cams F and G. These cams transmit motion to slide H, which is connected with slide E through a toggle and link mechanism. Cam F, acting upon roll T, moves the slide E for gripping the stock, whereas cam G, in engagement with roll V, withdraws the die after the forging operation is completed. The upper detail view to the right shows the relief mechanism in its normal position, and the lower view shows it after being tripped to relieve any abnormal pressure on the dies.

When the machine is operating normally, link J, which connects with link K of the main gripping toggle, oscillates link K about pivot L and, through link M, imparts a reciprocating motion to the gripping slide E. If a piece of stock or some other part is caught between the flat die faces, the gripping action continues until the strain exceeds a certain amount; then the backward thrust upon link N causes it to swing about pivot O (see lower detailed view) carrying with it the other links of the "by-pass toggle" and compressing the spring S which is shown in the plan view at the left.   As the result of this change in the position of the by-pass toggle, pressure on the gripping die is released.   Meanwhile the heading tool at-

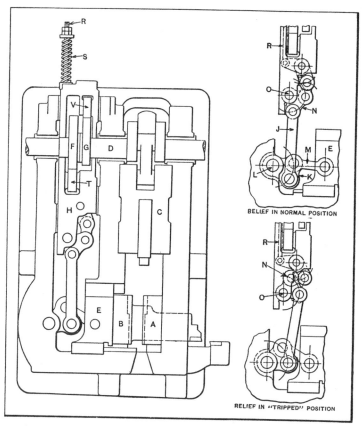

BELIEF IN NORMAL POSITION

RELIEF IN "TRIPPED" POSITION

Fig. 9. Plan and Detail Views of Forging Machine Showing Automatic Relief or Tripping Mechanism

tached to slide C completes its full stroke and, upon the return stroke, the by-pass toggle is re-set automatically by spring S which expands and, through rod R, swings the toggle links back to their normal position shown in the upper detailed view.   This automatic re-setting of the toggle makes it unnecessary to stop the machine, as is necessary with safety devices of the breaking-bolt type. There is no movement of the by-pass toggle, except when a "sticker"—to use the shop expression—is caught between the gripping dies. While this relief mechanism safeguards the working parts of the forging machine from excessive strains, it is capable of transmitting enormous pressures to the gripping dies.

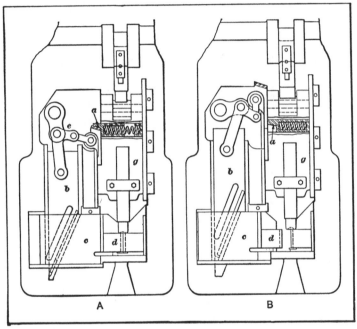

Fig. 10. Bevel Spring Plunger Type of Relief Mechanism on Bolt and Rivet Header

Beveled Spring-plunger Relief Mechanism. — The type of relief mechanism illustrated in Fig. 10 is applied to a wedge-grip bolt and rivet header. The movable die d is attached to slide c which is beveled to correspond with the tapering end of slide b. Slide b is given a reciprocating movement by the toggle mechanism at e, and, when slide c is pushed inward for closing the dies, the beveled end of slide b forms a solid metal backing, which securely locks the movable die during the heading oper-ation. When forming the heads on bolts or rivets, it is necessary to place the stock directly in the impression in the gripping dies, and not between their opposing faces, as these dies are intended to come together, so that the stock is firmly held in the impression between them while the rivet or bolt head is formed by the tool attached to slide g. The relief mechanism, which comes into action in case the stock is caught between the dies, consists of a spring plunger a, which has a beveled end and is held outward by the spring shown. The beveled end of this plunger bears against an angular projection on a slide for transmitting motion, through the toggle mechanism, to slide b and the movable die. If this die, however, is prevented from moving inward by a piece of stock that is not in the die impression, but caught between the faces, the increased pressure on plunger a forces it back against the tension of the spring and off of the beveled seat, as indicated by illustration B.

Overload Release with Positive Lock Acting During Short Period. — Often it is desirable, in feeding or other operations, to have a safety device that will function at any time except just before the end of the stroke, when the action must become positive as, for example, when a part is being locked or held solid while it is being operated on by a punch or other tool, after having been pushed into position with the automatic release ready to act in case of any obstruction. The device shown in Fig. 11 was designed to provide such an overload release and positive lock during the last 1/16 inch of the stroke. In this instance, the automatic release is used in connection with a slide that transmits a tensional or pulling force, but this type of release has also been applied with slight changes where a compressive or pushing force is utilized. Such a mechanism is applicable in conjunction with auxiliary rams or slides of punch presses, and it may also be applied to automatic machines or other mechanisms.

In the particular design illustrated, connection with the source of power is at A, and motion is transmitted through link B to slide C and through the safety device to the point of application at D. (In some cases, the operating stroke might be derived from an eccentric at the side of a press or

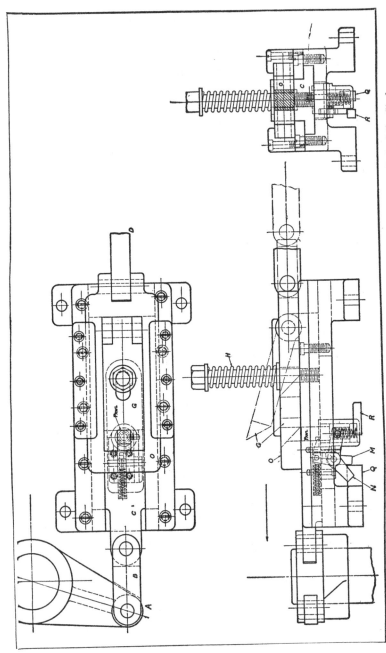

Fig. 11. Safety Overload Release which Changes to a Positive Drive Just Before the End of the Stroke

from a crankshaft.) Pivoted to slide C is a releasing plate or flap G, which is free to swing about its pivot or pin, as indicated by the dotted lines in the lower view. This releasing member G normally is held in the closed or driving position by spring H. As the end of plate G is beveled and engages an inclined surface on the driven part O, plate G tends to swing upward during the working stroke, but is prevented by spring H unless unusual resistance is encountered. The part O is in the form of a strap or loop which surrounds plate G, as indicated by the plan view, and O is positively connected to member D.

Unless some obstruction or abnormal resistance causes disengagement of the driving and driven part through the release mechanism described, movement of the slide continues in the direction of the arrow until lever M strikes projection Q, thus releasing the pawl, the location of which is indicated on the side and plan views. As the pawl is instantly forced upward by a spring beneath it, the drive is transferred from the beveled surfaces between parts O and G to vertical surfaces between parts O, G, and the pawl, so that the release mechanism is no longer effective and the stroke becomes positive.

During the return stroke, lever N strikes projection R and withdraws the pawl, so that it is in position for the next forward stroke. The illustration shows the relative positions of the parts just before the pawl is released or tripped. A spring is used to force the pawl upward when released, because this provides the instantaneous action desired.

If during the working stroke an obstruction had been encountered by the driven member, part O would have stopped as flap G disengaged from it and slide C would have traveled forward without injuring whatever tools or other members are connected at D. The action should, of course, be timed by properly locating the projections or stops for manipulating the pawl levers. The angle of the bevel between parts O and G depends upon the amount of force to be transmitted, and adjustment of spring H provides a more delicate means of regulating the releasing action.

Automatic Stop for Wire-winding Machine. — The diagram, Fig. 12, shows an automatic tripping device that is applied to a machine used for winding small wire onto spools. In this illustration, *A* represents the reel which contains the stock of wire, and *B* is the spool upon which the wire is wound. This spool is driven at a constant speed. If, for some reason, the wire should not uncoil easily from reel *A*, it might be broken or the mechanism damaged, assuming that the wire passed directly from the reel to the spool. In order to avoid trouble from any resistance to uncoiling which may occur, the wire, after leaving the reel, is guided by idler pulleys, so

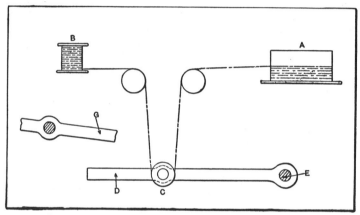

Fig. 12. Safety Tripping Device for Wire-winding Machine

as to form a loop; at the end of this loop, there is an idler pulley *C* mounted on a lever *D* which is free to swing about fulcrum *E*. When the uncoiling and winding is proceeding under normal conditions, the weight of lever *D* is sufficient to prevent the wire from lifting it; any abnormal resistance, however, such as might be caused by a kink on reel *A*, will result in swinging lever *D* upward into contact with trip *G*, which, by disengaging a clutch, stops the machine.

Automatic Stop Mechanisms for Textile Machines. — Some very ingenious tripping mechanisms or "stop motions" are applied to different classes of textile machines. The examples described illustrate the possibilities of the use of comparatively simple devices for automatically controlling the action of machines under conditions which might, at first, seem to be very complex and difficult.

The stop motion shown in Fig. 13 is applied to a machine used for twisting yarn. The yarn passes from the guide wire at A around the rolls B and C, through an eye in wire D and out through the guide at E. The wide D is attached to another wire F, which is normally held by the yarn in the position shown by the full lines. If the yarn or thread should break, the wires fall to the position shown by the dotted lines, thus bringing wire F into engagement with the lower roll C. Contact with this roll immediately moves the wires to the left until a tongue G enters between the rolls and raises B out of contact with C, which prevents it from revolving and stops the delivery of yarn.

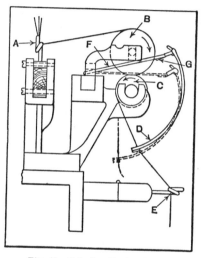

Fig. 13. Tripping Device for a Textile Machine

Another stop motion which acts when a thread is broken is shown in Fig. 14. This mechanism is applied to a machine used for winding thread on spools. It is designed to raise the spool out of contact with a flange which drives it by friction, if a thread breaks, thus arresting the motion of the spool without stopping the spindle on which the spool is mounted. The device is also arranged so that the wire which drops when a thread breaks is raised automatically to its normal position for re-threading. The thread A passes through the eye of a drop wire B and serves to hold this wire in its normal position. Attached to this wire there is a lever C pivoted at D and connected by link E with the catch F. The lever G is normally held in a horizontal position by catch F. If a thread breaks, however, the dropping of wire B releases catch F and lever G falls to the position shown in the illustration. This lever is connected by a rod H with a sleeve J pivoted at K. The downward movement of lever G swings the sleeve about its pivot and brings a pin under the flange R of the spool, thus raising it from the supporting disk L, as shown in the illustration; at the same time, the flange of the spool engages a rubber disk m which stops the rotation. Attached to the shaft of lever G, there is a small finger O which is given a partial turn when the catch lever falls. As the result of this movement, the finger engages lever C and swings it with the drop wire B back to the normal position ready for re-threading. As soon

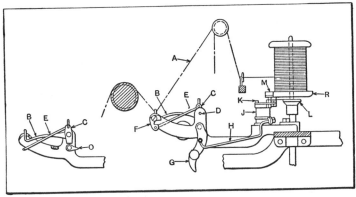

Fig. 14. Another Tripping Device or "Stop Motion" for a Textile Machine

as the catch lever has been re-engaged with the catch F, the spool drops into contact with its driving flange and again begins to wind the yarn.

Stop Mechanism That Operates When Work is not in Position. — A machine for placing covers on cans required a device to prevent a cover from being dropped when there was no can in place to receive it. The cans are pushed one at a time in the direction of the arrow (see plan view of the diagram, Fig. 15), and they slide along bridge B and stop over table A. This table is slotted to allow it to move upward until the recessed surface is raised above the surface of the bridge. As the table rises, the can is centered by a chamfered

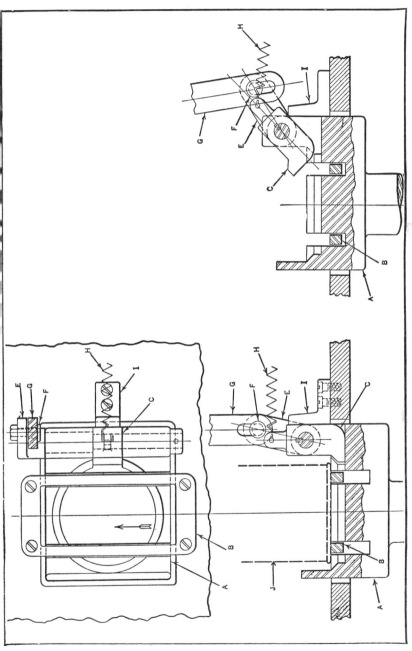

Fig. 15. A Machine which Places Covers on Cans is Equipped with this Mechanism which Stops Action cr Cover-dropping Mechanism if Can is not in Position

edge, and raised to receive a cover. The cover dropping mechanism is operated by link G which, in turn, is operated by stud F on arm E.

This device is most easily explained by first considering its action when table A rises without a can in place. Referring to the sectional view at the left, a detector foot is shown at C. Spring H is attached to the upper end of the detector foot and tends to swing it to the right, but this turning is prevented by block I. The detector foot is keyed to a shaft which is carried by the table, and when the center of the shaft passes the top of the block I, the foot is pulled over by the spring, as shown in the right-hand view. This also allows arm E to swing over without causing an upward movement of the link G. It will be noticed that this link is slotted so that stud F cannot move it until after the detector foot has reached a point where it can swing. Thus it is seen that if there is no can on the table, the detector foot, arm, and link will swing without actuating the cover-dropping mechanism.

Now suppose that a can J is in place on the bridge. As the table rises, the can settles into the recess, and when the shaft center passes the top of block I, the toe of the detector foot will come into contact with the rim of the can. The can prevents any further swinging movement of the foot, and therefore stud F pushes link G upward, thereby causing a cover to drop. The center of the stud in arm E is placed slightly to the right of the center of the shaft; so that while there is some tendency to swing the foot, it is not enough to crush the can.

Safeguard for Delicate Mechanism of Adding Machine. — The delicate and intricate mechanism of an adding machine is safeguarded from injury resulting from careless or rough operation, by the ingenious mechanism shown in Fig. 16. This controlling device is so arranged that the force or power exerted upon a hand lever by the operator is not transmitted directly to the mechanism, but the operation of the machine is subject to spring action at a certain known rate and with a known driving force. The operator is only allowed to supply the power for stretching certain springs and releasing their action, the arrangement being such that he cannot apply his strength directly to the mechanism.

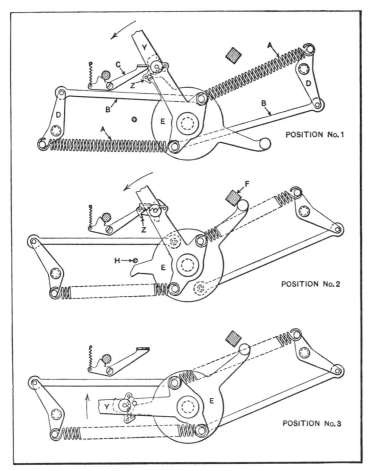

Fig. 16. Spring Controlling Device which Safeguards the Delicate Mechanism of an Adding Machine

The inner end of the operating lever is shown at Y. The upper view marked "position No. 1" shows the lever after it has been pulled forward for operating the machine. This lever revolves freely on its shaft and the operating parts are driven by the member E. As lever Y is pulled forward, catch

C releases pawl Z, thus allowing member E to fly forward as springs A contract and transmit motion to E through levers D and links B. This free movement of part E operates the forward stroke of the machine, and the rate of action is controlled by an oil by-pass governor (not shown). The movement of E is stopped by abutment F in position No. 2. The forward movement of operating lever Y, which is now free of the mechanism, is continued by the operator until the lever arrives at position No. 3. At this point, the end of pawl Z strikes pin H, throwing the pawl back into engagement with the projecting lug on part E. As the lever moves from position No. 2 to position No. 3, the springs A are extended, and acting through levers D and links B they return lever Y to the starting position, the lever carrying with it part E and the mechanism of the machine. This reverse movement is also under the control of the oil governor. Near the end of the return movement, latch C rises to permit pawl Z to pass, and the various parts return to position No. 1 ready for another stroke.

The provision of two springs A is simply for balancing the strain on the mechanism. The double action of these springs, which makes it possible for them to operate the mechanism on both the forward and backward strokes, is due to the fact that they are connected to movable members at each end. In one case, the connection is to levers D and in the other to operating lever Y. The contraction of these springs between positions No. 1 and No. 2 operates the forward motion, and their contraction from position No. 3 to No. 1 operates the backward motion.

Centrifugal Type of Speed-limiting Device. — The automatic speed-limiting device described in the following was designed for application to gas or gasoline engines. In case the speed becomes excessive, owing to the failure of the governor, this tripping mechanism, which is of the centrifugal type, operates by breaking the ignition circuit. It may be attached either to the secondary shaft or to the main shaft. The controlling element is a weight A (Fig. 17) which is attached to a rod connecting with a spring B on the opposite side of the hub. This weight is located within a casing C carried by a stud D screwed into the end of the shaft. Pivoted near the casing is a latch E which normally holds the weighted trip-lever F in the position shown. The ignition switch is located at G and, when the lever F is held up by latch E, the ignition circuit is closed. If the speed of the engine is increased to such an extent that the action of centrifugal force causes weight A to fly outward against the tension of spring B, the end of rod H, by striking catch E, releases lever F and allows it to fall, thus breaking the ignition circuit.

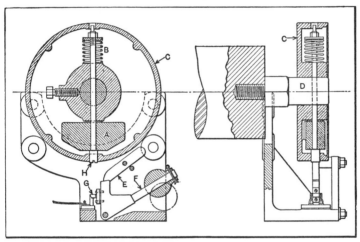

Fig. 17. Centrifugal Type of Speed-limiting Device Designed for Gas or Gasoline Engines

Over-speed Limiting Device for Electric Locomotives. — There exists, on certain railroads, the necessity of employing a reliable over-speed limiting device, which will prevent the application of further propelling power to an electric locomotive after a predetermined speed has been reached. The prevention of over-speeding protects the traction motors or other locomotive parts from being injured by speeds in excess of that for which they were designed. The device should be so designed that the engineer may again apply power as soon as the speed has decreased to some permissible value.

A device that meets these requirements has been developed for application to electric locomotives. It is, however, capable of alteration so that its field of usefulness can be extended to steam or other motive power, as the fundamental device is capable of various simple modifications which will perform different functions as desired. By mounting a Veeder counter on the contact box, a record can be obtained of the number of times an engineer has over-speeded his train. The system may also be interlocked with the air brake by the use of additional relays. One relay can be made to give the engineer a warning and at the same time cause a counter to register. Another relay can start a time element device which will, after the elapse of a definite time, cause the brake pipe to be vented, thus applying the brakes. In case the engineer reduces the speed of the train, when given warning that he is exceeding the prescribed speed, before the automatic feature has had time to work, nothing further will happen and the device will then return to its normal position.

The centrifugal member seen at the right in Fig. 18 is mounted on the center of an idle axle. One half of this member carries the centrifugal arm and adjusting spring, so that adjustments can be made on the test floor and left undisturbed when applying to the axle. The centrifugal arm has two cam surfaces A and B which are offset one from the other. These cams engage light arms A_1 and B_1 on the contact box, which is mounted close to the centrifugal member, as shown by the relative positions in the illustration.

Normally, a spring D holds the centrifugal arm in the "in" position. At some predetermined speed, the arm flies outward about the center C, compressing the spring, while cam A strikes arm A_1. This motion causes a toggle switch in the contact box E to break contact and thus interrupt the control current to the master controller. Thus further application of power to the locomotive is prevented until the speed is reduced. Upon a reduction of speed to the necessary value, the centrifugal arm moves inward about the center C and cam B strikes arm B_1, causing the toggle switch again to make the circuit in the contact box. This permits the engineer again to apply power to the locomotive. As developed at present, the range of speed between trip out and reset is from 39 to 37 miles per hour in one case and 65 to 60 in another.

In designing the centrifugal member, it was necessary to reduce friction to a minimum and also to produce bearings that would require almost no attention for maintenance. This arm, therefore, was mounted in a double-row, deep-

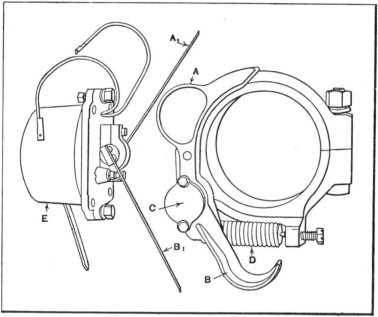

Fig. 18. Device for Electric Locomotives which Automatically Prevents Over-speeding

groove ball bearing, which insures minimum and constant friction over a long period of time. The spring D operates between hardened conical points or centers J and K, bearing in hardened pockets (see Fig. 19). It is restrained from buckling by hardened sliding guides F and G within the spring. The outside of guide G has an increased diameter at H to prevent the spring from buckling. This mounting resulted, on test, in an extremely sensitive centrifugal arm which, depending on the characteristic of the spring, could be made to fly out positively and return for a difference in speed of approximately 1 per cent.

The contact box is designed with special reference to withstanding the hard blows received from the cams. The striking arms are made of hardened steel springs, tapered toward the outer ends to reduce inertia and consequent rebound when the cam strikes. It was found necessary to provide a spring loaded friction drag in box E to prevent the striking arms

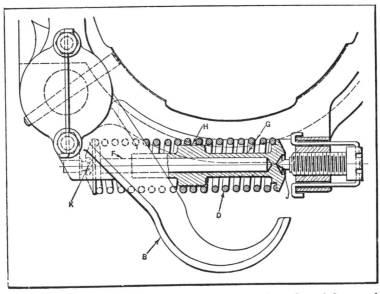

Fig. 19. Detail of Sensitive Spring Mounting for Centrifugal Arm of Over-speed Limiting Device

from rebounding and rubbing against the cams. The moving contact is a light weight element, which is connected to the striking arms only by means of two "over the center" springs, and is supported on knife-edge bearings. The contact does not move until after the striking arms have passed the center. It then snaps over quickly, regardless of the speed at which the striking arms are moving. The contacts are made of phosphor-bronze and in the form of flat springs which bear against copper-graphalloy buttons.

Speed-limiting Device for Steam Engines. — A speed-limiting device which is governed by the inertia of a weight and the tension of a spring is shown in Fig. 20. This automatic stop was designed for application to steam engines but devices operating on the same general principle could doubtless be applied to other classes of machinery. This mechanism is primarily a safety device and is intended to stop the engine and prevent damage such as might be caused by a bursting flywheel, in case the governor failed to operate. The lever A is pivoted at B to the engine cross-head and is normally pre-

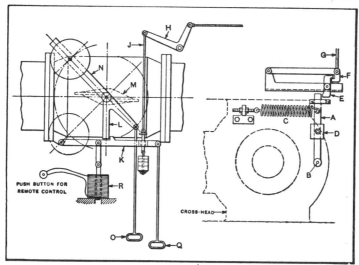

Fig. 20.   Automatic Speed-limiting Mechanism for Steam Engines

vented from swinging about pivot B by the spring C attached near the upper end. The inertia of weight D, which may be adjusted along the lever A, tends to swing the lever to the right when the motion is suddenly reversed and the cross-head moves to the left. When the cross-head is at one end of its stroke, the upper end of lever A is quite close to the catch E, which engages latch F. Rod G attached to this latch connects through whatever additional rods or levers may be needed, with the tripping mechanism used in conjunction with a quick-closing valve which controls the flow of steam to the engine cylinder. This valve and its operating mechanism is shown in detail at the left of the illustration. Rod G is connected in any convenient way with bellcrank lever H, from which rod J carrying weights at its lower end is suspended. This rod passes through trip-lever K, which normally engages lever L connected with the quick-closing valve M. If, for any reason, the speed of the engine becomes excessive, the lever A and its attached weight resists the sudden reversal of motion at the end of its stroke sufficiently to overcome the tension of spring C, and lever A strikes catch E, thus releasing latch F; as rod J drops, the flange on it strikes trip K and allows the steam valve to be closed by the weighted lever N. This speed-limiting device may be adjusted by varying the tension of spring C and also by changing the position of weight D. The greater the spring tension and the nearer the weight is to the pivot B, the faster the speed will have to be to overcome the tension of the spring at the point of reversal. The handle O is for resetting the steam valve and handle Q, for tripping the valve by hand. If remote control is required, this may be obtained by the use of rods or cables directly connected to latch K, or by the use of a solenoid R, as indicated by the illustration. This automatic safety stop is recommended as being simple, positive in action, adjustable, inexpensive, and easily applied to almost any engine.

CHAPTER VIII

INTERLOCKING DEVICES

THE primary purpose of an interlocking device is to prevent the simultaneous engagement of conflicting mechanisms, so as to eliminate, as far as possible, the danger of straining or breaking any of the machine parts. There are many examples of interlocking devices in which the component parts are much more costly to produce than is necessary or justifiable, considering the function of the device. Notwithstanding that the purpose of an interlocking device is to protect expensive parts of the machine from destruction, there is seldom any need to provide other than the simplest mechanism to accomplish the desired end. Elaboration or complication in this connection is unnecessary, and should be guarded against.

A separate movement should not be required to lock or unlock such a device, but the ordinary manipulation of the parts controlled by the interlocking mechanism should, without any further effort insure the correct functioning of the safety device. Otherwise, the operative time will be increased so that there will be much less available time for productive work. Over-elaboration in design as well as a tendency to make the parts much stronger than is necessary should be avoided. Although the latter fault may have very little influence on the cost, a more careful consideration of the design and the stresses imposed might result in much greater compactness and neatness.

Two-lever Type of Interlocking Mechanism. — Fig. 1 shows a typical interlocking device, which, in various forms, is extensively used. The two levers A and B usually control two pairs of sliding gears by means of which one shaft is driven

229

from another shaft at four different speeds. Since all the four pairs of gears are of different ratios, it is necessary to have some element introduced that will insure that one or the other of the two levers shall always be in the neutral position. Obviously, if some such provision were not made, there would always be a serious risk of one of the levers being accidentally engaged while the other was in engagement. This would, of course, be rather disastrous for the gears and other trans-

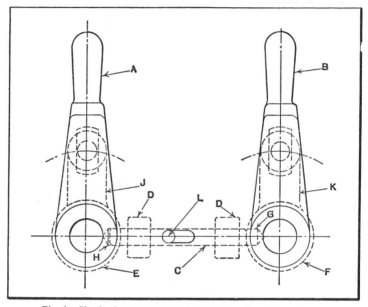

Fig. 1. Simple Arrangement of Interlocking Gear-change Levers

mission elements concerned, and it is to avoid such a possibility that the levers are designed to interlock so that one is always in the neutral position.

The locking element itself consists of a short piece of round steel C carried in bearings D, and engaging with holes H and G, in the bosses of the levers J and K. There is really no reason, except for the sake of appearance, why the sliding rod C should not be outside the gear-box housing and engage the bosses of the hand-levers A and B. A pin L, working in an elongated slot and fixed in the rod C, is the medium through which the rod is moved from left to right, or vice versa, as required. This arrangement is non-automatic and locking and unlocking are accomplished in a separate movement. As previously mentioned, this is not a desirable feature.

In the illustration, the lever B is shown unlocked ready for engagement. When it is swung either to the left or to the right, the hole G in the boss of lever K, is moved out of alignment with the rod C. It is therefore impossible for rod C to be moved, and since its left-hand end is engaged with the hole

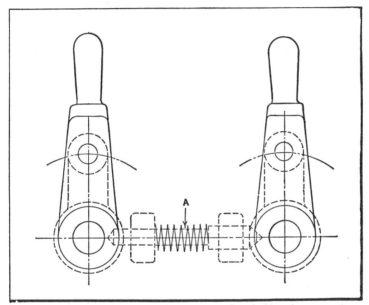

Fig. 2. Gear-change Levers with Automatic Interlocking Arrangement

in lever J, it is also impossible to move the lever A, and the gears that it controls. If it were desired to use one of the gears controlled by lever A, it would first be necessary to move lever B to the vertical, or neutral, position, then slide the rod C to the right by means of the pin L, after which lever A could be moved either to the right or left as required, and B would be definitely locked out of engagement. It will be seen that the interlocking device involves another movement by the operator, and, since this can be eliminated without adding to the cost of the mechanism, it should be done, if possible, as it insures easier and more convenient operation.

Automatic Interlock for Speed-changing Levers. — Fig. 2 shows an automatic interlocking mechanism in which the objectionable feature of the preceding design is overcome. Here a spring plunger *A* automatically performs the interlocking function. The plunger is supported in much the same way as in the previous example, but the spring tends to push it to the right. At the right-hand end it has a conical point which engages with a depression in the boss of the lever. Whenever the lever at the right is moved into the neutral position, the plunger is pushed into the hole in the boss by the action of spring *A*. This unlocks the lever at the left, and allows it to be immediately shifted as required. When the left-hand lever has been moved, however slightly, from the neutral position, the end of the plunger comes up against the boss and prevents the plunger from moving to the left.

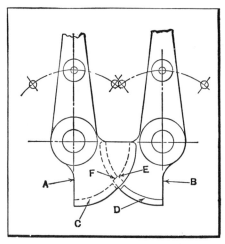

Fig. 3. Interlocking Levers that Can be Made Economically

Since the right-hand lever can be shifted only if the plunger is allowed to move to the left, this lever is positively locked in the neutral position whenever the left-hand lever is in any position other than neutral. The arrangement shown in Fig. 2 is therefore to be preferred to that shown in Fig. 1. It not only works automatically, without requiring any attention or effort on the part of the operator, but it has also the additional advantage that it is rather cheaper to produce.

Levers having Interlocking Segments. — Fig. 3 shows another arrangement for use under conditions similar to the two previous ones. The two levers are each fitted with a fan-shaped projection A and B. On the front side of the latter and the rear side of the former are two ribs D and C, respectively, which interfere with each other. Grooves E and F are cut across the center of the ribs to allow either of the levers to be moved as required. When the two levers are in the position shown in the illustration, the gears are in the neutral position, and either lever may be shifted. Whichever lever is moved, and no matter whether it is moved to the right or to the left, one or the other of the ribs is moved across the

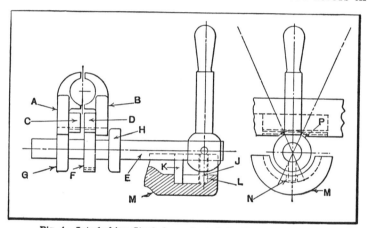

Fig. 4.   Interlocking Single-lever Control for Two Sets of Gears

gap in the other rib, thus positively locking it in the out-of-gear position. The arrangement is quite an effective one and very cheap to produce, the only criticism that might be made against it being that it takes up more space than some of the other types. Good designers take a pride in concealing such devices, as a rule, but whether the practice is worthwhile when it involves greater expense is a matter for individual decision.

Single-lever Interlocking Device. — While the three previous examples are much used in machine tool practice, none of them is ideal because there are two levers which must be manipulated in order to operate the sliding gears. The design shown in Fig 4 is much superior in this respect, since only one hand-lever is required, and this lever is so arranged that it not only operates both sets of gears but also functions as an interlocking device. This combination permits quicker operation and, incidentally, gives greater compactness to the design.

The two sets of sliding gears are engaged by forks carried on bars A and B to which are secured racks C and D. The hand-lever shaft E has a pinion F and two plain disks G and H formed solid with it, and is so arranged that it can be shifted endwise as well as rotated on its own axis. The hand-lever is pinned or keyed to the shaft, and is fitted, in addition, with a pin J which projects into two segmental grooves K and L in a hollow semi-cylindrical casting M.

Connecting the two grooves and corresponding with the vertical position of the hand-lever, is a cross-groove N, approximately equal in width to the diameter of the pin J. With the hand-lever in the position shown, the pinion F is in mesh with rack D, and consequently the gears controlled by slide-bar B will be operated if there is any angular movement of the hand-lever. If the lever and its shaft are pushed bodily to the left, the rack C will be engaged by the pinion F, and the second set of gears will thus be brought under control. The function of the two disks G and H is to lock the slide-bar that controls one set of gears in a central position when the other set is in mesh. This is done by the disk being moved into suitably shaped hollows formed in each of the slide-bars, as shown at P.

In the illustration, the disk G is shown engaged with slide-bar A, so that the gears controlled by this bar are definitely locked out of engagement. If it is required to operate slide-bar A, the hand-lever must first be moved to the vertical position (all gears then being out of mesh), after which the lever is pushed to the left, thus engaging the pinion with rack C, and the disk H with slide-bar B. In this position, the gears controlled by slide-bar B are positively locked out of engagement.

The purpose of the grooves in casting M is to insure that the lever-shaft will not be moved axially, except when it is in the neutral position. This part of the device and the pin J

could, if necessary, be carried inside the gear case, and the pin *J* could be fitted to either of the disks *G* or *H*. On these lines the design would be somewhat neater, though it is possible that it would not be quite so convenient in operation, as the grooves would then be out of sight and the exact vertical position of the hand-lever would have to be judged, more or less, by the "feel."

Interlocking Device for Lathe Apron. — Lathe aprons are usually fitted with some arrangement for interlocking the conflicting gear mechanisms. In some cases the cross and

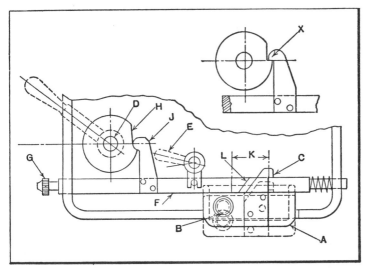

Fig. 5. Interlocking Device for Lathe Apron

longitudinal feeds are not interlocking but provision is made to prevent the engagement of the lead-screw nut when either of the other two feeds is in use. The design shown in Fig. 5 is of this type, the only interlocking action being between the lead-screw nut and the feed combinations.

The feeds are engaged by a drop-worm, carried in box *A* and engaged by handle *B,* the trigger *C* holding the drop-worm box in the engaged position. The square rod *F* has a slight endwise movement, and is cut away near the worm-box to provide a support for trigger *C.* To release the feed, the rod is moved to the right by means of handle E, while a knurled-head screw G, for fine adjustment, forms the abutment for a dog fixed to the bed of the lathe which trips the feed automatically at any predetermined point. At D is the camshaft for operating the lead-screw nut. A counter-clockwise rotation of this shaft from the position shown closes the nut on the screw. Fixed to the camshaft is a cam H which, when the lead-screw nut is engaged, pushes the rod F to the right through the medium of lever J which is firmly secured to the rod. This action releases the trigger C and allows the drop-worm to fall out of engagement with the wormgear, thus disengaging both the cross and longitudinal feeds before the lead-screw can be used. There are, however, several details in this type of design that are open to criticism, as will be pointed out in the following.

The arrangement illustrated does not meet all the requirements of a satisfactory interlocking device. Suppose a machine were fitted with a device of this type, and the work required that the lead-screw be geared up, and running, the whole time, as would be the case on almost any job that involved both turning and threading.

Suppose now, that the tool had been set up for, say, facing a moderately deep shoulder and that the cut had nearly reached the body of the shaft. If at this time the handle on the camshaft that operates the lead-screw nut happened to be accidentally depressed, the cross-feed would be disengaged, but the lead-screw would come into action· and there would be the possibility of a smash-up. There are also similar possibilities with the longitudinal feed and the lead-screw, though in this case the damage would probably not be quite so extensive.

The trouble is due to the fact that the feeds are not really interlocked, because this term means, that when one feed is engaged it is impossible to engage another; in any case it does not mean that the accidental engagement of one particular feed merely disengages the one that is in use, as in the type shown in Fig. 5. If this illustration is carefully examined, it will be seen that the portion of the rod F that is cut away at K to form the trigger is much wider than it need be. If the width were reduced and the left-hand side were beveled as at L to match the back of the trigger piece, the whole principle of the device would be changed, and it would become an effective interlock, provided cam H and lever J were made of such a shape that the former could not be made to impart any movement to the latter. To meet these requirements, cam D and lever H should be designed as shown at X.

When the trigger is engaged with the rod, that is, when the worm is in mesh, the lead-screw nut cannot be closed, because lever X is held up by the cam, so, obviously, it is not possible to engage the nut when the worm and gear are in mesh. Further, when the nut is engaged, the beveled face L prevents the engagement of the worm, as it interferes with the back side of the trigger when the rod F is positively prevented from movement toward the left. This simple alteration has converted an indifferent and imperfectly operating device into one that is correct.

Lathe Apron Interlocking Mechanism for Three Feeding Movements. — Fig. 6 shows another type of lathe apron in which all three feeds are interlocked to prevent conflicting engagement. In this design there is no drop-worm, the actual starting and stopping of the feeds being effected by a hand friction knob A, which couples gear B to the worm-gear C. At D is the rack pinion on which is keyed the large gear E, while at G is the cross-feed screw pinion which permanently engages with an intermediate gear H. Approximately midway between the intermediate gear H and the rack gear E is an eccentric shaft J carrying a gear F and a pinion Y. The eccentric shaft is controlled by a small lever K, which swings into three positions—the one shown, one to the right, and another to the left. In the position shown, the feed gears are entirely disconnected; if the lever is swung over to the right, pinion Y engages gear E, and gear F engages friction gear B. If swung the opposite direction—to the left—the eccentric shaft gear F connects the intermediate gear H to gear B.

In the first case the longitudinal feed gears are engaged,

while in the latter case the cross-feed gears are connected. Lever K is used only for engaging the required feed gears; that is to say, it does not actually start or stop the feed, this being done by the hand friction knob A. Up to this stage the mechanism is perfectly interlocked, because when lever K is swung over, either to the right or to the left, only the particular feed represented by the given location of the lever can be engaged, the other being totally inoperative—that is, of course, as an automatic feed.

Since the lever cannot be in two positions at once, only one feed can be engaged at a time, the change from longi-

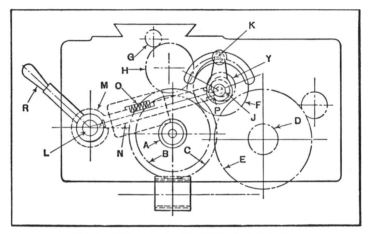

Fig. 6. Lathe Apron with All Feeds Interlocking

tudinal to cross, or vice versa, necessitating, first, the disconnection of the feed by the hand-knob; second, the correct placing of the lever for the feed required; and third, the actual starting up of the feed by the hand-knob. So far as the cross and longitudinal feeds are concerned, then, we may assume that fairly complete interlocking is secured. However, there still remains the screw-cutting mechanism.

The lead-screw nut is operated by a camshaft L, and on this shaft is a collar M in which is cut a V-shaped groove. On the inside of the apron casting is a rectangular groove running from camshaft L to the eccentric shaft J, and fitting into the groove is a plunger N which is pressed always in the direction of the camshaft by a spring O. The lower or left-hand end of the plunger is pointed to fit into the V-shaped groove in collar M so that, in effect, the plunger acts on the camshaft as a snap plunger, thus reducing the risk of its being moved by vibration. At the opposite end, the plunger is narrowed down so as to enter a cross-groove P, cut in the boss of lever K. When the lever is in the position shown (the feed-gears being out of mesh), the camshaft handle R can be depressed and the nut connected with the lead-screw. The plunger N, meanwhile, is pushed to the right into the groove P in lever K, thus effectually locking the latter in the central position.

Suppose, now, that the lever K had been moved to either the cross or longitudinal feed position. The groove P in the boss on the lever would not then be in alignment with the plunger, so the latter could not move toward the right. This being the case, the nut-operating handle R could not be depressed, as it would be locked by the V-point of the plunger. Thus, only when the lever K is in the central or out-of-gear position can the lead-screw nut be engaged, and on the other hand, only when the lead-screw nut is disengaged can the feed-lever be moved, all three feeds being by this simple means, interconnected so that no two can possibly be in simultaneous engagement.

Simple Interlock for Lead-screw and Longitudinal Feed. — An example of effective design combined with low production costs is shown in Fig. 7, the interesting feature of this particular arrangement being that a perfect form of interlock is provided without the introduction of a single extra part. The device is fitted to a lathe apron and is of the type that applies only to the lengthwise feed and screw-cutting movement, the cross-feed not being interlocked in any way. The eccentric shaft A operates a sliding clutch which engages the longitudinal feed when handle B is correctly manipulated. When the handle is pulled outward, the clutch is engaged and the feed started. At C is the cam-plate which controls the lead-screw nut; in the position shown the nut is disengaged.

On the right-hand side of the cam-plate is cast a projection *E*, while on the boss of handle *B* is a lug *D*, the lower side of which is roughly located at the same height as the upper end of the cam-plate projection *E* when the lead-screw nut is disengaged. If the handle *B* is pulled outward to engage the feed, lug *D* moves over the end of projection *E* and prevents the cam-plate from being rotated to engage the nut. On the other hand, should the nut be first engaged, projection *E* moves upward into the position shown by the dotted lines at *F* and prevents the engagement of the feed by interfering with

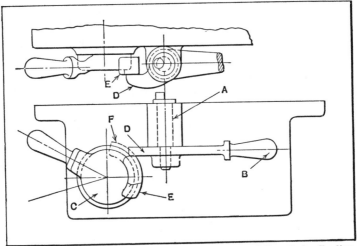

Fig. 7. Simple Interlocking Device for Lead-screw and Lengthwise Feeding
Movement

lug *D* on the hand-lever. So far as it goes, the design is perfectly effective and a good example of simplicity and cheapness, though, of course, it is rather lacking in completeness in that it does not also interlock the cross-feeding movement of the tool slide.

Interlocking Device for a Back-shaft Type of Lathe. —
Fig. 8 shows still another example of a feed interlocking device as applied to the saddle of a back-shaft type of lathe. Ordinarily, in the old style of back-shaft lathe, the three feeds, cross, longitudinal, and screw-cutting, are almost always entirely disconnected, so there is nothing to prevent all three be-

ing simultaneously engaged. To overcome the risk from this possibility, the arrangement illustrated in Fig. 8 was patented some years ago by an English firm. At the back of the saddle are two gears *A* and *B,* the former mounted on the cross-feed

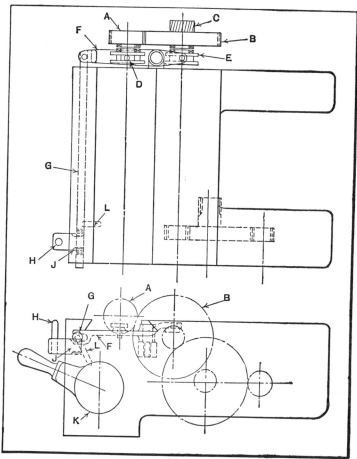

Fig. 8. Back-shaft Type of Lathe Saddle with All Feeds Interlocking

screw and the latter on the feed-shaft for the longitudinal feed. Gear *B* is driven from a worm *C* which is keyed to the back-shaft and free to slide on it, and both gears *A* and *B* run loosely on their shafts and are provided with clutch teeth on their front faces, which are engaged by clutches *D* and *E,* which are feathered to the cross-feed screw and the feed-shaft, respectively. Both clutches are operated by a single lever F through a small handle H and rack and pinion G and J.

When lever H is pushed toward the rear, clutch D is engaged with gear A, the other clutch E being meanwhile moved further away from its mating gear B. If the handle H is pulled toward the front, clutch E is moved into engagement, thus starting up the longitudinal feed. When the handle is vertical or in the mid-position, both clutches are out of engagement, and the cross and longitudinal feeds are inoperative. Obviously, then, it is impossible for the two feeds to be engaged at the same time.

Referring now to the screw-cutting motion, the nut is engaged through the usual type of cam-plate K by depressing the handle, so that the two half-nuts are engaged with the lead-screw. Immediately above the cam-plate is a short length of round rod L, the lower end of which is pointed to fit into a conical depression in the periphery of the cam-plate. The upper end just touches the round rack G when rod L is in its lowest position. Corresponding with the neutral position of the two feed clutches is a hole in rack G opposite rod L. When handle H is in the mid-position, the cam-plate handle can be depressed, rod L meanwhile being slightly raised until its upper end enters the hole in rack G, thus preventing the latter (and with it the feed clutches) from being moved. When the cam-plate handle is in the upper position so that the half-nuts are disengaged from the lead-screw, either of the feed clutches can be engaged as required, but no matter which is engaged, the hole in rack G is moved out of line with rod L, so it is impossible for the rod to be raised, and consequently for the lead-screw nuts to be operated. It will be seen, therefore, that only one of the feeds can be in operation at any given time, so that the engagement of any of the three feeds definitely locks the remaining two out of engagement.

Interlocking Device for a Geared Drive. — Fig. 9 shows a more complicated system of interlocking, as utilized in the design of a six-speed, all-geared drive machine tool. In the example to be considered the transmission consists of nine gears and three shafts. The drive is from a pair of tight and loose pulleys. The speed of the pulleys is about 500 revolutions per minute, and it is this comparatively high speed that makes it necessary that all the speed-changing levers be interlocking, so as to minimize the risk of damage by preventing the engagement of the gears while under load. The correct relative location of the shafts is shown in the end view at the left. The plan view at the right merely shows the three sets of gears as if they were all in the same plane. At *A* is the

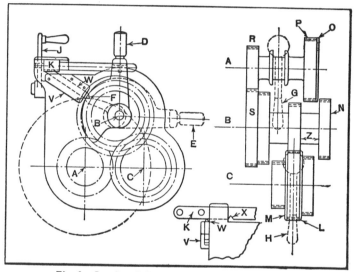

Fig. 9. Gear-box Drive with All Controls Interlocked

pulley shaft which carries twin gears having an axial movement that permits them to be engaged with either of two gears on shaft *B*. The latter is an eccentric shaft carrying four gears, all of which are fixed together, but run loose (as a whole) on the shaft. Shaft *C* carries three gears which may be engaged at will with the three mating gears on shaft *B*.

The gears on shaft *B* give three speeds, which are doubled by the twin gears on shaft *A*, thus giving six speeds in all. It might be mentioned, at the outset, that this design of gearbox is not presented as an ideal one, but it is an excellent example of interlocking, and it is only with this point that we are immediately concerned. The eccentric shaft B is really the foundation of the whole system. This shaft is controlled by a lever D, fitted with a spring plunger which holds it in the vertical position. Slightly more than a 90-degree angular movement is required to bring the lever into the position indicated at E, which causes the gears on the eccentric shaft to be thrown entirely out of mesh both with the gears on shaft A and with those on shaft C.

Referring now to shaft A, the larger of the two gears carried by this shaft has a washer O secured to it, which is of a diameter equal to the external diameter of the gear itself. This washer must lie either on the right or on the left of gear N; that is, it cannot be located anywhere in the width of face of gear N, because it would interfere with the teeth. The washer O thus insures that either the gears R and S or P and N are in mesh. Were there no washer, it would be possible for the twin gears R and P to be in a mid-position, with two pairs of different ratio gears in mesh at the same time. Of course, this trouble could be obviated by increasing the distance between the gears, but such a course would affect the interlocking properties of the design as originally conceived. The washers M and L on the largest of the three gears on shaft C serve a similar purpose to washer O. These washers prevent the half and half engagement of the gears, and necessitate that the eccentric shaft be thrown into the out-of-mesh position before the sliding gears can be shifted. It is really the object of the washers to insure that the eccentric-shaft gears are disconnected before a change of speed is made.

The next step is the connecting of the eccentric shaft to the belt-shifting gear, and this is provided for quite simply. The belt forks are mounted in a fairly substantial belt bar K, which is moved by the lever J. On the end of the gear-box is bolted a short slide V provided with a plunger W which has pointed ends. The lower end enters a V-shaped groove F in the eccentric-shaft lever, and the plunger is held down against the lever by gravity. In the belt bar K, is another V-shaped groove which receives the opposite end of the plunger.

When the belt is on the fast pulley, the belt bar is in its extreme right-hand position, and the V-groove X is out of line with the plunger, the plunger end coming snugly up against the body of the bar. In this position, the eccentric-shaft lever is locked by the plunger and is engaged with groove F in the quadrant of the lever. Since none of the sliding gears can be moved (because of the washers previously referred to) until the lever has been swung to position E, it follows that no change of speed can be made until the belt is first shipped on the pulley and the power thus shut off. Further, when the belt bar is moved to the left until groove X is opposite the upper end of the plunger—the belt then, of course, being on the loose pulley—the act of swinging down the eccentric-shaft lever raises the plunger into groove X, thus locking the belt on the loose pulley until the eccentric shaft is again returned to its "in-gear" position.

With the lever in position E, the two sets of sliding gears may be manipulated as occasion requires, and when the desired change has been made, the gears are again engaged by returning the lever to its original position, after which the belt is shipped to the fast pulley. If the sliding gears are not correctly positioned sidewise, the washers O, M, or L, by interfering with their respective gears, will not allow the eccentric shaft to go fully back into its normal "in-gear" position, so that, here again, the operator has little chance to blunder .

To alter the speed, the sequence of movements is as follows: (1) Move the driving belt on the loose pulley; (2) swing down the eccentric-shaft lever; (3) move the two sets of sliding gears to the required positions by means of levers G and H; (4) swing back the eccentric-shaft lever; and (5) ship the belt on the fast pulley. The interlocking arrangements insure that this sequence is always followed, and though it may sound a little complicated and involved, it is much more quickly done than described, and the arrangement certainly removes practically all risk of injury to the gear.

Regarding the introduction of an eccentric shaft, there may

be some difference of opinion as to the efficacy of this arrangement. It is open to the objection that it tends to slow down the operation of the machine, because of the necessity for allowing the machine to come to rest before the eccentric shaft can be safely swung back into the "in-gear" position. When ordinary sliding gears are used, the power is first shut off, the necessary speed changes made, and the belt reshipped, before the machine has fairly come to rest, thus saving a certain amount of time. On the other hand, there is one advantage to be derived from the use of the eccentric shaft, and that is the fact that double helical gears can be used for the driving gears, if desired. If this is done, the drive will be much smoother and less noisy — two features of importance when the speeds are so high.

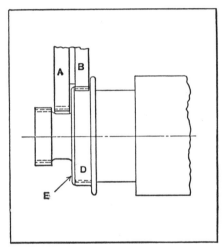

Fig. 10. Interlocking Device for Double Back-gears

Another advantage of the eccentric shaft is that it reduces the over-all length of the gear-box, because it does away with the necessity of having a space of two face-widths between the gears, as at Z With ordinary sliding gears, this distance would have to be slightly greater than twice the face width of the gear so as to allow one gear to slide out of engagement before the other entered. By swinging the gears out of mesh, as is done by the eccentric shaft, the distance Z between the fixed gears is reduced to slightly over half, thus giving greater compactness.

Interlocking Arrangement for Back-gears. — Another application of the washer or shroud idea is illustrated in Fig. 10, which shows a simple device by means of which the double back-gears of a lathe or other machine are prevented from being engaged so as to cause damage. As ordinarily made, there is nothing to prevent double back-gears from being engaged with both pairs of teeth in mesh, in which case serious damage would probably result. If, however, the gears are arranged as shown, with a washer, or shroud, equal to the external diameter secured to the larger cone gear D, there is not the slightest possibility of the gears being otherwise than correctly engaged. In some cases the two sliding gears A and B on the quill are not provided with washers and are pushed across from one position to the other by hand, so that should they happen to stick in some intermediate position, both pairs of gears would be engaged at once if the eccentric gear - meshing shaft were manipulated.

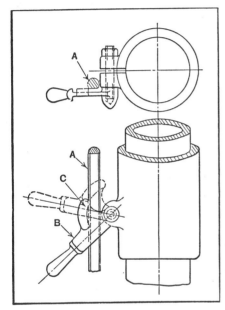

Fig. 11. Interlocking Device for Power Elevating Mechanism

With the arrangement shown, the gears must be correctly located sidewise; if they are not, the washer E will prevent their engagement by coming in contact with the top of the teeth of sliding gear B. Incidentally, it might be mentioned that double helical gears in this connection would do away with the necessity for any interfering device whatever, as the V-shaped teeth would prevent any axial movement unless the gears were first thrown out of mesh.

Interlocking Device for Power-driven Machine Tool Elevating Mechanism. — Power elevating mechanisms, which are often used in machine tool design, should be equipped with some form of interlocking device which would insure that the parts to be elevated are unclamped before the power is applied. The arrangement illustrated in Fig. 11 is designed to prevent such accidents. The shaft A, which by partial rotation controls the elevating gears, has a wide flat on one side, while the clamping lever B has a fan-shaped projection C which, when the lever is in the upper position, fits snugly against the flat of the shaft and thus prevents the shaft from being rotated. When the lever is pulled into the lower position (as shown by the full lines), the parts to be elevated are unclamped, the tail of the lever moves clear of shaft A, and the elevating gears may then be engaged as required, in the certainty that everything is in order and that there is no risk of breakage.

In some cases a different method from the one just described is followed. Somewhere in the elevating mechanism—usually in the connection between the elevating screw and its driving gear—a slipping device is fitted, this device being so adjusted that it is quite capable of performing its normal duty; but should there be any accidental over-running, or should the operator have forgotten to unclamp, or any other abnormal circumstances arise which unduly increase the load, there would be a slippage and not a breakage, thus saving the machine from serious damage and breakdown.

CHAPTER IX

DRIVING MECHANISMS FOR RECIPROCATING PARTS

MACHINES of many different types are equipped with some form of mechanism for changing a rotary motion to a rectilinear or straight-line motion, or *vice versa*. The design of such a mechanism may depend upon the kind of motion required, the amount of power to be transmitted, or other considerations. In this chapter, various mechanisms, especially of the more unusual designs, are described.

Relative Motions of Crankpin and Cross-head. — In some cases, especially in connection with steam engine work, it is important to note the relative motions of the crankpin and cross-head, or whatever part has a straight-line movement. The crankpin has a practically uniform velocity, but the sliding member, which in the case of a steam engine consists of the cross-head and piston, has a variable velocity. Each time the cross-head reaches the end of its stroke, it starts from a state of rest and the velocity increases during approximately one-half of its stroke and then decreases until the cross-head again comes to a state of rest at the opposite end of the stroke. The relative positions of the crankpin and cross-head also vary at every point of the stroke. The position of the crank when the cross-head has traversed one-half its stroke is indicated by the diagram Fig. 1. If the crank were rotating in the direction indicated by the arrow, it would turn through some arc a less than 90 degrees, to bring the cross-head to its mid-position, and through a greater arc b for the remaining half of the stroke of the cross-head. It will thus be seen that the relative motion between the cross-head and crank during the first half of the stroke is different from that of the second half. This variation in movement is further illustrated by locating the distances that the cross-head moves for equal movements of the crank; for example, if the crank is moved through an arc c, from the dead-center position, the cross-head will move a distance y, but if the crank is placed on the opposite dead center and then moved through an arc d, which is equal to c, the cross-head will move a distance z, which is less than y. This is due to the fact that one-half of the crankpin circle curves toward the cross-head, whereas the other half curves away from it. This variation of motion has an important effect on the design of steam-engine valve-gears, and it is objectionable in some types of mechanisms. The length of the connecting-rod from the center of the cross-head wrist-

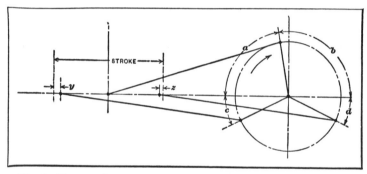

Fig. 1. Diagrams Showing Relative Motions of Crankpin and Cross-head

pin to the center of the crankpin is usually equal to from 4½ to 6½ times the crank radius, on steam engines.

Crank and Slotted Cross-head or "Scotch Yoke." — The irregularity in the motion of a cross-head relative to the crank with the ordinary form of crank mechanism depends upon the length of the connecting-rod. The greater the length of the connecting-rod, the less the irregularity of motion. If it were practicable to use a connecting-rod of very great length, the horizontal movement of the cross-head would be practically the same as the movement of the crankpin measured horizontally. If the connecting-rod were of infinite length, theoretically the movement of the cross-head and crankpin in a horizontal direction would be alike.

A simple form of mechanism for eliminating the irregularity of motion common to all ordinary crank drives, is known as a "crank and slotted cross-head" or the *Scotch yoke*. The cross-head *a* (see Fig. 2) has a slot which is at right angles to the center-line *xx* representing the direction of rectilinear movement. The crankpin carries a block, which is a sliding fit in this slot, and is free to revolve about the pin. As the crank revolves, the distance which the crankpin moves, as measured

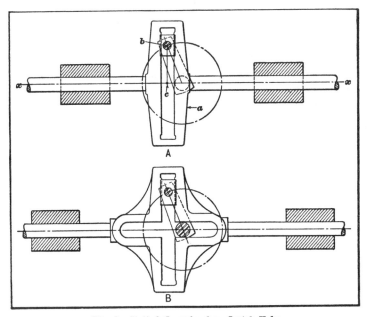

Fig. 2.   Slotted Cross-head or Scotch Yoke

in a horizontal direction, will be the same as the movement of the cross-head. This mechanism is sometimes called a *harmonic motion,* because if the crank rotates uniformly, the cross-head will be given a harmonic motion. When a point, as at *b,* moves with uniform velocity along a circular path, point *c* will have a harmonic motion along the center-line *xx;* hence, harmonic motion may be defined as the movement of a point along the diameter of a circle, which is projected from a point moving with uniform velocity along the circumference.

The crank and slotted cross-head has been applied to some types of steam pumps. One of the rods extending from the slotted cross-head carries the steam piston and the other, the water piston. The crank is a driven member, and its radius regulates the length of the stroke. By mounting a flywheel on the crankshaft, steam may be cut off before the end of the stroke and used expansively, because of the energy stored in the flywheel. The crank and slotted cross-head is a very compact form of mechanism, although the sliding motion of the block in the slotted member causes more friction and wear than the ordinary crank and connecting-rod. The latter is also simpler in construction and is, therefore, used almost exclusively as an engine connection, as well as for many other classes of machinery.

The diagram B, Fig. 2, shows a modification of the crank and slotted cross-head or Scotch yoke. This mechanism gives the same motion as the one illustrated at A, but the cross-head has two slots at right angles to each other, so that it can be placed on a continuous shaft. The vertical slot is for the sliding crank block, whereas the horizontal slot forms a clearance space for the shaft. With this design, the crank could be placed at any intermediate point on the shaft without using a center crank. It is not as compact, however, as form A, and the vertical slot is not continuous, which is an objectionable feature.

Scotch Yoke Modified to Give 60-degree Dwell at Each End of Stroke. —With the design of Scotch yoke shown in Fig. 3, the driven cross-head or slide has a dwell at each end of the stroke equivalent to about 60 degrees of crank rotation. This mechanism is part of a double flanging press used in the manufacture of certain cans. The flanging is done in three operations at the rate of eighty cans per minute, and as a slight amount of time is required for the cans to drop from one working position to another, the dwell obtained with this mechanism allows for this. The drive is from pinion A to gear B. The eccentric crank C has a bearing in gear B, and carries at one end (see plan view) a pinion D which meshes with a stationary gear E, the ratio of these two gears being 2 to 1.

As gear B revolves, the planetary pinion D rotates around fixed gear E; consequently, crankpin C turns about the axis F (see Fig. 4) of its bearing in gear B, while this axis fol-

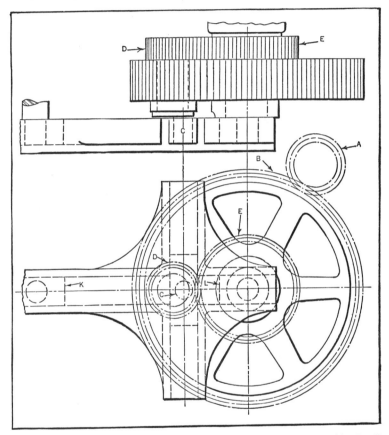

Fig. 3. Scotch Yoke Having an Eccentric Planetary Crankpin which Provides Dwell at Each End of Stroke

lows a circular path G. These combined rotary motions of C about F and of F around path G cause the axis of eccentric pin C to describe an oblong path, as indicated by line H. The straight sides of this oblong path represent the dwelling periods at each end of the stroke. This approximately straight-line

movement of the center of pin C during nearly 60 degrees of
crank rotation on each side is due to the fact that the center of
crankpin C moves inward toward center J, so as to offset,
during this period, the circular movement. The driven slide
or cross-head is supported on pivoted guides K and L, Fig. 3.
This mechanism doubtless can be applied to various classes
of machinery requiring dwell at the ends of the stroke.

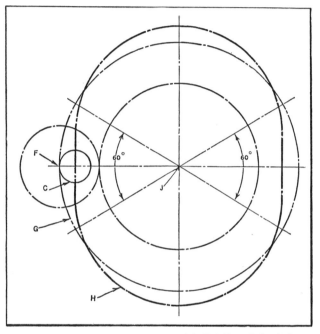

**Fig. 4. Diagram Showing How Eccentric Rotation of Crankpin
Causes Dwell at Ends of Stroke**

Crank Designed to Traverse Slide at Uniform Rate. —
An interesting and unique mechanism for controlling the
traverse motion of the carriage on a grinding machine is
shown in Fig. 5. This mechanism is designed to feed the
carriage by means of a crank motion. The accelerated motion
given to the crank at the ends of the stroke is the principal
feature of the device. This acceleration gives the carriage
a uniform rate of travel the full length of the stroke which

prevents under-grinding at each end of the spindles. The illustration shows a plan view and two positions of the mechanism.

In Position 1, the parts are shown as they slide the carriage toward the end of the first traverse, while in Position 2 the return traverse has started. The driving rod, which is not shown, is connected to the connecting arm *K,* by an adjusting

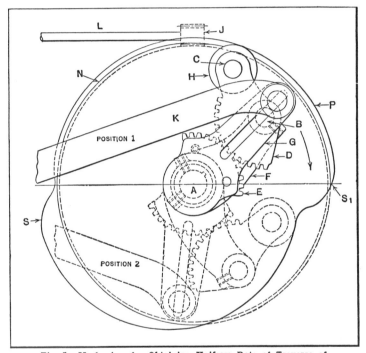

Fig. 5. Mechanism for Obtaining Uniform Rate of Traverse of Grinder Carriage, with Crank Motion

screw and nut. Worm-shaft *L,* which receives its motion from the main drive of the machine by means of bevel gears, carries worm *J,* which drives the upright shaft *A* through the worm-wheel *N* in the direction shown by the arrow. Driving arm *F,* is keyed and fastened by set-screws to *A,* as shown, and carries, at its outer end, a stud which is fastened by set-screws in hole *B.* Segment *D* has a forked or clevised end, in which the roll *H* is carried, revolving freely on stud *C.* The segment and roll are capable of pivoting on the stud at *B*. Segment *E* is loose on shaft *A*, and carries crank-arm *G* fastened to it by means of a screw and dowel-pin. The segment is prevented from lifting off by a head on the upper end of the shaft.

The grinding starts with roll *H* at point *S*, one full traverse being from *S* to S_1, where the return motion starts, ending automatically at *S*, after the necessary number of traverses has been completed. As the arm *F* swings around on *A*, segment *D*, which meshes with segment *E*, forces roll *H* against cam *P*.

When *H* reaches point S_1, and one traverse has been completed, the heel of cam *P* throws *H* toward the center, forcing the segment *E*, which is free to turn clockwise on *A*, to suddenly revolve about five teeth. This action throws arms *G* and *K* forward quickly, and incidentally the carriage as well, preventing a dwell at the end of the spindle. The resulting relation of the parts is shown in Position 2. Arm *G* has a T-slot, in which the connection *K* may be adjusted, in or out, to obtain the desired throw, that is, the proper length of the traverse.

The feed of the carriage being constant, the same amount of time is consumed on either long or short traverses. Since the speed of *A* is always the same, as the length of the spindles increase, the number of traverses required is also increased, by means of a knock-off device. It is obvious that the depth of the cut on short spindles is greater than that on long spindles, and that fewer traverses are required to grind short spindles to the desired size, and vice versa.

Crank Mechanism for Doubling the Stroke. — A crank and link mechanism is shown in Fig. 6 which makes it possible to obtain a rectilinear motion approximately equal to twice the throw of the driving crank. This mechanism is shown applied to an air pump for use on automobiles, either for the inflation of tires or in connection with engine starting apparatus requiring compressed air. The crank proper is of the center type with a bearing on each side. The connecting-rod is attached to the yoke A which is mounted on the main crank-pin. The opposite end of this yoke is pivoted to link B which is suspended from a pin attached to the compressor casing. As the crankshaft rotates, this link oscillates and so controls the position of yoke A that the stroke of the piston is approximately doubled. The view to the left shows the piston at the lower end of its stroke. As the crank turns in a counterclockwise direction, link B swings to the right so that the right-hand end of yoke A is forced downward and the left-hand

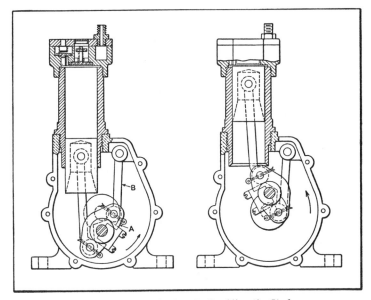

Fig. 6. Crank Mechanism for Doubling the Stroke

end upward, as indicated by the right-hand illustration which shows the piston at the top of its stroke. The advantage of this crank mechanism is that it enables a comparatively large capacity to be obtained from a small compact pump.

Pinion and Rack Mechanism for Doubling Stroke. — Another method of doubling the stroke when a crank of relatively small size is necessary, owing to a limited space, or desirable, in order to obtain a compact design, is by means of a fixed and a movable rack having a crank-driven pinion

interposed between them.   The pinion is pivoted to the end
of the crank connecting-rod so that it is free to roll along
the stationary rack when the crank revolves.   As the result of
this rolling movement of the pinion, the movable rack is given
a rectilinear motion equal to twice the stroke of the crank,
or twice the diameter of the path described by the crankpin.
This mechanism has been used for driving the beds of cylinder
presses.

A modification of the plain gear-driven crank is shown in
Fig. 7 which illustrates the bed motion of a two-revolution
pony press.   The driving and driven gears *A* and *B* are of
the elliptical form in order to compensate for the motion

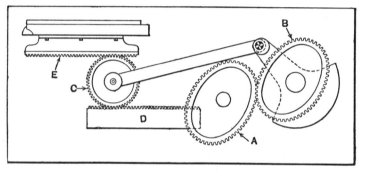

Fig. 7. Crank-driven Pinion Engaging Stationary and Movable Rack for
Doubling Stroke

derived from a crank rotating at uniform velocity.   The driven
gear *B* revolves the crank which, in turn, transmits motion
to pinion *C* by means of the connecting-rod shown.   This
pinion is rolled in first one direction and then the other along
the stationary rack *D*, and imparts a rectilinear motion to rack
E and the press bed.   The press bed moves a distance equal
to twice the distance that the axis of gear *C* moves, or four
times the radius of the driving crank.   The elliptical gears
are so proportioned and located relative to the crank as to
give a more uniform motion to the press bed than could be
obtained with a crank rotating at uniform velocity.   With
an ordinary crank, whatever part is given a rectilinear motion

starts from a state of rest, and the velocity gradually increases toward the center of the stroke and then decreases until it again becomes zero at the opposite end of the stroke. With the elliptical gearing shown, as the pinion C approaches either end of its stroke and the crank advances toward the "dead-center" position, the long side or radius of the driving gear comes into engagement with the driven gear and increases its velocity, and also the velocity of the crank. As the return stroke begins, the velocity of the driven gear and crank gradually decreases, because the radius of the working side of the driving gear gradually diminishes; the result is that, when the crank is at right angles to the line along which the axis of pinion C moves and is in a position to impart the maximum velocity to pinion C, the speed of the crank is slowest, because it is then driven by the shortest radius of the driving gear. As the crank moves away from this central position at right angles to the center-line of motion, the speed is gradually accelerated again so that pinion C does not slow down as it would with a crank rotating at uniform speed. The reversal of the heavy press bed is assisted by means of "air springs" or cushions, the same as on cylinder presses in general. This mechanism is intended for small presses.

Compact Long-stroke Mechanism. — The mechanism shown in Fig. 8 was designed to give a long stroke within a small space. A harmonic reciprocating motion is obtained. The device consists of two gears A and B fitted on eccentrically positioned shafts C and D. Two concentric surfaces turned on each gear act as eccentrics with respect to the shafts. Two yokes E and F with two stays secured by bolts G, keep the two gears in mesh as shown, the gears being meshed in such a manner that the shafts are nearest each other at one end of the stroke. The drive shaft C is mounted in bearings on the machine frame, and the other shaft D is attached to the reciprocating part of the machine with its gear free to turn on the shaft or with the shaft.

If desired, the machine can be equipped with ball or roller bearings throughout, and it can be made fairly light by using hollow shafts and gears. The full lines show the mechanism in the position it occupies at the upper end of the stroke, the dotted lines at H show the mechanism in the position it occupies when the drive shaft has rotated through an angle of 90 degrees, while the dotted lines at J show the mechanism at the end of the down stroke, after the drive shaft has rotated through an angle of 180 degrees. If the driven gear is located 180 degrees from the position shown, rotation of the driving

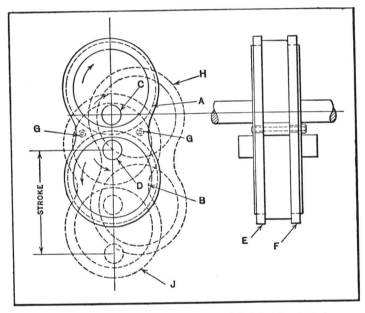

Fig. 8. Long-stroke Mechanism so Designed that Length of Stroke Depends upon Relative Positions of Eccentric Gears

gear will not result in reciprocating motion; therefore, a stroke from zero to the maximum can be obtained by varying the relative positions of the driven and driving gears. For any relative position except that which gives maximum stroke length the stroke will not be a simple harmonic motion.

Long-stroke Mechanism for Windmill Pump. — An ingenious mechanism used in the head of a windmill in order to obtain a long stroke for the pump rod and still allow the

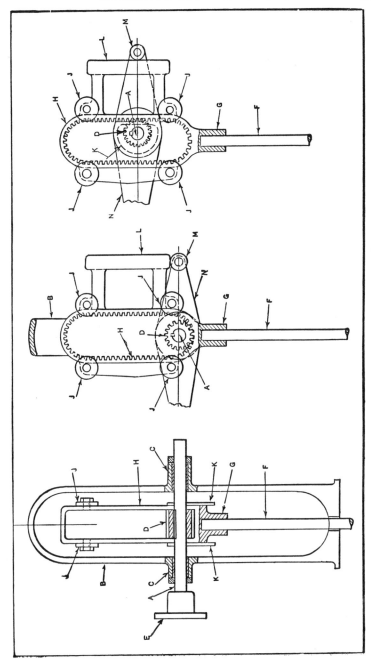

Fig. 9. Windmill Pump Drive for Obtaining Long Stroke and Efficient Windmill Speed

wind wheel to run at an efficient speed is shown in Fig. 9. A vertical section through the gearing is shown at the left, a front view in the center, and a front view at approximately midstroke at the right. In these different views, each part referred to is indicated by the same reference letter.

The main shaft *A* is supported in the main bearings *C* (left-hand view) which are held in housings in the head casting *B* of the mill head. The actual wind wheel is connected to the shaft at *E,* a coupling only being shown. On the main shaft is keyed a pinion *D* which meshes with an internal rack *H.* This rack consists of two vertical sections which the pinion *D* engages alternately, and connecting semicircular sections at top and bottom. The rack is provided with four steel rollers *J,* which engage with the cam *K* at the bottom and top of the pump stroke. These cams throw over the rack at the end of each stroke and thus reverse the direction of travel of the pump rod *F,* which is shown connected to the bottom of the rack at *G.* An extension of the rack at *L* consists of a steel guide plate which is in contact with a flanged roller *M,* and this keeps the pinion *D* in mesh with the rack *H.* This is clearly illustrated in the right-hand view, which shows the pump at approximately midstroke. The roller *M* is shown in the central view, passing across the bottom portion of the guide plate; it will make contact next with the inside face of the guide plate, and thus keep the pinion *D* in mesh with the opposite side of the rack *H.* The swing of the rack is only about 1¾ inches; consequently there is very little angular motion of the pump connecting-rod.

With the gearing shown there is one stroke of the double-acting pump to two and a half revolutions of the wheel. On account of the slow motion of the gear, the wear of the working parts is slight and the mechanism as a unit is efficient.

Reciprocating Motion from Double Rack and Shifting Spur Gear. — If a gear rotating continuously in one direction is between parallel racks, so that it can be engaged with first one rack and then the other, these racks will be moved in opposite directions. For instance, if the top side of the gear moves one rack to the right, the lower side will move the other rack toward the left. Some flat-bed printing presses are equipped with this double-rack and shifting-gear mechanism for driving the bed in first one direction and then the other. With mechanisms of this class, one rack is first traversed past the gear; when the gear and rack are entirely disengaged, the gear is shifted axially far enough to align it with the other rack. While this shifting movement takes place, the motion of the bed is arrested, and it is reversed by some auxiliary mechanism which moves it far enough to bring the other rack into engagement with the driving gear. Press bed motions of this general type differ principally in regard to the method of moving the press bed at the ends of the stroke, at the time when the driving gear and rack are disengaged.

Napier Motion for Press Beds. — When a gear or pinion is in mesh with a single rack and rotates in one position, obviously both the gear and rack must reverse their direction of motion at the end of each stroke. The gear, however, may rotate continuously in one direction if it is arranged to engage the upper and lower sides of a rack designed especially to permit such engagement. A mechanism of this type, known as the Napier motion and also as "mangle gearing," has been used for imparting a rectilinear motion to the tables of flat-bed printing presses. The principle of the Napier motion will be apparent by referring to Fig. 10. The rack A is attached to a frame B which is secured to the table of the printing press. The rack teeth are of such a form that the gear C may mesh with the rack on either the upper or lower sides. The shaft D, upon which the gear C is mounted, is rotated through a universal coupling, which permits it to swing in a vertical plane so that the gear may pass from the upper side of the rack to the lower side, and *vice versa*. The gear shaft is made to move in a vertical plane by a stationary slotted guide E having a vertical slot that is engaged by a sliding block mounted on the shaft. Spherical-shaped rollers F are mounted at each end of the rack, and the gear has a socket or spherical depression formed in it for engaging the rollers, each time the gear moves around the end of the rack when passing from one side to the other. Opposite each end of the rack, there are guide plates G having curved surfaces which are concentric with the rollers at the ends of the rack. The gear C also carries a roller H which engages these curved guides as the gear moves upward or downward at the points of reversal.

The action of the mechanism is as follows: If the gear is on the upper side of the rack, as shown in the illustration, and it is revolving to the left or counter-clockwise, the rack will be driven to the right with a velocity equal to the motion at

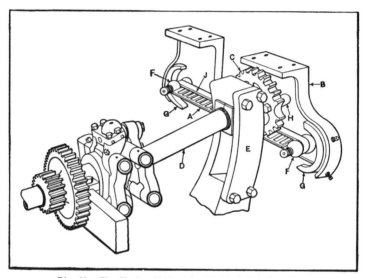

Fig. 10. The Napier Motion for Flat-bed Printing Press

the pitch circle of the gear. As soon as the gear engages the roller F on the end of the rack, it begins to move downward in a vertical plane, because its motion is constrained by guide E. When the gear is in mid-position so that its axis coincides with the center-line of the rack, it will have made a quarter turn, thus moving the center of roller F farther to the right, a distance equal to the radius of the pitch circle. Farther movement of the gear downward causes the rack to reverse and move toward the left; the gear then operates on the under side of the rack until the roller at the right-hand end of the rack is engaged, when the upward movement of the gear takes place and there is another reversal of motion.

The total length of the stroke is equal to the distance between the centers of the rollers on the rack, plus the pitch diameter of the gear. The length of the rack must equal the pitch circumference of the gear or some multiple of it, so that the rollers at the end will engage the socket or depression in the gear at the points of reversal. If a gear is used having two roller spaces located 180 degrees apart, the length of the rack or the center to center distance between the rollers may be some multiple of half the pitch circumference. The teeth on each side of the rack incline from the horizontal at the same angle as the gear axis when in its upper and lower positions, to obtain a full contact of the gear teeth. The gear also has a plain cylindrical shoulder on the inner side, which rolls upon a plane surface J at the base of the rack, to give a smoother action than would be obtained from a gear supported entirely by tooth contact. This arrangement of gearing imparts a uniform motion to the press table, excepting any variable movement resulting from a universal joint, and gives a gradual reversal of motion at the ends of the stroke. The Napier motion may be designed for any length of stroke, although the stroke remains constant, as there is no way of making an adjustment.

Modified Napier Motion for Saw-filing Machine. — An interesting mechanical movement for obtaining the motion required in filing the teeth of handsaws is shown in Fig. 11. As one file is being drawn across the saw, the other file, which has been raised to clear the saw, is returning. When the file in the raised position reaches the end of the return stroke, it is lowered to the saw, and during the filing stroke, the first file mentioned is being returned in the raised position, the files operating alternately.

The mechanical motion for operating the two files is derived from the combination of a special internal gear driven by a pinion having a projecting shaft which engages a slotted cam guide that keeps the pinion and internal gear in the proper

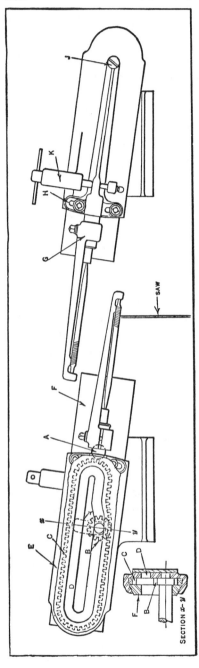

Fig. 11. Mechanism of a Duplex Handsaw-filing Machine, which Raises Each File During the Return Stroke. Left-hand View Shows File Near the End of its Cut and About to Lift Away From the Saw

relative positions. The illustration shows a front view of the two filing heads and a section x-y through the left-hand head. The outside of one filing head is shown at the right of the illustration, and the inside mechanism is shown by the left-hand view of the opposite head.

Each slide is pivoted on a shaft A, and pinion B revolves in mesh with the internal gear C. The pinion is held in contact with the internal gear teeth by an extension of the pinion shaft beyond the face of the pinion gear. The end of this shaft revolves in the slotted guide D which controls the motion. The file-holding head E operates on the slide F during the filing and return strokes. On the filing stroke, the internal gear teeth are being driven from the bottom of the pinion gear, whereas on the return stroke the internal gear is being driven from the top of the pinion gear. The file-holding arm G is under the pressure of a coil spring in K. The file-holding arm pivots from J, there being slots at H to allow for the necessary swinging movement.

The left-hand view shows the file nearly at the end of its cut and about to leave contact with the saw and travel a short distance to clear the end of the file-holder; then as it is being raised to its highest point, the file of the right-hand head is lowered to start its cut. A plan view would show the files working at opposite angles, to file the proper bevel into the

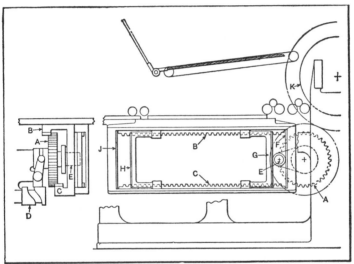

Fig. 12. Double-rack Shifting-gear and Crank Combination for Traversing Bed of a Printing Press

handsaw teeth. An interesting feature is a 5 to 1 gear ratio, but due to a loss of one-half revolution at each end of the stroke, the ratio is 4 to 1 for a complete filing cycle.

Crank Type of Reversal for Press Bed Motion. — An ingenious mechanism of the double-rack and shifting-gear type is shown diagrammatically in Fig. 12. This design is applied to some flat-bed or cylinder presses. In the operation of presses of this general type, the sheets to be printed are carried around by a revolving cylinder K so that contact is made with a flat form on the press bed which moves horizontally beneath the cylinder. This cylinder makes one revolution during the printing stroke and a second revolution while the press bed is being returned. In order to avoid contact between the cylinder and the bed or form during the return stroke, the cylinder is raised slightly by a suitable mechanism. The rotation of the cylinder is continuous in one direction and it is imperative that the cylinder and press bed move exactly in unison. The circumferential velocity of the cylinder should equal the linear velocity of the bed, because any relative motion would cause slurring on the printed sheet and it would be impossible to obtain sharp clean-cut impressions. As the cylinder revolves at a uniform speed, obviously the mechanism for driving the bed must be designed to give a uniform motion while the impression is being made. In order to properly time the motion of the cylinder and bed, the cylinder is connected by gearing and suitable shafts with gear A, which transmits motion to the bed; therefore, the press bed motion must be designed to reverse the movement of the bed without reversing the motion of gear A, since this gear rotates in unison with the cylinder or continuously in one direction.

This driving gear A is mounted between parallel racks B and C, both of which are attached to and travel with the bed. The distance between the pitch lines of these racks corresponds to the pitch diameter of the driving gear A. The racks are not directly in line, but are offset as shown by the end view, so that, when the gear is in mesh with one rack, it will clear the other one. The lateral movement of gear A for aligning it alternately with racks B and C is derived from cam D, which transmits motion by means of a lever and yoke engaging the gear hub.

When the press is in operation, the bed is moved in one direction by the engagement of gear A with rack B and in the opposite direction by meshing gear A with rack C. If gear A is revolving in a clockwise direction while in mesh with rack C, the latter and the press bed (the motion of which is constrained by guides) will move toward the left. When the press is in motion, this movement toward the left continues until the rack is entirely out of mesh with gear *A;* just before the disengagement of gear *A* and rack *C,* the crankpin *E,* which is provided with rollers, comes around and enters between the parallel faces of a fixed reversing shoe *F* and a swinging or movable reversing shoe *G.* The fixed shoe is rigidly attached to the press bed and rack frame, whereas the movable shoe is pivoted and free to swivel. This swinging reversing shoe has a pin on its lower side (not shown) which engages a slot or cam that controls its swinging movements. As soon as rack *C* has moved far enough to the left for shoe *G* to clear the crankpin, the cam swings the shoe inward so that crankpin *E* is confined temporarily between the faces of shoes *G* and *F,* which form a vertical guide or slot. As the crankpin passes its lowest position and begins to move upward, the roller on it bears against the face of *G* and "picks up" the load as gear *A* moves out of mesh with rack *C.*

When crankpin *E* arrives at the position shown in the illustration, the motion of the press bed is reversed, because a roller on the crankpin then engages the face of shoe *F* thus moving the driven member toward the right. The motion continues to be derived from the crank independently of the disengaged gear and rack, until the crankpin has passed the top quarter or highest position; then gear *A* enters the upper rack *B* and the motion is transmitted entirely through the gear and rack until the crank again comes into action at the opposite end of the stroke. At this end, the crankpin is again confined between a swinging shoe *H* and a fixed shoe *J.* After rack *B* has moved out of engagement with gear *A,* crankpin *E,* which is now in its highest position, comes into contact with shoe *H* and continues the movement toward the right while making a quarter turn, and then reverses the motion as it swings downward against the face of shoe *J.* While crankpin *E* is controlling the motion and gear *A* is entirely out of mesh, this gear is shifted by cam *D* out of line with the rack *B* which it just left, and into line with rack *C.*

An ingenious feature of this mechanism lies in the provision of two rollers for crankpin E and locating the fixed and swinging shoes in different vertical planes. With this arrangement, each roller is free to revolve in opposite directions as the crankpin moves along the vertical faces of the shoes. The momentum of the bed is gradually checked at the points of reversal, by air cushions or "air springs." A plunger enters a cylinder at each end of the stroke and air is compressed to arrest the movement, and, by expanding, this air assists in accelerating the heavy bed when its motion is reversed. Provision is made for regulating the air cushion or pressure according to the speed of the press. The air cushion is a feature common to flat-bed or cylinder presses in general.

Reversal of Motion by Reciprocating Pinions. — The mechanism illustrated in Fig. 13 is similar, in some respects, to the press bed motion just described, in that the parallel-rack and shifting-gear construction is employed. The method of operating the press bed at the ends of the stroke, however, is entirely different from that shown in Fig. 12, as reciprocating pinions are used to pick up the load and reverse the motion. The uniform motion of the bed is derived from pinion A which is constantly in mesh with gear D carried on the main driving shaft. Pinion A is located between parallel racks B and C which are attached to the press bed. These racks are offset, as in the design shown in Fig. 12, so that the pinion will clear one rack while in engagement with the other one. The shifting of the pinion is controlled by cam E which transmits motion to the pinion by means of lever F. The pinions for reversing the motion of the bed are located at G and H. The shafts upon which these pinions are mounted are connected to a heavy yoke J which has a vertical slot or groove in which a swiveling block attached to the crank K operates. This crank is rotated by the main driving shaft, and transmits to yoke J and pinions G and H a rectilinear motion equal to the throw of the crank. This is a harmonic motion, as yoke J and the sliding crank-block operate on the same principle as the well-known Scotch yoke. The outer ends of yoke J are supported by horizontal guides, and the pinions G and H

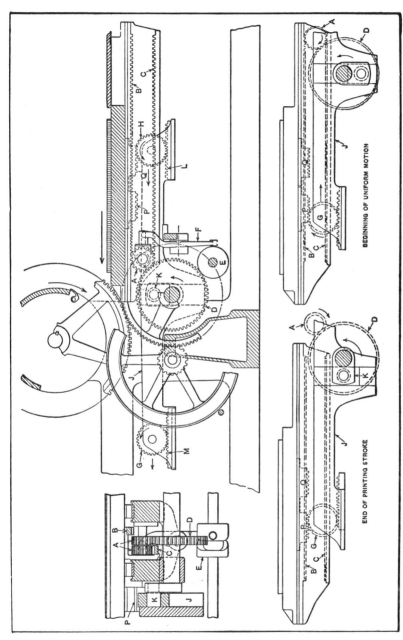

Fig. 13. Double-rack and Shifting-gear Mechanism for Press Bed having Reciprocating Pinions for Controlling Motion at Ends of Stroke

are constantly in mesh with short racks M and L along which the pinions roll as the crank moves them to and fro.

The action of the mechanism will be apparent by considering the various movements which occur during a forward and return stroke. The side view of the assembled mechanism shows the press bed in the position where the driving pinion A has just come into engagement with the lower rack C. As this pinion rotates in a clockwise direction, the bed will be driven to the left with a uniform motion. (The relative positions of pinion A and racks B and C are clearly shown by the end view.) When the bed has moved so far to the left that pinion A is about to roll out of mesh at the right-hand end of rack C, pinion G, which, meanwhile, has been moving along its rack M, comes into engagement with another short rack P (see also end view) attached to the bed. To insure the proper engagement of pinion G with rack P, the action of crank K relative to the motion of the bed is so timed that pinion G is rolling to the left when rack P which is also moving to the left comes into engagement with it. As pinion A leaves rack C, pinion G, which is then in mesh with P, continues the movement of the bed toward the left until crank K is in the position shown by the diagram in the lower left-hand corner of the illustration, which represents the end of the printing stroke. Further rotation of crank K in the direction indicated by the arrow causes a reversal of the rolling motion of pinion G and starts the press bed toward the right, motion being transmitted from G to rack P. While this reversal of movement occurs, pinion A is being shifted by cam E into alignment with the upper rack B.

When crank K has moved a quarter revolution from the position it occupies at the extreme end of the stroke, pinion A comes into mesh with the upper rack B and the short rack P leaves pinion G. The view at the lower right-hand corner of Fig. 13 shows pinion A about to enter rack B and pinion G leaving rack P. As the rectilinear motion of yoke J is harmonic, the movement of the bed is uniformly retarded as it approaches the point of reversal and is then accelerated until pinion A engages its rack, when the motion is uniform. When pinion A enters at the end of either rack, the velocity of the movement derived from crank K and the reciprocating pinion corresponds to the velocity obtained from the driving pinion A, so that there is no abrupt change of motion as the load is being transferred from the reversing pinion to the driving pinion A. As the press bed approaches the opposite end of its stroke, pinion H comes into engagement with rack Q and continues the movement for a short distance each side of the point of reversal or while pinion A is out of mesh with either rack and is being shifted, the action being the same as previously described.

Variable Reciprocating Motion. — The fly frames used in the manufacture of cotton goods are equipped with a mechanism for traversing the rovings or slightly twisted slivers of cotton as they pass between the rolls of the fly frame, which is used to make the rovings more slender and give them a twist. The reason for traversing the roving as it passes between a steel and a leather-covered roll is to prevent wearing the leather covering at one place. On some machines, this reciprocating or traversing motion is obtained from a crank or a cam. This simple arrangement distributes the wear but, if the length of traverse is uniform, the tendency is for the leather covering to wear the most at the points of reversal. In order to distribute the wear more evenly, the mechanism shown in Fig. 14 was designed. With this arrangement, the length of traverse gradually increases until it reaches a maximum and then decreases until the shortest length of traverse is obtained; the gradual increasing and decreasing of the stroke are then repeated.

The diagram A illustrates graphically the action obtained with a crank motion, and diagram B illustrates the variable stroke derived from the mechanism to be described. The guide-bar C, which extends the full length of the rolls, has small holes opposite each roll section through which the rovings pass, and it is this guide-bar which receives the reciprocating motion. The automatic variation of the traversing movement is derived from two eccentrics D and E, which revolve at different rates of speed. These eccentrics are formed on the hubs of gears F and G, which are adjacent to each other, and are both driven by one worm H as shown by the end view. The motion of the eccentrics is transmitted to guide-bar C through rods J and K and the bracket L. One of the gears meshing with worm H has one more tooth than the other, which causes the gears to rotate at a varying speed. The result is that the eccentrics formed on the two gear hubs are

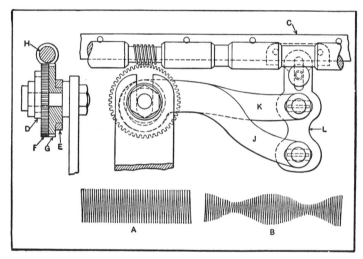

Fig. 14.   Double Gear and Shifting Eccentric Combination for Automatically Varying Traversing Movements

continually changing their position relative to each other, which automatically varies the length of traverse for guide-bar C. For instance, at one period during the cycle of movements, both eccentrics will move rods J and K in the same direction, and, at another period, one eccentric rod will be moving backward while the other is moving forward, thus reducing the stroke of the guide-bar. The connections between the eccentric rods and the bracket are adjustable; an adjustment is also provided where the bracket is attached to the guide-bar, so that the maximum traversing movement may be varied to suit the requirements.

Another Design of Variable Reciprocating Motion. —
A mechanism used on cotton fabric machinery for varying
the traverse of rolls is designed to give a variable reciprocating motion to slide *A* (see Fig. 15), which has a movement varying from zero up to the maximum of 1 inch with
a gradual reduction back to zero.   The slide is traversed

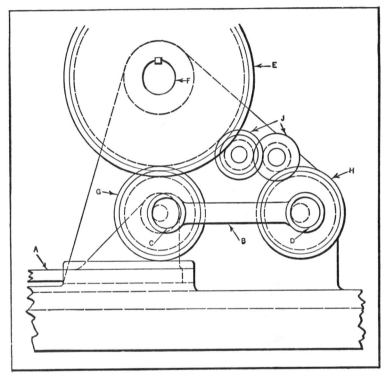

Fig. 15. Mechanism for Alternately Increasing and Decreasing Movement of Slide

through link *B,* and the variation in stroke is obtained by
eccentrics *C* and *D,* the relative positions of which are changed,
thus lengthening or shortening the effective length of link *B.*

The driving gear *E* is keyed to shaft *F* which receives its
motion from the machine proper.   Gear *E* meshes with gear
G directly, and with gear *H* through two idlers *J.*   Gears *G*
and *H* turn freely on studs.   The stud for gear *G* is fastened

into the driven slide, and the stud for gear H into a bracket. The eccentrics are integral with their respective gears, and there is a variation in the eccentricity due to a difference of one tooth in the gears G and H. When the movement of one eccentric neutralizes that of the other, the driven slide remains stationary, and this movement gradually increases from zero up to the maximum as the relative positions of the two eccentrics change. This traversing movement causes a change in the center-to-center distance between gears F and G, but since involute teeth are used, this variation does not interfere with the tooth action.

Rapid Reciprocating Motion from Epicyclic Gearing. — What is known as a "wabble" gear is used on mowing machines for imparting a rapid reciprocating motion to the cutter bar. The arrangement of this gearing and the other parts of the mechanism is shown in Fig. 16. The internal gear C is so mounted that it cannot rotate but is free to oscillate on a universal gimbal joint D. The gear B which meshes with one side of C is mounted on the main shaft which connects with the driving wheels. The frame J is rigidly connected to gear C and is pivoted in the revolving part H. By this means, gear C is given an oscillating or wabbling movement, so that the entire gear describes or follows a circular path. This circular motion causes the teeth of gear C to mesh with those of gear B all around the circumference for each rotation of the part H. This part H turns on a fixed shaft E and acts somewhat as a flywheel to maintain steadiness of action besides constraining gear C to follow a circular path.

In this case, gear C has forty-eight teeth and gear B, forty-six teeth; therefore, if gear B were free to turn on its shaft, it would be displaced two teeth for each rotation of part H or each time gear C completed a circular movement. Consequently, twenty-three revolutions of part H and a like number of oscillations of frame J would be required to turn B one revolution. Tracing the motion in the opposite direction, it will be noted that one rotation of gear B, which acts as the driver when the mechanism is in operation, will cause twenty-three oscillations or wabbling movements of gear C and a like number of rotations for part H. The frame J is connected to the cutter bar by the ball joint at K, so that one turn of the driving wheels which are mounted on shaft A will traverse the cutter bar twenty-three times. This combination of gearing makes it possible to use a gear B having only two teeth less than the number in gear C, which would be practically impossible with gears having teeth parallel to the axis of the shaft. With the usual forms of epicyclic gearing, in which a high velocity ratio is obtained, the efficiency of transmis-

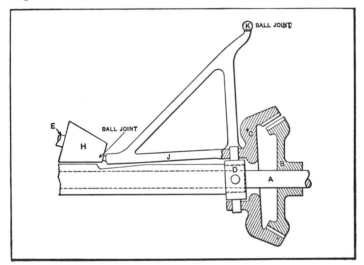

Fig. 16. Epicyclic or "Wabble" Gearing for Producing a Rapid Reciprocating Motion

sion is low on account of the excessive tooth friction, but, in this case, the efficiency is said to be nearly as high as that obtained with a train of spur gears having the same velocity ratio.

Epicyclic Gear and Crank Combination. — The mechanism illustrated in Fig. 17 is applied to an electric coal-puncher. One of the difficulties encountered in designing coal-punchers, excepting the solenoid type, has been in changing the rotation of the motor into a reciprocating motion for the drill. If the blow is directly dependent upon the motor, the latter causes trouble, owing to the vibrations and strains incident to the blows of the pick, and if springs are utilized they are liable to break. Types having separate motors and flexible shaft connections have also been tried in order to avoid some of these difficulties, but complications were introduced which at least partially offset the benefits derived.

The coal-puncher of which the mechanism shown in Fig. 17 forms a part uses both compressed air and electricity.

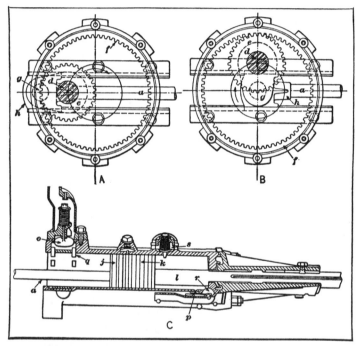

Fig. 17. Epicyclic Gear and Crank Combination from which Reciprocating Motion is Derived

Power for operating the coal-puncher is obtained from a motor and the compressed air gives the blow. There is no direct connection between the motor and striking pick, so that the vibrations are cushioned. The illustration shows the mechanical means by which the rotation of the motor armature is changed to a reciprocating motion for driving the air-compressing piston. A small pinion attached to the armature shaft engages a large driving gear (not shown) which has a solid web carrying the stud d upon which the crank pinion e is mounted. This crank pinion has 33 teeth and meshes with internal gear f which is rigidly fastened to the frame of the machine and is concentric with the main driving gear which surrounds it. The pitch diameter of the crank pinion e is just one-half that of the internal gear f which has 66 teeth. The crankpin g is attached to the pinion e and engages cross-head h which is mounted in guides and receives a rectilinear motion as pinion e revolves around the internal gear. Attached to the cross-head, there is a piston-rod a which enters the air-compressing cylinder and has a piston secured to its forward end.

When the main driving gear is revolved by the motor, the crank pinion stud d describes a circular path, as indicated by the arrows, thus causing pinion e to revolve about the stud and around the internal gear. When the pin d has moved one-quarter of a revolution, it will be in the position shown by the illustration to the right, and pin g attached to cross-head h will be in the center of the internal gear. At the completion of one-half a revolution, pin g will have moved in a straight line a distance equal to the pitch diameter of the internal gear, and will be at the right-hand end of its stroke. Similarly, at three-quarters of a revolution, the pin will again be in mid-position, and at the completion of a full revolution, it will be at the starting point, as shown by the view to the left. In this way, the crank pinion, as it revolves around the internal gear, transmits to pin g and the attached cross-head h a rectilinear forward and backward movement. The cross-head is mounted in guides, but pin g would follow a straight line even though guides were not used.

The way in which the air is compressed and utilized to impel the pick-carrying piston forward, all in one cylinder, will be described. A sectional view of the air cylinder is shown at C in Fig. 17. The air cylinder contains two pistons j and k. The rear piston j is attached to rod a connecting with the cross-head. The front piston k has no connection with j, but it is attached to the drill or pick socket by the rod l. The first stroke of the pick is purely mechanical. The rear piston j moves forward, pushing the front piston k. During this stroke, air is drawn into the cylinder behind the piston j, through the main inlet valve o. On the return stroke, this air is compressed and at the same time the front piston k is drawn back by the partial vacuum created by the piston j, air being admitted in front of k through a port p. When the return stroke is completed, the rear piston has passed the by-pass opening q in the cylinder, which opening is between the two pistons at the time. This allows the compressed air to force the front piston forward, exactly as in any compressed air drill. In this way, the first real stroke of the machine is made; that is, the mechanical stroke previously mentioned is made only once or when starting from rest. On the forward stroke of the piston k, the air in front escapes through the port p, but after the piston has passed and, therefore, closed this port, a sufficient amount of air remains to cushion the blow and prevent damage to the front cylinder head. This cushion of air may leak somewhat, and to prevent an insufficient supply remaining, which would have the effect of creating a partial vacuum in this space and holding the piston on the return stroke, a small inlet valve r is placed in the forward part of the cylinder. This allows air to flow in under these conditions before the open port is passed. When the front piston k has made its forward stroke, the rear piston follows, mechanically driven as before, and would compress the air which has just made the stroke of the front piston, were it not for the so-called vacuum valve s which allows all air between the pistons above a certain pressure to escape to the atmosphere. This action prevents the two piston faces from coming together.

A mechanism operating on the same general principle as the one shown in Fig. 17 has been applied to printing presses of the flat-bed type, for imparting a rectilinear motion to the bed. This mechanism has the advantage of giving a long, gradually increasing and decreasing motion with a short crank and without the use of a connecting-rod or a slotted cross-head; therefore, it can be applied to some classes of mechanisms when there would not be sufficient room for a connecting-rod or in preference to the slotted yoke, because of mechanical objections to the latter. In designing this mechanism, the center of pin g should exactly coincide with the pitch circle of the internal gear; then, if the internal gear has twice as many teeth as the revolving gear, the center of g will move in a straight line, even though its motion is not constrained by means of guides.

Converting Fast Rotary Speed to Slow Reciprocating Motion. — A novel method of converting a fast rotary speed into a very slow straight-line motion consists in using two

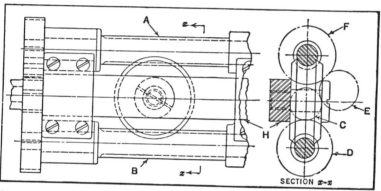

Fig. 18. Mechanism for Reducing a High Rotary Speed to a Slow Straight-line Motion

worms A and B (see Fig. 18) which differ very slightly in pitch and which mesh with opposite sides of a worm-wheel mounted on the part that is to receive the slow straight-line motion. Worm B is driven directly by pinion C and gear D, and worm A is revolved in the opposite direction through the provision of an idler gear E between pinion C and gear F. The worm-wheel turns freely on a shoulder stud which is held in slide H.

Worm B has six threads per inch, and worm A, 5 31/32 threads per inch, all threads being right-hand. Now if the pitch of both worms were exactly the same, no motion would be transmitted to slide H, but as there is a difference in pitch equal to 1/1146 inch, the center of the worm-wheel and slide will move one-half this amount or about 0.0004 inch per revolution of the worms, or 0.0002 inch per revolution of the driving pinion C, as the latter has 15 teeth, whereas gears D and F each have 30 teeth. When the slide H reaches the end of its stroke, engagement of a dog with a suitable trip

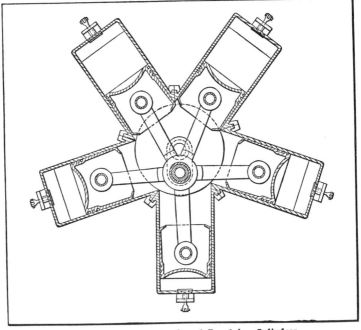

19.   Stationary Crank and Revolving Cylinders

operates a clutch and the traversing movement of the slide is reversed.

Cylinders which Revolve about a Stationary Crank. — A crank which is connected to a piston or other reciprocating part ordinarily revolves, but a piston may be given a rectilinear motion relative to a cylinder by holding the crank in a fixed position and revolving the cylinder, connecting-rod, and piston about the crank. An example illustrating this method of utilizing the crank is shown by the diagram, Fig. 19, which illustrates the general arrangement of an early type of airplane motor. With this form of motor, as the cylinders revolve about the stationary crank, the pistons move in and out relative to the cylinders, the same as though the latter were stationary and the crank revolved. The cylinders form the flywheel and drive the propeller and, as they revolve rapidly, the temperature is reduced sufficiently by air cooling and without any auxiliary cooling device.

Combined Reciprocating and Rotary Movements. — The piston of the pump shown in Fig. 20 has, in addition to a rectilinear movement, a rotary motion. This pump was de-

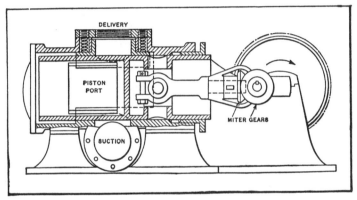

Fig. 20. Piston Having Combined Rectilinear and Rotary Movements

signed for pumping water or other liquids containing foreign materials, such as weeds, pieces of rope, paper, etc., which might enter the pump cylinder. Instead of using suction or discharge valves which would become clogged and cause trouble, the opening and closing of the ports is controlled by the rotary movement of the piston, and any foreign materials of the kinds mentioned are sheared off by the edges of the ports. The rectilinear motion of the piston is obtained from a crank. A miter gear keyed to the end of the crankpin meshes with a mating gear keyed to the end of the connecting-rod, so that, as the piston is moved in and out, it is also given a rotary motion. The piston is of the trunk type with an opening at both ends and a partition in the center. The head end at the left of the partition contains a port which alternately registers with the suction and delivery ports. When the piston is in the position shown, both ports are closed, but, as soon as the pump rotates in the direction indicated by the arrow, the suction port begins to open. When the crank has moved 90 degrees, the piston port will be exactly over the suction

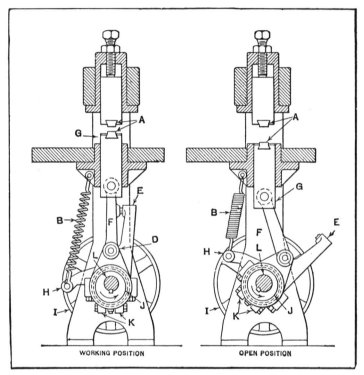

Fig. 21. Mechanism for Shifting Reciprocating Part from Working Position
Automatically

port and, when the opposite dead center is reached, both ports will again be closed. When the crank is on the bottom quarter or at the center of the return stroke, the piston port will be opposite the delivery port.

Shifting Reciprocating Part from Working Position. — The machine shown in Fig. 21 is used in a certain branch of the leather business to press a leather product between a pair of dies A by a series of reciprocating motions given to the lower die, which is afterwards withdrawn to the "open position" shown at the right, to allow the removal and insertion of the work. The mechanism to be described serves to automatically locate the lower die in the open position when the driving belt is shifted to the loose pulley, and into the "working" position as eccentric J, which imparts motion to ram G, is rotated by shifting the belt to the tight pulley. The shaft to which this eccentric is keyed turns in the direction shown by the arrow. The upper half of the eccentric strap is pivoted to the connecting-rod at D and carries an arm H to which is attached the long spring B. If there were nothing to prevent it, the spring would evidently tend to pull the joint D over, as shown by the right-hand view. In this position, with the belt on the loose pulley, the machine is ready to receive the work. The lugs K are attached to a leather band friction, bearing on an extension of the eccentric surface, and shown in dotted lines behind the eccentric strap. A finger screwed to the lower half of the strap and projecting between the lugs serves to keep the brake in position. If the machine is started by throwing the belt onto the tight pulley, the brake grips the eccentric with sufficient force to overcome the slight tension of spring B, and joint D is moved back to the central working position, where buffer E has reached its seat on the connecting-rod. As the shaft continues to turn, the brake slips on its seat and the eccentric gives the desired movement to the ram. When the operation is completed, the belt is shifted to the loose pulley, and spring B turns the shaft, eccentric, and strap backward until the machine is again in the open position with the ram lowered to allow a change of work.

Rectilinear Motion from Revolving Pawls. — The mechanism for driving a conveyor is shown in Fig. 22. This conveyor consists of a pair of endless chains between which the conveyor buckets are carried. These buckets are hung on pivots, so that they are kept in an upright position by gravity. The chains are equipped with wheels which run on tracks. The chain and buckets are propelled along the tracks as indi-

cated by the arrow, by a system of rotating pawls which
receive their motion from a large gear D. Each pawl, in turn,
engages one of a series of pins on the chain and, after having
pushed the conveyor ahead, the pawl is raised by cam C and
the next pawl repeats the operation. When a pawl, as at A,
is passing through the lowest arc of its travel, the conveyor
is propelled forward. The pawl shown at B has passed the
lowest point, and it gradually lags behind the conveyor, so
that the end of the pawl is readily lifted out of engagement

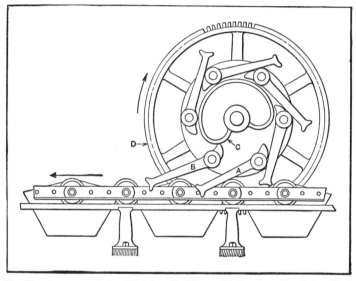

Fig. 22. Arrangement for Obtaining Rectilinear Motion from Revolving Pawls

without interference. As will be seen, the inner end of pawl
B is in contact with the cam surface which controls its position.

Compact Reciprocating Mechanism of Air Compressor. —
A sectional view of the air compressor of an oil burner is
shown in Fig. 23. In this equipment, four pistons A are
reciprocated by a revolving thrust plate B, which is located
at an angle of 12 degrees. As this plate is revolved by worm-
gearing C, the pistons will, of course, be forced upward due
to the angular location of plate B, and the return of each piston is insured by the action of a spring D which keeps the

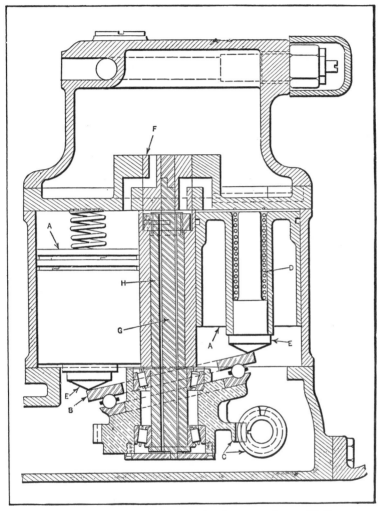

Fig. 23. Reciprocating Mechanism of Air Compressor for an Oil Burner

hardened steel plug E in contact with plate B. Every revolution of the worm-wheel causes each of the four pistons to make a forward and return stroke. This worm-wheel revolves on Timken roller bearings, and the thrust plate is supported on a ball bearing. The rotary valve F, which opens and closes the air ports, is driven from the worm-wheel by a shaft G.

In the illustration, the right-hand piston has delivered its quota of air and is ready for the downward stroke, the valve being timed to allow air to be drawn through shaft H as the piston descends. The valve, as shown, has connected the compressed air reservoir with the left-hand cylinder, although actually the port does not open until the piston has completed nearly three-fifths of its stroke, thus preventing, as far as possible, a pulsating action of the air due to the small capacity or space above valve F. The four cylinders, each of which is 3 inches in diameter, are so located that their center lines form a $3\frac{1}{4}$-inch square, as seen on a plan view of the compressing mechanism.

In this oil burner, the oil is atomized by the use of compressed air generated by this mechanism, and a surplus of low velocity air for complete combustion of the fuel is furnished by a fan, not shown. The worm-gearing is continually submerged in oil to half its depth, and the splash from the worm lubricates the Timken bearings and the pistons. The oil spray, carried along by the air being compressed, is sufficient to lubricate the rotary valve. A relief valve in the cover above valve F allows any surplus air not needed for atomizing the oil to be returned to the intake side so that it can be compressed again.

Drop-hammer Lifting Mechanism. — The drop-hammers used for making drop-forgings are so designed that the hammer head is raised by rolls which run in opposite directions and bear against opposite sides of a board attached to the hammer head. Front and side elevations of a drop-hammer lifting mechanism are shown in Fig. 24. The board A passes between the rolls B and C. One roll rotates in a fixed position and the other one is alternately pressed against the board and then withdrawn from it, when the hammer is in operation. The pressure of the movable roll is applied for raising the hammer head and released for allowing it to drop upon the work. The roll that is withdrawn is usually the front one which has an eccentric bearing so that a slight rotary movement will cause the roll to release the board. As the hammer

drops and approaches the bottom of its stroke, it engages some
form of trip or latch which holds the eccentric roll in the outward position so that the roll moves in against the board; the
hammer is then immediately elevated preparatory to striking
another blow. As the hammer approaches the top of its stroke,
the eccentric roll is again automatically withdrawn, thus stopping any further upward movement. The hammer will then
fall and repeat the cycle of movements and will continue to
run automatically, provided the board clamps at D are not
allowed to grip the board. The position of these clamps is
controlled by a foot-treadle. When this treadle is released,

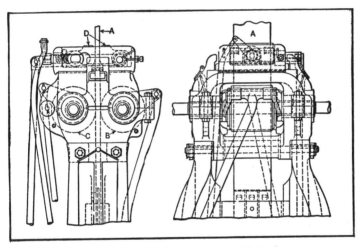

Fig. 24. Board Drop-hammer Lifting Mechanism

the clamps grip the board as it reaches the top of its stroke
and starts to move downward, so that the hammering action
discontinues until the foot-treadle is again depressed. This
mechanism for transmitting the rotary motion of the rolls to
board A, which has a rectilinear movement, is similar in principle to the rack and pinion, except that motion is transmitted
entirely by frictional contact instead of by means of teeth
which give a positive drive.

   Toggle Joint. — A link mechanism commonly known as a
toggle joint is applied to machines of different types, such as

drawing and embossing presses, stone crushers, etc., for increasing pressure. The principle of the toggle joint is shown by the diagrams A and B, Fig. 25. There are two links, b and c, which are connected at the center. Link b is free to swivel about a fixed pin or bearing at d, and link c is connected to a sliding member e. Rod f joins links b and c at the central connection. When force is applied to rod f in a direction at right angles to center-line xx, along which the driven member e moves, this force is greatly multiplied at e, because a movement at the joint g produces a relatively slight movement at

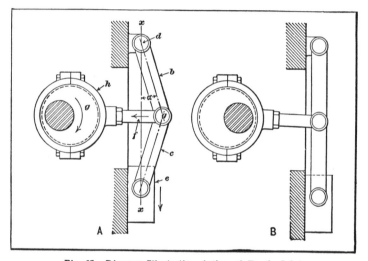

Fig. 25.　Diagram Illustrating Action of Toggle Joint

e. As the angle a becomes less, motion at e decreases and the force increases until the links are in line, as at B. If $R =$ the resistance at e; $P =$ the applied power or force; and $a =$ the angle between each link and a line xx passing through the axes of the pins, then:

$$2\,R\,\sin a = P\,\cos a.$$

Single- and Double-stroke Toggle Mechanism. — A toggle mechanism is often utilized for changing a rotary to a rectilinear motion, especially when a powerful squeezing action is required. An arrangement of this kind is used on some cold-heading machines, such as are employed for forming heads on bolts, rivets, etc. The diagram *A,* Fig. 26, illustrates a crank-driven toggle mechanism which gives a forward and return stroke for each revolution of the crank. When the links of the toggle are straightened, as indicated by the heavy lines, the punch which forms the head on the work is at the end of its stroke, and it is then withdrawn as the crank makes another half revolution. This form of drive as applied to a cold-header is known as the "two-cycle type," because two revolutions of the crankshaft are necessary to complete a rivet or bolt requiring two blows of the punch.

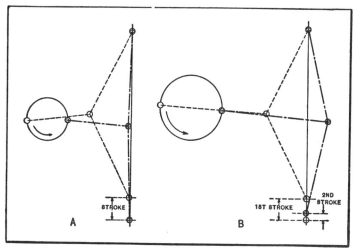

Fig. 26. Action of Single- and Double-stroke Toggle Mechanisms

Many classes of work cannot be done satisfactorily with a single stroke, owing to the amount of metal that must be upset in order to form the head of a bolt or rivet. A design of toggle mechanism which is extensively used on double-stroke machines is illustrated by the diagram *B,* Fig. 26. With this arrangement, two blows are obtained for each revolution of the crank connecting with the toggle. The location of the crank is such that the links of the toggle are straightened before the crank has made one-half revolution; consequently, when the half revolution is completed, the links of the toggle are carried beyond the center-line, as indicated by the diagram, which causes the ram and die to be withdrawn preparatory to making a second stroke. As the crank continues to revolve and the toggle is again straightened, a second working stroke is made and then the ram and die are withdrawn; this cycle of operations is repeated for each revolution of the crank. The two strokes which are obtained for each revolution of the crank may be of unequal length, as shown by the diagram, or of equal length, depending upon the position of the crank relative to the line of the straightened toggle. A cold-header having this form of drive is known as a "one-cycle" machine, since it will impart two blows to the work for each revolution of the crankshaft.

Toggle Mechanism of Drawing Press. — One way of arranging the toggle mechanism of a drawing press is illustrated in Fig. 27. When a press of this kind is in operation, the sheet of metal to be drawn is pressed firmly down upon the die face by a blank-holder, while the drawing punch forces the metal into or through the die. The blank-holder prevents the sheet stock from buckling, and it should remain in the downward position while the drawing punch is at work. Toggle mechanisms are employed on large drawing presses to operate the blank-holder. The toggle mechanism illustrated is operated from crank H on the main crankshaft. This crank connects with link A, the lower end of which is attached to yoke B. The upper end of yoke B is guided by link E, which is pivoted to the frame of the press, and the lower end is guided by another link C pivoted at D. These two links C and D compel the yoke to move in practically a vertical straight line when it is traversed by the action of crank H. Attached to the yoke are two other links connecting with bellcranks F and G which, in turn, are pivoted to the side of the press frame. The outer arms of these bellcranks are connected by long links or rods with cranks on the ends of two rockshafts J and K, at the front and rear of the press, respectively. From these rockshafts, motion is transmitted to the blank-holder by means of arms L and links M. The dotted lines on **one**

side indicate the action of the rockshaft and its connecting link when in the extreme upper position. The bellcrank levers F and G, together with the links connecting them with the rock-shafts J and K, form a toggle mechanism which is straightened out at the same time that the driving crank H is passing its center and the arms L and links M are in line. This central or straight-line position for the toggles occurs while the blank is being held for the drawing operations; the

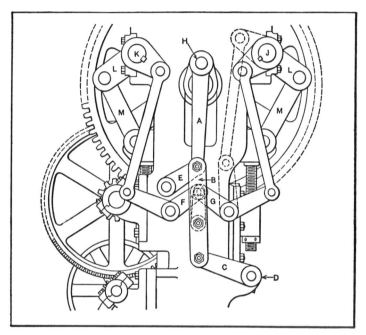

Fig. 27. Application of Toggle Mechanism to a Drawing Press

blank-holder dwells or remains down long enough to enable the drawing punch to complete its work before the sheet metal stock is released by the blank-holder. The slide to which the drawing punch is attached receives its motion from the main crankshaft.

Adjustable Double-motion Eccentrics. — The diagram Fig. 28 represents a disk at A encircled by a strap in which the disk is free to run. If this disk is mounted on a concentric shaft perpendicular to it, no motion will be imparted to the strap. However, if the shaft passes through the geometrical center of the disk, but the disk is located at some angle a (see diagram B), then as the shaft makes one complete revolution, the disk and its strap will oscillate harmonically through angle $2a$ in a direction parallel to the shaft.

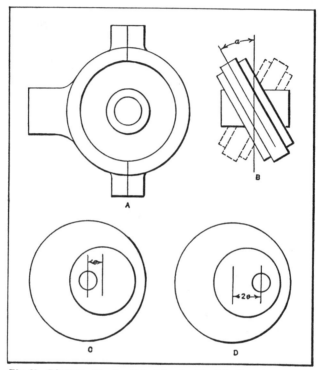

Fig. 28. Diagrams Showing Principles Combined in Eccentric, Fig. 29

Referring now to diagram C, two eccentrics are shown in outline, each having an eccentricity equal to e. The outer eccentric is adjustable relative to the inner one, and it is assumed that they can be fastened in any angular relation to each other. When clamped in the position shown, one eccentric offsets the other, so that the resultant throw is zero. When the positions are changed, as shown by diagram D, the eccentricities are added, and intermediate positions will, of course, vary the throw from zero to a maximum distance 2e.

Fig. 29 illustrates how the principles shown separately in Fig. 28 are combined. The inner eccentric E is made spherical, so that the outer eccentric F may be set at an angle to the shaft, as indicated by diagram B, Fig. 28. These inner and outer eccentrics, Fig. 29, have the same eccentricity, so that any throw from zero up to the maximum may be secured by adjustment, as represented by diagrams C and D, Fig. 28. Finally, the mechanism as a whole may be arranged for a circumferential adjustment about the shaft. This mechanism or some modification embodying the same principle may be

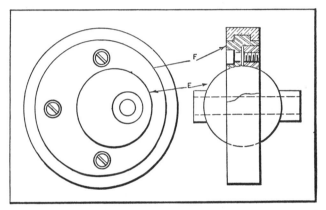

Fig. 29. Eccentric Adjustable as to Throw and Angle Relative to Shaft

used to obtain a harmonic motion in perpendicular planes with adjustment as to amplitude and phase.

Reversing Screw. — When a relatively slow but powerful reciprocating movement is required, a reversing screw may be employed. Many broaching machines of the horizontal type, which operate by pulling long broaches through holes in castings and forgings, are equipped with the reversing screw type of drive. As the broaching is done by a series of cutting teeth which gradually increase in size in order to produce a hole of the required shape progressively, considerable power is required for pulling the broach through the work, especially when cutting hard tough metal. Therefore, the draw-head to which the broach is attached is given a rectilinear movement by means of a screw which does not revolve but is moved in a lengthwise direction by a nut. The screw passes through this nut which is held against endwise movement, and, with one design, is rotated from the driving shaft through suitable gearing. This gearing is so proportioned that a comparatively slow motion is imparted to the nut and screw for the cutting stroke and a faster movement for the return or idle stroke. The nut which engages the screw is alternately connected with these two combinations of gearing by means of a clutch that is shifted by adjustable tappets or dogs that control the length of the stroke. Some of the smaller broaching machines intended for lighter work have belt pulleys that revolve about the screw in opposite directions, and are alternately engaged with a central clutch which transmits motion to the draw nut on the screw.

Screws for Power Transmission. — When screws are used for power transmission, multiple-threaded screws are preferred ordinarily, as they are much more efficient than single-threaded screws, the efficiency being decidedly affected by the helix angle of the thread. Single-threaded screws are sometimes preferable, owing to their mechanical advantage as compared with the multiple-threaded form. Thus a heavier load can be moved by a single-threaded screw than by the multiple-threaded type, assuming, for example, that the screw diameters and the force applied are the same in each case. The single-threaded screw, in this instance, moves the load only one-half as far as a double-threaded screw, so that it is capable of overcoming greater resistance for a given applied force, although the mechanical efficiency is lower than that of the multiple-threaded screw.

Force Required to Turn Screw. — In determining the force that must be applied at the end of a given lever arm in order to turn a screw (or nut surrounding it), there are two conditions to be considered: (1) When rotation is such that the load *resists* the movement of the screw, as in raising a load with a screw jack; (2) when rotation is such that the load *assists* the movement of the screw, as in lowering a load, assume that:

$F =$ force applied at end of lever arm;

$L =$ load moved by screw;

$R =$ length of lever arm;

$l =$ lead of screw thread;

$r =$ mean or pitch radius of screw $=$ outside radius minus one-half of thread depth;

$\mu =$ coefficient of friction.

When load resists screw movement:

$$F = L \times \frac{l + 2r \times 3.1416\,\mu}{2r \times 3.1416 - \mu l} \times \frac{r}{R}$$

When load assists screw movement:

$$F = L \times \frac{2r \times 3.1416\,\mu - l}{2r \times 3.1416 + \mu l} \times \frac{r}{R}$$

If lead l is large in proportion to the diameter so that the helix angle is large, F will have a negative value, which indicates that the screw will turn due to the load alone, unless prevented by a force F which is great enough to prevent rotation of a non-locking screw.

Coefficients of Friction. — According to experiments by Professor Kingsbury made with square-threaded screws, a coefficient of 0.10 is about right for pressure less than 3000 pounds per square inch and velocities above 50 feet per minute, assuming that fair lubrication is maintained. If the pressures vary from 3000 to 10,000 pounds per square inch, a coefficient of 0.15 is recommended for low velocities. The coefficient of friction varies according to lubrication and the materials used for the screw and nut. For pressures of 3000 pounds per square inch, using heavy machinery oil as a lubricant, the coefficients were as follows:

Mild steel screw and cast-iron nut, 0.132; mild steel nut, 0.147; cast-brass nut, 0.127.

For pressures of 10,000 pounds per square inch, using a mild steel screw, the coefficients were, for a cast-iron nut,

0.136; for a mild steel nut, 0.141; for a cast-brass nut, 0.136. For dry screws, the coefficient may be 0.3 to 0.4 or higher.

Coefficient of Friction for Angular Thread Forms. — Frictional resistance is proportional to the normal pressure, and for a thread of angular form the increase in the coefficient of friction is equivalent practically to μ sec β, in which β equals one-half the included thread angle; hence, for a U. S. standard thread, a coefficient of 1.155μ may be used.

Effect of Helix Angle on Efficiency. — The efficiency between a screw and nut increases quite rapidly for helix angles up to 10 or 15 degrees (measured from a plane perpendicular to the screw axis). The efficiency remains nearly constant for angles between about 25 and 65 degrees, and the angle of maximum efficiency is between 40 and 50 degrees.

In determining the efficiency of a screw and a nut, the helix angle of the thread and the coefficient of friction are the important factors. If E equals the efficiency, A equals the helix angle, measured from a plane perpendicular to the screw axis, and μ equals the coefficient of friction between the screw thread and nut, then the efficiency may be determined by the following formula, which does not take into account any additional frictional losses, such as may occur between a thrust collar and its bearing surfaces:

$$E = \frac{\tan A \ (1 - \mu \tan A)}{\tan A + \mu}$$

This formula would be suitable for a screw having ball-bearing thrust collars. Where collar friction should be taken into account, a fair approximation may be obtained by changing the denominator of the foregoing formula to $\tan A + 2\mu$. Otherwise the formula remains the same.

The square form of thread has been used chiefly for power transmission, because it has a somewhat higher efficiency than threads with sloping sides. However, when the inclination is comparatively small, as in the case of an Acme thread, the preceding formula, which was deduced for a square thread, may be applied without serious errors. The Acme thread has practical advantages in regard to cutting and also in compensating for wear between the screw and the nut.

It is evident that the screw of a jack, or of any lifting or hoisting appliance, will not be self-locking if the efficiency exceeds 50 per cent, as higher values would mean that the screw would turn in its nut under the action of the load. In actual designing practice, it will be understood that the maximum efficiency, even for a screw intended solely for power transmission, may not be practicable, as, for example, when it is necessary to employ a helix angle smaller than that representing maximum efficiency, in order to permit moving a given load by the application of a smaller turning moment.

CHAPTER X

QUICK-RETURN MOTIONS FOR TOOL SLIDES

MANY machines, especially of the type used for cutting metals, are equipped with a driving mechanism which gives a rapid return movement after a working or cutting stroke, in order to reduce the idle period. For instance, shapers and slotters are so arranged that the tool, after making the cutting stroke, is returned at a greater velocity, thus increasing the efficiency and productive capacity of the machine. The method of obtaining this rapid return varies with different types of machine tools. In some cases, motion for the return movement is obtained by using two belts which alternately come into the driving position and rotate the driven member at two rates of speed. This method is employed with belt-driven planers, the belt for the return movement of the table connecting with pulleys having a higher speed ratio. The rapid return movement for some other types of machines is obtained by transmitting motion through a different combination of gearing which is automatically engaged at the end of the working stroke. The term "quick-return motion," however, as applied to machine tools, generally relates to a driving mechanism so designed that the increased rate of speed for the return movement is obtained through the same combination of parts which actuate the driven member during the forward or working stroke, and the quick-return feature is due to the arrangement of the mechanism itself.

Crank and Oscillating Link. — A simple form of quick-return mechanism which has been applied extensively to shapers is shown diagrammatically in Fig. 1. The pinion A drives gear C at a uniform speed, and this gear carries a swiveling block B which engages slotted link L. The lower end of this link is pivoted at D and the upper end connects by means of a link with the ram of the shaper. As the crankpin or swiveling block B revolves with gear C, it slides up and down in the slot of link L and causes the latter to oscillate about the fixed pivot D at its lower end. The ram of the shaper is mounted in guides or ways so that it is given a rectilinear movement.

A quick-return movement is obtained with this form of drive owing to the fact that the crankpin B moves through an arc a during the cutting stroke, whereas, for the return stroke, it moves through a much shorter arc b. As gear C

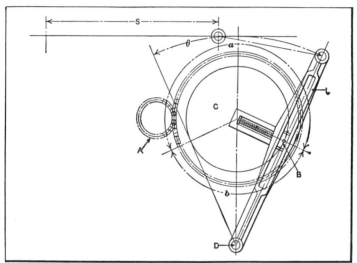

Fig. 1. Quick-return Motion from a Revolving Crank and
Oscillating Slotted Lever

rotates at a uniform speed, obviously the time required for the return stroke, as compared with the cutting stroke, is in the same proportion as the lengths of the arcs a and b. The radial position of block B may be varied in order to change the length S of the stroke. This mechanism imparts a variable speed to the ram, the speed increasing toward the center of the stroke and then diminishing. The angle made by the crankpin for the forward stroke equals 180 degrees + the angle θ through which slotted link L moves; for the return stroke, the crankpin moves through an angle equal to 180 degrees — the angu-

lar movement θ of the slotted link. The sine of one-half
angle θ equals the radius of the crank divided by the distance
from pivot D to the center of the gear C.

Whitworth Quick-return Motion. — A type of quick-return
motion that has been widely used in slotter construction is
illustrated in Fig. 2. This mechanism, which is known as
the "Whitworth quick-return," is similar in principle to the
crank and oscillating link combination previously referred to,
although the construction is entirely different. The pinion A
drives gear C at a uniform velocity, and this gear carries a
block B which engages a slot or groove in part D, which is

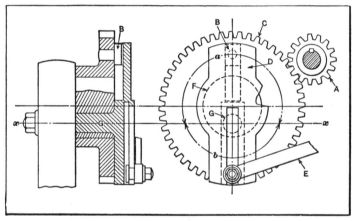

Fig. 2.   Whitworth Quick-return Motion

connected by a link E with the tool-slide of the machine. The
line xx represents the center-line of motion for the tool-slide.
The gear C revolves upon a large bearing F which is a part of
the machine frame. The slotted member D has a bearing G,
within F, and the center about which D rotates is offset with
relation to the center of driving gear C; consequently, the
crankpin or block B moves through an arc a during the cutting stroke and through a shorter arc b for the return stroke,
so that the latter requires less time in proportion to the respective lengths of arcs a and b. The stroke is varied by changing
the radial position of the pin which connects with link E.

So far as the principle of operation is concerned, the chief difference between the Whitworth motion and the crank and slotted link is that, in the former case, the bearing for the slotted or driven member is inside of the crankpin circle, whereas, with the crank and slotted link combination, the pivot is outside of the crankpin circle. As the result of this difference in arrangement, part D in Fig. 2 has a continuous rotary motion, whereas the slotted link L, in Fig. 1, swings through a definite angle. With the Whitworth quick-return, the ratio of the time required for the forward and return strokes is not varied by changing the length of the stroke. With the crank and oscillating link, a change of stroke does affect this ratio, the latter increasing as the length of the stroke is increased.

Modification of Whitworth Motion. — A quick-return mechanism that is a modification of the Whitworth motion combined with the slotted link and rotating crank is illustrated by the sectional view, Fig. 3. This form of drive has been applied to a shaper in order to secure in addition to a quick return a cutting speed that is practically constant throughout the working stroke. The driving gear F transmits its rotary movement through a swiveling block A to a ring E which turns about an eccentric C. On the opposite side of this ring there is a second swiveling block B, which drives the crank-disk G, on which is mounted the main crankpin block H, engaging the vibrating arm or link L that, in turn, is connected with the ram. The eccentric C is offset with relation to the center of the driving gear F, and it remains permanently in a fixed position; therefore, the circular path of the eccentric ring blocks A and B is not concentric with the path described by the main crankpin H. In other words, the circle which these blocks describe as they are driven around by gear F has a constantly varying radius from the center of the gear, which compensates for the irregularity of speed obtained by a plain slotted link, and gives a practically constant movement during the working stroke.

Quick Return from Elliptical Gearing. — Elliptical gearing has been used to obtain a quick-return motion, although such gearing is difficult to cut without special attachments, and comparatively few mechanisms requiring a quick-return motion have this type of drive. The driving and driven gears are of the same proportions and size as shown in Fig. 4, and each gear revolves about one of its foci as a fixed center. The distance between the shaft centers is made equal to the length

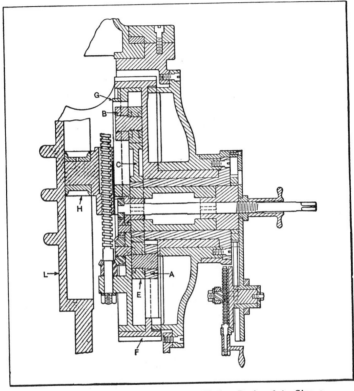

Fig. 3. Crank-operated Quick-return Motion Designed to Give a Uniform Forward Speed

of the common major axis. The angular velocity ratio varies according to the respective radii of the driving and driven gears at the point of contact. If *A* is the driving shaft and it rotates at a uniform speed, the angular velocity of shaft *B* will increase during the first half revolution from the position shown in the illustration, and then decrease during the remaining half revolution. When the gears are in the position shown, the angular velocity of the driven shaft B is minimum, because that side of the driver having the shortest radius is in contact with it; as the driver revolves, the radius at the point of contact gradually increases, and, consequently, the angular velocity increases until tooth C is in mesh, when the angular velocity is maximum. When point C representing the longest radius of the driving gear has passed the point of contact, the angular velocity gradually diminishes until it is again at a minimum.

The actual number of revolutions made by each shaft in a given time is, of course, the same, and the driving and

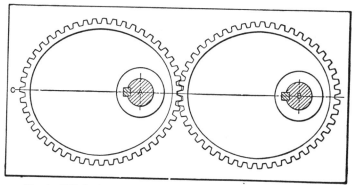

Fig. 4. Elliptical Gearing Arranged to Return Driven Part Quickly

driven gears both require the same time to complete the half revolution between the two positions representing the minimum and maximum angular velocities. The variable motion of the driven gear, however, may be utilized to give a quick-return movement to a driven tool-slide or other part.

This type of quick-return motion has been applied to shapers in order to return the tool quickly after the cutting stroke. The driven gear is connected to the ram by a link. The bolt or crankpin on the gear which connects with the link may be adjusted along a groove for varying the distance from the center of the driven shaft and the length of the stroke. Elliptical gearing has also been used for operating the slide valve of a steam stamp, such as is used for crushing rock. In this case, the variable motion obtained from the gearing is utilized to so control the motion of the valve as to admit steam above the piston throughout almost the entire downward stroke, whereas, on the upward stroke, just enough steam is used to return the stamp shaft, in order to reduce steam consumption.

Eccentric Pinion and Elliptical Gearing for Quick-return Motion. — The eccentric and elliptical gear combination, in conjunction with gears mounted concentrically, as shown in Fig. 5, has been utilized to secure a quick-return motion. The pinions A and B are keyed to the driving shaft. The smaller pinion A is concentric with the shaft and meshes with a half spur gear F. The larger pinion B is eccentrically mounted on the shaft and is in line with a half elliptical gear H, the two gear segments on the driven shaft being offset as shown by the end and plan views.

In the operation of this gearing, the semi-circular gear F is driven by the small pinion A and the elliptical gear by the eccentric pinion B. The elliptical gear makes one-half revolution to each complete revolution of its eccentric driving pinion. If the driven shaft is revolving in a counter-clockwise direction, the eccentric pinion will be the driver from C to D. At the latter point, the elliptical gear segment leaves the eccentric pinion and the smaller pinion A comes into mesh with the half spur gear and continues to be the driver through the remaining half revolution of the driven shaft, or until the elliptical gear again comes around into mesh with the eccentric pinion. Owing to the difference in the diameters of the half spur gear and its pinion A, the latter must make two revolutions before the eccentric pinion can again engage the teeth of the elliptical gear.

At the point C where the eccentric pinion again becomes the driver, the radius of pinions A and B is equal, and the transfer of the load from A to B does not cause an abrupt change of speed for the driven member. As the eccentric pinion, however, begins to swing the elliptical gear around, the speed of the driven shaft is increased until the maximum radius of the eccentric pinion is opposite the minor axis of the elliptical gear. The speed is then at maximum and, as the movement continues, the speed gradually decreases until the load is transferred to the concentric pinion A which imparts a uniform velocity to the driven member.

With the eccentric-elliptical combination of gearing just described, one revolution of the driven shaft is obtained for every three revolutions of the pinion driving shaft, two revolutions of the concentric pinion A being required for a half

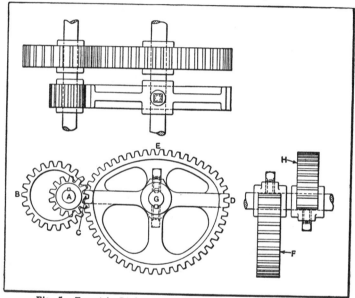

Fig. 5.   Eccentric Pinion and Elliptical Gear for Accelerating Return Movement

revolution, and one revolution of the eccentric pinion B for the remaining half revolution. If this mechanism is applied to a slotter or other machine requiring a similar movement, the cutting stroke will occur while pinion A is the driver, because a relatively slow and uniform speed is imparted to the driven shaft. As the eccentric pinion starts the drive, the speed of the driven shaft is gradually accelerated and, after reaching the maximum, is reduced to the cutting speed, so that the tool-slide is rapidly returned to the starting position ready for the next cutting stroke. The ratio of the quick return to the cutting speed should not be too great, because a jerky motion and excessive vibrations in the machine will result. It has been found, by experiment, that a ratio of 2 to 1 is about the highest that will give satisfactory operation.

When laying out gearing of this kind, there are a few fundamental points which must be observed in all cases: 1. The long radius AB of the eccentric pinion from the shaft center to the pitch line should equal one-half the distance between the centers of the driving and driven shafts. 2. The short radius AC of the eccentric pinion should equal one-half the diameter of the concentric pinion. 3. The major axis CD of the elliptical gear should equal twice the distance between the shaft centers, minus twice the short radius AC of the eccentric pinion. 4. The minor axis of the elliptical gear, or twice the distance EG, should equal the distance between the centers of the shafts. 5. The elliptical gear, assuming that it were complete, should have twice the number of teeth that there are in its eccentric driving pinion, and the number of teeth that there are in its eccentric driving pinion, and the number of teeth in both the elliptical gear and eccentric pinion should be even. 6. The shaft hole for the elliptical gear should always be located at the intersection of the major and minor axes, or in the center of the gear. This type of gearing is employed when it is especially desirable to secure a uniform motion during the entire cutting stroke.

Quick-return Movement which Operates Independently. — On one design of automatic screw machine, the quick-return and advance movements of the turret-slide are controlled independently of the turret-slide feed cam by means of a crank. The turret A (Fig. 6) is carried by a slide that moves horizontally along the machine bed. The movements of the turret-slide are derived from two different sources. When the turret tools are at work, the slide is operated by a lead cam through lever B, which has teeth at its upper end meshing with rack C. While the turret is being indexed, it is withdrawn rapidly and then quickly advanced to the working position again, by the action of crank E which is revolved once for each indexing movement. The rack C transmits motion to the turret-slide through connecting-rod F, which is pivoted to crank E on the turret-slide. This crank is on the "dead center," as shown in the illustration, while the tools are cutting; when the turret is to be indexed to bring the next successive tool in position, it is first withdrawn far enough for the tool to clear the work, and then the shaft carrying crank E is turned one revolution, through suitable gearing, by the

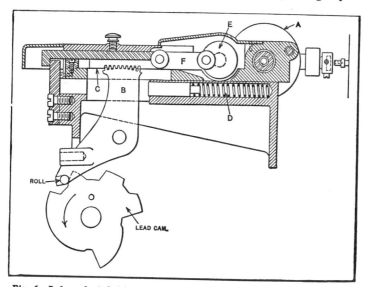

Fig. 6. Independent Quick-return Movement for Screw Machine Turret-slide

engagement of a clutch the action of which is controlled by a trip dog. When the crank revolves, it allows spring D to draw back the turret-slide without rack C, while making one-half turn, and then advance it during the remaining half turn, the rate of movement being increased by the motion derived from the cam, which is laid out to suit the work. This quick-acting crank operates while the roll on the lower end of lever B is passing from the highest point of the cam lobe to the point for starting the next cut.

CHAPTER XI

SPEED-CHANGING MECHANISMS

MECHANISMS for changing the speeds of rotating parts may be divided into two general classes. Those in one class provide convenient means of varying speeds to suit operating conditions. Usually a range of several speeds may be obtained from mechanisms of this class and the change from one speed to another may only require the shifting of a lever or some other form of controlling device.

Speed-changing mechanisms in the second class are designed either for reducing or for increasing the speed a fixed amount, the mechanism being designed for one speed ratio only. For example, the speed between driving and driven shafts may be reduced in the ratio of, say, 10 to 1, or to secure whatever speed change is required. Ordinarily, mechanisms in this class are for reducing the speed of a driving shaft, but in some cases an increase of speed is required. A speed-changing mechanism of this general class may be incorporated in the design of a machine for changing the relative speeds of certain parts, or such a mechanism may be located between a driving motor and the machine to reduce the relatively high motor speed down to the normal working speed of the machine.

General Methods of Speed Regulation. — When speed variations are essential to the operation of machines such, for example, as are used for some kinds of manufacturing work, the changes are usually obtained by hand-controlled speed-changing devices. If such variations are seldom required, it may be necessary to stop the machine and make an adjustment, or replace one or more gears with others of different diameters. When changes of speed are frequently needed, the machine is generally equipped with some mechanical device enabling one or more variations to be obtained rapidly, by sim-

310

ply moving a wheel, lever, or rod which controls the combination or velocity ratio of the mechanism through which the motion is transmitted. If the machine is of the automatic type, the speed may be regulated according to varying conditions, by the mechanism of the machine itself, which is constructed or adjusted beforehand to give the proper changes. The exact arrangement of the details depends, in any case, upon conditions such as the speed variation required, the importance of rapid changes, the relation of the speed-controlling mechanism to other parts of the machine, etc.

Mechanical devices for varying the speed are of special importance on machine tools. In fact, most machine tools are so constructed that the speed of the cutting tool or of the part being operated upon can be varied, the range or extent of the variation depending upon the type of machine. These changes are desirable in order to cut different kinds of metal at the most efficient speed; for example, soft brass may be turned, drilled, or planed at a much higher speed than cast iron or steel, and, by using the fastest speed that is practicable, obviously the rate of production is increased. Another important reason for speed variation is to secure the proper surface speed for revolving parts, regardless of the diameter, and the correct cutting speeds for rotating tools of different sizes. In the case of lathes or other turning machines, the speed of the work is increased as the diameter decreases, in order to maintain a cutting or surface speed which is considered suitable for the kind of metal being machined. Similarly, drilling or boring machines are so designed that the speed of the drill or boring bar can be varied in accordance with the diameter of the hole being drilled or bored. The design of this part of any machine tool involves determining the minimum and maximum speeds that would ordinarily be required, the total number of variations, the amount of increment by which each step or change varies, and the design of the mechanical device for securing speed changes and transmitting them to the work-spindle or tool. These speed-changing devices usually consist of different combinations of gearing, although belt-driven pulleys and friction gearing are often utilized.

Types of Mechanical Speed-changing Mechanisms. — When a variation of speed is obtained by changing the velocity ratio of two or more parts forming a train of mechanism, one of the following methods is generally employed: (1) By means of conical pulleys connected by a belt or cone-pulleys having "steps" of different diameters upon which a connecting belt may be shifted; (2) by the use of cone-pulleys in conjunction with one or more sets of gears; (3) by means of toothed gears exclusively, with an arrangement that enables the motion to be transmitted through different ratios or combinations of gearing; (4) by employing a friction transmission consisting of driving and driven disks, pulleys, or wheels, so arranged that one member (or an intermediate connecting device) can be shifted relative to the axis of the other for varying the speed. These different types or classes of speed-changing mechanisms are constructed in various ways.

Combination of Cone-pulley and Gearing. — One method of changing speeds by using a cone-pulley in conjunction with gearing is illustrated by diagram *A*, Fig. 1. This particular arrangement is commonly employed on engine lathes and is known as "back-gearing." When the pulley is driving the spindle direct, it is usually locked to the spindle by means of a bolt which connects it with the "face gear" *d*. For the direct drive, the back-gears are disengaged and the main spindle and cone-pulley revolve together. By disengaging the cone-pulley from gear *d* so that it rotates freely about the spindle and engaging the back-gears, motion is transmitted from the "cone gear" *a* to gear *b,* and from *c* to *d;* in this way, the range of speeds obtained by the direct drive is doubled. With a four-step cone-pulley, there would be four direct speeds and four slower speeds with the back-gears engaged, the drive being so proportioned that a gradual increase of speeds from the minimum to the maximum, or *vice versa,* may be obtained. The sleeve which carries the two back-gears revolves about a shaft having eccentric bearings at the ends, so that, by turning this shaft, the back-gears are engaged or disengaged.

Many modern engine lathes have double back-gears, one arrangement being shown at *B*. There are two cone gears *a* and *b* and two mating gears *c* and *d* on the rear shaft, so that a double range of geared speeds may be obtained, in addition to variations secured with the direct drive; thus, with a three-step cone-pulley, there would be a total of nine speeds. The gears *c* and *d* are shifted along the rear shaft for changing their position relative to the cone gears. A modification of

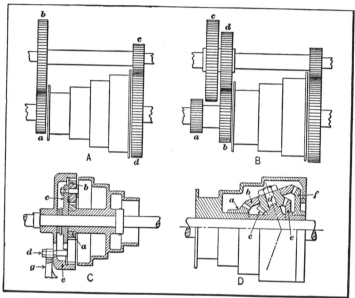

Fig. 1. Gearing and Cone-pulley Combinations for Varying Speed

the double back-geared drive is so arranged that the two gears on the rear shaft are connected by a friction clutch controlled by a conveniently located lever. Another design of lathe headstock gearing is commonly known as "triple gearing," although this term is not always applied to the same form of drive by machine-tool builders. Ordinarily, however, a lathe is said to be triple-geared when there are two gear shafts. The cone-pulley speeds are doubled by driving through one combination of gears, and a third range of speeds is obtained by transmitting the motion through the other combination, the pinion of the second shaft being engaged directly with a large internal gear on the faceplate. Triple gearing is used on large lathes and the direct drive to the face-plate provides a very powerful turning movement, such as is required for taking heavy cuts on castings or forgings of large diameter.

Cone-pulley and Epicyclic Gearing. — The use of a cone-pulley and planetary or epicyclic gearing is shown at C, Fig. 1. The cone-pulley has a pinion a, which meshes with pinion b, mounted on a stud carried by plate c. Pinion b also meshes with an internal gear forming part of casting e. This casting and the cone-pulley are both loose upon the shaft, but plate c is keyed to it. When lock-pin d engages a notch in plate c, the gears are locked together and the shaft is driven directly by the cone, the entire mechanism revolving as a unit. When lock-pin d is engaged with a stationary arm g, the internal gear is prevented from rotating and motion is transmitted to the spindle of the machine from the cone-pulley, as pinion a causes pinion b to revolve about the stationary internal gear and carry with it plate c, which transmits a slower speed to the spindle than is obtained with the direct drive. This design, which has been applied to some upright drilling machines, is sometimes known as a "differential back-gear."

Another cone-pulley containing epicyclic gearing is shown by the diagram D, Fig. 1. Bevel gears are employed in this case, instead of spur gears, and the combination is known as "Humpage's gear." This gearing was designed originally to replace the back-gearing of a lathe, but it has been applied to various classes of machinery. When used in conjunction with a cone-pulley, the arrangement is as follows: The cone-pulley is loosely mounted on its shaft and carries a pinion a which meshes with gear b. This gear is locked to pinion c, thus forming a double gear that is free to turn about arm d, the hub of which is also loosely mounted on the spindle or shaft. Gear b meshes with gear f, whereas pinion c meshes with gear e. Diametrically opposite arm d, there is another arm which carries gears corresponding to b and c. This additional gearing is included because of its balancing effect and need not be considered in studying the action of the gearing. The gear e is keyed to the spindle, and, except when a direct drive is employed, gear f is stationary. With the fulcrum gear f stationary and gear a revolving, gear e and the spindle are rotated at a much slower speed, as the arm d and the intermediate connecting gears roll around gear f. The direction in which gear e rotates for a given movement of gear a depends upon the ratio of the gearing, and the direction may be reversed by changing the relative sizes of the gears. When the ratio $\dfrac{f \times c}{b \times e}$ is less than 1, gears a and e will revolve in the same direction, whereas, if this ratio is greater than 1, they will revolve in opposite directions. This is compact gearing and the velocity ratio may be varied considerably by a slight change in the relative sizes of the gears.

The velocity ratio when $\dfrac{f \times c}{b \times e}$ is less than 1 may be determined by the following formula, in which the letters represent the numbers of teeth in the gears marked with corresponding reference letters in the illustration:

$$\text{Ratio} = \frac{\dfrac{f}{a} + 1}{1 - \dfrac{f \times c}{b \times e}}$$

If gear a has 12 teeth, b, 40 teeth, c, 16 teeth, e, 34 teeth, and f, 46 teeth, then,

$$\text{Ratio} = \frac{{}^{46}/_{12} + 1}{1 - \dfrac{46 \times 16}{40 \times 34}} = \frac{4\tfrac{5}{6}}{\dfrac{39}{85}} = 10.53$$

Therefore, gear a will revolve 10.53 times while gear e is making one revolution. If the expression $\dfrac{f \times c}{b \times e}$ is greater than 1, the formula may be changed as follows:

$$\text{Ratio} = \frac{\dfrac{f}{a} + 1}{\dfrac{f \times c}{b \times e} - 1}$$

Geared Speed-changing Mechanisms. — When toothed gearing is used exclusively in a speed-changing mechanism, the most common arrangements may be defined as the (1) sliding-gear type; (2) the clutch-controlled type; (3) the gear-cone and sliding-key type; (4) the gear-cone and expanding-clutch type; (5) the gear-cone and tumbler-gear type; and (6) the multiple crown-gear and shifting-pinion type. Diagram A, Fig. 2, illustrates the principle of the sliding-gear design. One of the parallel shafts carries two fixed gears, a and c; the gears b and d on the other shaft are free to slide axially so that motion may be transmitted either through gears a and b or c and d. The first combination gives a faster speed than the latter, because driving gear a is larger than gear c. For obtaining a greater range of speeds, two or more sets of sliding gears are used in many cases.

Clutch Method of Control. — Diagram B, Fig. 2, illustrates the use of a clutch for controlling speed changes. This clutch is located between the two driven gears and it can be engaged with either of these gears by a lengthwise movement effected usually by a lever. While this clutch is free to slide axially, it is prevented from revolving about the shaft by a spline or key. The driven gears, however, turn freely about the shaft unless engaged by the clutch. A positive clutch is shown in the diagram, or one having teeth which engage corresponding notches in the hubs of the gears; many of the clutches for speed-changing mechanisms, however, are of the friction type.

In the diagrams A and B, single-belt pulleys are shown upon the driving shafts. This is a common method of rotating the initial driving shaft of speed-changing mechanisms of the all-geared type, the shaft rotating at a constant speed and all of the changes being obtained by the shifting of gears or clutches. On many machines, however, the single constant-speed belt pulley is replaced either by a motor of the constant-speed type or one of the variable-speed type.

Intermeshing Gear Cones and Sliding Key. — The use of intermeshing gear cones and a sliding key for changing speeds is represented by diagram C, Fig. 2. Two cones of gears are mounted upon parallel shafts so that they intermesh, one shaft being the driver and the other, the driven member. All of the gears on shaft *a* are attached to it, whereas those on shaft *b* are free to revolve around the shaft, except when engaged by the key *c,* which can be shifted from one gear to another by moving rod *d.* If the key were in the position shown by the diagram, the drive would be through gears *g* and *e;* if *a* were the driving shaft, the speed of shaft *b* could be increased by

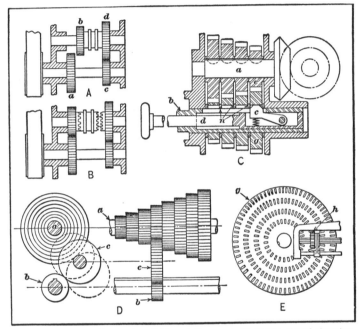

Fig. 2.   Diagrams Illustrating Different Types of All-geared Speed-changing Mechanisms

engaging the key with gears to the left. Obviously, the number of speed changes corresponds to the number of gears in the cone.

The driving end of the key projects through a slot in the shaft and the edges are beveled to an angle of about 45 degrees, so that, as the key is moved in a lengthwise direction, it will be depressed by the action of the beveled edge against a steel washer or guard *n* placed between each pair of gears.

With this arrangement, the key is completely disengaged from one gear before meshing with the next one, which is essential with a drive of this kind. The key is forced upward into engagement with the keyways of the different gears, by means of a spring beneath it. A modification of the mechanism just described is so arranged that, instead of locking the gears in the upper cone by means of a sliding key, each gear is fitted with a ring which may be expanded by means of a wedge, the action of which is controlled by suitable means. The gear-cone and sliding-key mechanism is applied to many different types of machine tools, although this form of mechanism is usually installed either for transmitting feeding motion or in connection with spindle drives which require a relatively small amount of power.

Gear-cone and Tumbler-gear Mechanism. — The arrangement of a gear-cone and tumbler-gear mechanism is represented by diagram D, Fig. 2. There is a cone of gears on shaft a and a pinion b which is free to slide on a splined shaft and is connected with cone gears of different diameters, by means of the tumbler gear c. The tumbler gear is carried by an arm which can be shifted parallel to the axis of the gear cone for aligning the tumbler gear with any one of the cone gears; this arm can also be moved at right angles to the axis of the gear cone for bringing the tumbler gear into mesh with the various sizes of gears composing the cone (as shown by the dotted circles), and provision is made for locking the arm in its different positions. Cone-and-tumbler gearing is not always arranged as shown by diagram D; for instance, the tumbler gear, instead of engaging with a pinion mounted upon a splined shaft, may mesh with a long pinion, or the tumbler gear may be carried by a frame which is adjusted to bring the tumbler gear into mesh with the different cone gears. Another modification consists of a cone of gears which are adjusted axially for alignment with the tumbler gear which is only moved in a radial direction.

Multiple Crown-gear and Shifting Pinion. — The multiple crown-gear type of speed-changing mechanism is represented by diagram *E*, Fig. 2. The crown gear *g* has several concentric rows of teeth, and the speed is varied by shifting the pinion *h* so that it engages a row of larger or smaller diameter. This mechanism has been applied to drilling machines for varying the feeding movements of the drill.

The design and application of the various kinds of speed-changing mechanisms previously described, and the exact arrangement of the gears or other parts are governed very largely by the type of machine and the general nature of the work which it does. Mechanisms of the same general type are often constructed along different lines.

Gear Ratios and Speed Variations. — Proportioning a train of gears to obtain a given velocity ratio, or possibly a given series of speeds, is frequently encountered in the design of geared transmissions. When the problem is simply that of obtaining a given velocity ratio, and when the latter is so large that more than one pair of gears should be used, a uniform reduction between the different pairs is conducive to the highest efficiency. Whenever this arrangement is practicable, the ratio of each pair in a train may be determined by extracting the root of the total ratio. If there are two pairs of gears, extract the square root; for three pairs, extract the cube root, etc. For example, if the total ratio between the first driving and the last driven gear is to be 125 to 1 and three pairs of gears are to be used, the ratio of each pair should preferably equal 5 to 1, since $\sqrt[3]{125} = 5$.

Speeds in Geometrical Progression. — In designing gear combinations for varying spindle speeds or feeding movements, it is general practice, among machine tool builders particularly, to vary the speeds in geometrical progression, successive speeds being obtained by multiplying each preceding term by a ratio or constant multiplier. Thus, if the slowest speed is 50 revolutions per minute and the ratio is 1.3, the succeeding speeds will equal

$$50 \times 1.3 = 65$$
$$65 \times 1.3 = 84.5$$
$$84.5 \times 1.3 = 109.8$$

When the fastest speed f and the slowest speed s in a series are known and also the total number of speeds n, the ratio may be determined by the well-known formula:

$$\text{Ratio} = \sqrt[n-1]{\frac{f}{s}}$$

Since logarithms would ordinarily be used for the extraction of this root, the ratio may be obtained as follows:

Rule. Subtract the logarithm of the slowest speed from the logarithm of the fastest speed and divide the difference by the total number of speeds minus 1. The result will equal the logarithm of the ratio.

Ratios for Machine Tool Drives. — In actual practice, the exact progression obtained may be modified slightly to permit using gears of a certain diametral pitch. For machine tool transmissions, the ratio of a geometrical progression should, as a general rule, be between 1.3 and 1.5, as otherwise there will be either too small or too great a difference between successive speeds. There would be no practical advantage in a series of speeds varying by small increments equivalent to a ratio of say, 1.1, whereas, if the ratio were 1.7 or possibly 2, the changes from one speed to the next would be excessive. Feeding mechanisms may be designed for ratios of 1.2 or less, depending on the type of machine.

Speeds of machine tool drives and especially feed changes are sometimes varied according to "chromatic scale progression," with a ratio of either 1.4142 or 1.189 in case a lower ratio is required. The first ratio is the square root of 2, and the second the fourth root of 2. The object of using these particular ratios is to obtain a series of speeds or feeds containing the even ratios, 2, 4, 8, 16, etc.

Speed Calculations for Lathe Headstock. — As an example of gear designing, for obtaining speed changes, assume that the problem is to design a lathe headstock with a stepped cone pulley. The headstock is to have a five-step cone with back-gears, giving ten speeds in all, and the speeds are to range from 300 down to 10 revolutions per minute.

The first thing to determine is the geometrical progression

of ten numbers from 10 to 300. As the maximum and minimum speeds and the number of speed changes are known, the ratio of the progression may be determined by using the formula previously given. Inserting the values in this formula we have,

$$R = \sqrt[10-1]{\frac{300}{10}} = 1.46$$

Therefore to have speeds from 10 to 300 revolutions per minute, ten in number, we must use a ratio of 1.46, which on multiplying gives the results shown in the first column of Table 1. Five of these speeds are obtained from the five-step cone. The function of the back-gear is to double the number of speeds, making ten in all. We have decided, from observation or experience, that the least practicable diameter for the

Table I

Speeds Varying in Geometrical Progression	Speeds Actually Obtained
10......... = 10.0	9.7
10 × 1.46 = 14.6	14.5
14.6 × 1.46 = 21.3	21.0
21.3 × 1.46 = 31.0	30.0
31 × 1.46 = 45.3	45.0
45.3 × 1.46 = 66.0	65.0
66 × 1.46 = 96.4	97.0
96.4 × 1.46 = 140.5	140.0
140.5 × 1.46 = 205.0	202.0
205 × 1.46 = 299.3	300.0

smallest cone step is 5 inches. This gives a starting point from which all the other cone steps may be calculated, as their ratios must be to one another as the ratios of their speeds.

Cone Pulley Diameters For Crossed Belts. — The highest speed is obtained, of course, when the belt is driving from the largest step of the countershaft cone to the smallest step of the headstock cone and the back-gear is out of mesh. Countershaft and headstock cones are usually made alike; consequently the two middle steps will be the same diameter, which makes the countershaft speed the same as the middle speed of the cone, without the back-gears, or 140.5 revolutions per minute.

The size of the largest step may be obtained by the formula:

$$\text{Diam. largest step} = \frac{\text{Max. spindle speed} \times \text{Diam. smallest step}}{\text{Speed of countershaft}}$$

$$\text{Diameter of largest step} = \frac{300 \times 5}{140.5} = 10.7 \text{ inches}$$

The middle step should next be determined. It is the sum of the large and the small diameters divided by 2:

$$5 + 10.7 \div 2 = 7.85 \text{ inches}$$

The second largest step will be the mean between the middle and the largest step:

$$7.85 + 10.7 \div 2 = 9.27 \text{ inches}$$

The second smallest step in like manner will be:

$$5 + 7.85 \div 2 = 6.43 \text{ inches}$$

Calculating the Back-gears. — The back-gears obviously must be of such ratio as to reduce the first five speeds to a second or slower five. The ratio between the slowest of the first five and the slowest of the second five is as 10 to 66 revolutions per minute or 6.6 to 1. As the back-gear consists of two pairs of gears, this ratio is the product of their respective ratios, and we must extract the square root to find what these respective ratios may be. The square root of 6.6 is 2.56. Therefore the ratio of each pair of gears should be 2.56 to 1. We may now proceed to find the number of teeth in the gears. It is customary to make the gears next to the faceplate of heavier pitch, as they are the slowest running and consequently have the greatest tooth pressure. Suppose we find by calculation, experience, or comparison with other makers that the faceplate gears should be 4 diametral pitch and the back-gear shaft admits of using a 15-tooth pinion. Then:

$$15 \times 2.56 = 39 \text{ teeth, approximately}$$
$$15 \div 4 = 3\tfrac{3}{4} \text{ inches pitch diameter}$$
$$39 \div 4 = 9\tfrac{3}{4} \text{ inches pitch diameter}$$

One-half of the sum of the pitch diameters, or $6\tfrac{3}{4}$ inches, equals the center-to-center distance. Now we want to make the other pair of gears 5 diametral pitch, and 5 diametral pitch gears will not fit into $6\tfrac{3}{4}$-inch centers. The solution of the problem is to make the faceplate gears with 16 and 40 teeth, respectively, which will make the centers an even 7 inches.

$16 \div 4 = 4$ inches pitch diameter

$40 \div 4 = 10$ inches pitch diameter

Therefore the center distance equals $\dfrac{10 + 4}{2} = 7$ inches.

This alters the ratio slightly, as the ratio of 40 to 16 is 2.5. The difference must be made up in the other pair of gears as follows:

$6.6 \div 2.5 = 2.64$ ratio for 5 diametral pitch gears

Twice the center distance or 14 inches is the combined pitch diameters of the gears, and with 5 diametral pitch the combined number of teeth is 70. The desired ratio is 2.64. Add 1 to this ratio and divide it into the total number of teeth. Thus:

$70 \div 3.64 = 19$, approximately

Therefore we have 19 teeth in the pinion and $70 - 19 = 51$ teeth in the gear. The spindle speeds will now be checked as

Table II

Counter-shaft Speed, R.P.M.		Counter-cone Diam. Inches		Headstock Cone Diameter, Inches	Back-gears		Spindle Speeds, R.P.M.
140	×	10.7	÷	5.00	Out	=	300
140	×	9.27	÷	6.43	Out	=	202
140	×	7.85	÷	7.85	Out	=	140
140	×	6.43	÷	9.27	Out	=	97
140	×	5.00	÷	10.7	Out	=	65
140	×	10.7	÷	5.00 × (19 × 16 ÷ 51 × 40)	=	45
140	×	9.27	÷	6.43 × (19 × 16 ÷ 51 × 40)	=	30
140	×	7.85	÷	7.85 × (19 × 16 ÷ 51 × 40)	=	21
140	×	6.43	÷	9.27 × (19 × 16 ÷ 51 × 40)	=	14.5
140	×	5.00	÷	10.7 × (19 × 16 ÷ 51 × 40)	=	9.7

shown in Table 2 to see how close we have come to the desired result. We will make the countershaft speed 140 revolutions per minute even. The results shown in Table 2 are also compared with the ideal speeds in Table 1. It is possible to get even closer results by a little juggling, but such accurate speeds are seldom necessary.

Planetary or Epicyclic Gear Trains. — If one of the gears in a train is fixed or stationary, and another gear (or gears) revolves about the stationary gear in addition to rotating relative to its own axis, the mechanism is known as an *epicyclic train of gearing,* because points on the revolving gears describe

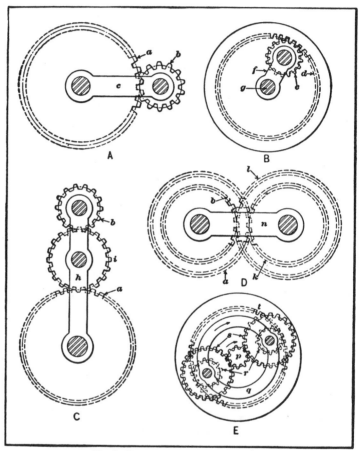

Fig. 3. Epicyclic or Planetary Gearing

epicycloidal curves. The two gears *a* and *b* (see diagram *A*, Fig. 3) are held in mesh by a link *c*. If this link remains stationary and gear *a* makes one revolution, the number of revolutions made by gear *b* will equal the number of teeth in *a* divided by the number of teeth in *b*, or the pitch diameter of *a* divided by the pitch diameter of b. If a and b represent either the pitch diameters of the gears or numbers of teeth, the revolutions of b to one turn of a equal $\frac{a}{b}$. If gear a is held stationary and link c is given one turn about the axis of a, then the revolutions of gear b, relative to arm c, will also equal $\frac{a}{b}$, the same as when gear a was revolved once with the arm held stationary. Since a rotation of arm c will cause a rotation of gear b in the same direction about its axis, the total number of revolutions of gear b, relative to a fixed plane, for one turn of c, will equal 1 (the turn of c) plus the revolutions of b relative to c or $1 + \frac{a}{b}$. For example, if gear a has 60 teeth and gear b, 20 teeth, one turn of arm c would cause b to rotate $\frac{60}{20}$, or 3 times about its own axis; gear b, however, also makes one turn about the axis of gear a, so that the total number of revolutions relative to a fixed plane equals $1 + \frac{60}{20} = 4$ revolutions.

In order to illustrate the distinction between the rotation of b around its own axis and its rotation relative to a fixed plane, assume that b is in mesh with a fixed gear a and also with an outer internal gear that is free to revolve. If the speed of the internal gear is required, it will be necessary, in calculating this speed, to consider not only the rotation of b about its own axis, but also its motion around a, because the effect of this latter motion on the internal gear, for each turn of link c, is equivalent to an additional revolution of b.

Diagram B, Fig. 3, represents an internal gear d in mesh with gear e on arm f. If arm f is held stationary, the revolutions of e for one turn of d equal $\frac{d}{e}$, d and e representing the numbers of teeth or pitch diameters of the respective gears. If the internal gear is held stationary and arm f is turned about axis g, the rotation of e about its axis will be clockwise when f is turned counter-clockwise, and *vice versa;* hence, the revolutions of gear e, relative to a fixed plane, for one turn of f about g, will equal the difference between 1 (representing the turn of f) and the revolutions equal to $\dfrac{d}{e}$.

Method of Analyzing Epicyclic Gear Trains. — A simple method of analyzing epicyclic gearing is to consider the actions separately. For instance, with the gearing shown at A, Fig. 3, the results obtained when link c is fixed and the gear a (which normally would be fixed) is revolved are noted; if gear a is revolved in a clockwise direction, then, in order to reproduce the action of the gearing, the entire mechanism, locked together as a unit, is assumed to be given one turn counterclockwise. The results are then tabulated, using plus and minus signs to indicate directions of rotation. Assume that gear a has 60 teeth and gear b, 20 teeth, and that $+$ signs represent counter-clockwise movements and $-$ signs clockwise movements. If link c is held stationary and gear a is turned clockwise $(-)$ one revolution, gear b will revolve counter-clockwise $(+)$ $^{60}/_{20}$ revolution. Next consider all of the gears locked together so that the entire combination is revolved one turn in a counter-clockwise $(+)$ direction, thus returning gear a to its original position. The practical effect of these separate motions is the same as though link c were revolved once about the axis of a fixed gear a which is the way in which the gearing operates normally. By tabulating these results as follows, the motion of each part of the mechanism may readily be determined:

	Gear a	Link c	Gear b
Link Stationary	−1 turn	0 turn	$+\frac{60}{20}$ turn
Gears Locked	+1 turn	+1 turn	+1 turn
Number of Turns	0	+1	+4

The algebraic sums in line headed "Number of Turns" indicate that, when gear a is held stationary and link c is given one turn about the axis of a, gear b will make 4 revolutions relative to a fixed plane in a counter-clockwise or $+$ direction, when link c is turned in the same direction.

The application of this method to the arrangement of gearing shown at B, Fig. 3, will now be considered. Assume that gear d has 60 teeth and gear e, 20 teeth. Then, if gear d is turned clockwise with link f stationary, and the entire mechanism with the gears locked is turned counter-clockwise, an analysis of the separate motions previously referred to will give the following results:

	Gear d	Link f	Gear e
Link Stationary............	-1 turn	0 turn	$-\frac{60}{20}$ turn
Gears Locked.............	$+1$ turn	$+1$ turn	$+1$ turn
Number of Turns..........	0	$+1$	-2

Effect of Idler in Epicyclic Gear Train. — If an idler gear i is placed between gears a and b (diagram C, Fig. 3), the latter will rotate about its axis in a direction opposite to that of the link (the same as with the arrangement shown at B), and the revolutions of gear b, relative to a fixed plane, for one turn of link h about the axis of a, will equal the difference between 1 (representing the turn of h) and the revolutions equal to $\frac{a}{b}$.

Assume that gear a has 60 teeth, idler gear i, 30 teeth, and gear b, 20 teeth. Then the turns of b, relative to a fixed member for one turn of h about the axis of a, are shown by the following analysis:

	Gear a	Idler i	Link h	Gear b
Link Stationary.........	-1 turn	$+\frac{60}{30}$ turn	0 turn	$-\frac{60}{20}$ turn
Gears Locked...........	$+1$ turn	$+1$ turn	$+1$ turn	$+1$ turn
Number of Turns.......	0	$+3$	$+1$	-2

The direction of rotation of b, relative to a fixed member, may or may not be in the same direction as that of link h, depending upon the velocity ratio between gears a and b. If gears a and b are of the same size, one turn of link h will cause b to revolve once about its own axis, but, as this rotation is in a direction opposite to that of h, one motion neutralizes the other, so that b has a simple motion of circular translation relative to a fixed member. If gear b were twice as large as a, it would then revolve, for each complete turn of link h, one-half revolution about its own axis, in a direction opposite to the motion of h; this half turn subtracted from the complete turn of link h gives a half turn in the same direction as h, relative to a fixed member.

Compound Train of Epicyclic Gearing. — Diagram D, Fig. 3, illustrates a compound train of epicyclic gearing. This arrangement modified to suit different conditions is commonly employed. Gear a represents the fixed member and meshes with gear k, which is attached to the same shaft as gear l. Gear l meshes with gear b the axis of which coincides with that of fixed gear a. Assume that gear a has 36 teeth, gear k, 34 teeth, gear l, 35 teeth, and gear b, 35 teeth. Then one turn of link n about the axis of gear a would give the following results:

	Gear a	Link n	Gears k and l	Gear b
Link Stationary...	-1 turn	0 turn	$+\frac{36}{34}$ turn	$-(\frac{36}{34} \times \frac{35}{35})$ turn
Gears Locked.....	$+1$ turn	$+1$ turn	$+1$ turn	$+1$ turn
Number of Turns.	0	$+1$	$+2\frac{1}{17}$	$-\frac{1}{17}$

From this analysis, it will be seen that, for each counter-clockwise turn of link n, the rotation of gear b equals $1-\dfrac{a}{k} \times \dfrac{l}{b}$ in which the letters correspond either to the pitch diameters or numbers of teeth in the respective gears shown at D in Fig. 3. If the value of $\dfrac{a}{k} \times \dfrac{l}{b}$ is less than 1, gear b will revolve in the same direction as link n, whereas, if this value is greater than 1, gear b will revolve in the opposite direction.

Compound epicyclic gearing may be used for obtaining a very great reduction in velocity between the link n and the last gear b in the train. As an extreme example, suppose gear a has 99 teeth, gear k, 100 teeth, gear l, 101 teeth, and gear b, 100 teeth. The speed of gear b will equal $1 - \dfrac{99 \times 101}{100 \times 100} = \dfrac{1}{10,000}$ revolution; hence link n would have to make 10,000 revolutions for each revolution of gear b. The arrangement of epicyclic gearing shown at D is known as a *reverted train.*

Diagram E shows another arrangement of reverted train. An internal gear t forms part of the mechanism, and either this gear, frame q, or pinion p may be the stationary member, depending upon the application of the mechanism. In this case, instead of a single set of gears between p and t, there is a double set located diametrically opposite and connected by a suitable frame q. This arrangement is similar to the mechanism of a certain type of geared hoist. The central pinion p is the driving member, internal gear t is stationary, and the frame q is the driven member and imparts motion to the hoisting sheave.

Sun and Planet Motion. — A mechanism of the general type illustrated by diagram A, Fig. 3, was employed by Watt for transmitting motion from the connecting-rod to the engine shaft, because the crank motion had been patented previously. This mechanism is known as a "sun and planet" motion, the fixed gear a representing the sun and the revolving gear b, the planet. In applying this mechanism to an engine, one gear was keyed on the shaft and the other was fixed to the connecting-rod. The connecting link between the gears was loose on both shafts. A forward and return stroke of the piston caused the connecting-rod gear to pass once around the shaft gear, but without revolving on its own axis, as it was attached to the connecting-rod. With this arrangement, if both gears are of the same diameter, the shaft gear will make two revolutions for one turn of the connecting link between the gears or one revolution for each stroke.

Differential Gearing for Large Speed Reduction. — When it is necessary to obtain a speed reduction of large magnitude, some arrangement of differential gearing may be preferable. Belting may not be suitable, either because it does not give a positive drive from the driving to the driven shaft, or because there is not sufficient room for the pulleys. Lack of room may also prevent the use of chains and sprockets. Spur gear trains often require too many gears, thus introducing high costs and an undue amount of power loss through friction. Differential gearing as a means of obtaining a satisfactory speed reduction mechanism of compact form is utilized on many classes of machines. With the form shown in Fig. 4, the speed reduction is made from the shaft A to the shaft B. As may be seen, the gears C and C_1 are fixed on shaft A. Gear E is merely an idler meshing with gears C and D. Gear

D is fixed on the hub of bevel gear F, which is bushed and free to revolve on the shaft B. Gear C_1 meshes directly with gear D_1. Gear D_1 is mounted on the hub of the bevel gear F_1 which is bushed and free to revolve on shaft B, the same as gear F. The crank H is keyed to the shaft B, and on its two arms are mounted the bevel idler gears G and G_1 which are bushed and free to revolve on the arms.

Now as C and C_1 are keyed to the shaft A, it is evident that

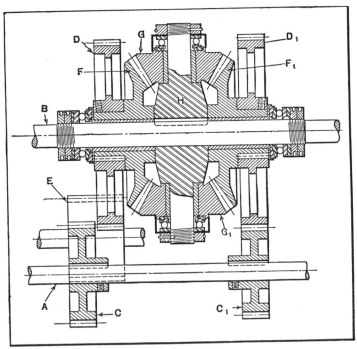

Fig. 4. Speed Reduction Mechanism Employing Differential Gearing

they must always turn or rotate in the same direction. It should be noted further that C_1 drives direct to D_1 and that C drives D through the idler gear E. Therefore D and D_1 must revolve in opposite directions. Their speeds are also different, as gear C has more teeth than gear C_1. Gears F and F_1 must, of course, revolve with gears D and D_1 which are mounted on their respective hubs. The bevel gears G and G_1

meshing with both F and F_1 must therefore revolve on their own axis. If F and F_1 should revolve at the same speed but in opposite directions, gears G and G_1 would be stationary with respect to the axis of the shaft B, that is, they would revolve on their own axis, but they would not revolve about shaft B. Now if F and F_1 should revolve at different speeds, G and G_1 must revolve about the axis of the shaft B, and thus revolve shaft B through crank H.

The number of revolutions per minute of shaft B, which is driven by crank H will be one-half the algebraic sum of the number of revolutions per minute of the speeds of the bevel gears F and F_1. In order to make this clear, let it be assumed that a point on the pitch circle of gear F_1 is traveling at a rate of $x + y$ feet per minute, and that gear F is stationary. It is obvious that a point the same distance from the center of shaft B on the axis of G and G_1 will travel at a speed of $\frac{1}{2}$ $(x + y)$ feet per minute. Assume that a corresponding point on gear F is traveling x feet per minute in the opposite direction or—x feet per minute, and consider gear F_1 to be stationary. The corresponding point on the axis of gears G and G_1 will now travel at a rate of $\frac{1}{2}$ $(-x)$ feet per minute. Next assume that both gears F_1 and F are traveling at their respective speeds of plus $x + y$ and minus x feet per minute. Now adding the speeds of F_1 and F we obtained $x + y - x = y$ feet per minute. Then a speed of one-half y feet per minute equals the speed of the point under consideration on the axis of gears G and G_1.

As the pitch diameters of gears F and F_1 are equal and cannot be otherwise, and as these gears mesh with gears G and G_1, it is evident that the number of revolutions per minute can be readily employed to designate the speed. Therefore, the speed of shaft B can be expressed as one-half the algebraic sum of the revolutions per minute of gears F and F_1. The selection of the bevel gears is merely a matter of choosing such sizes as can be used in the available space, and still be of sufficient size to give the necessary tooth strength for the material used and the load imposed. The diameters of gears

G and G_1 should, of course, be made as large as permissible so that they will not revolve on the crank H at a higher speed than necessary. In determining the sizes of the spur gears to be used, it is only necessary (not considering the available space) to select such sizes as will give the desired difference in speed between F and F_1, together with a suitable surface speed.

The following example will serve to make clear the procedure followed in laying out or designing a speed reduction device of the type shown in the illustration. We have shaft A running at a speed of — 625 revolutions per minute (anti-clockwise), and it is desired to drive shaft B from shaft A so that shaft B will revolve in a clockwise direction at a speed of approximately 3 revolutions per minute. The algebraic sum of the speeds of gears F and F_1, must, therefore, be about + 6 revolutions per minute. Taking 300 revolutions per minute as an approximate speed for F_1, we may proceed to determine the pitch and size of the gears to be used.

Let it be assumed that we select an 8-pitch gear having a pitch diameter of $7\frac{1}{2}$ inches for D_1 and an 8-pitch gear having a pitch diameter of $3\frac{1}{2}$ inches for gear C_1. These gears will give bevel gear F_1 a speed of + 291.67 revolutions per minute. The speed of gear F should therefore equal approximately — 291.67 + 6 or — 285.67 revolutions per minute. Now if gear D has a pitch diameter of $7\frac{3}{8}$ inches and gear C has a pitch diameter of $3\frac{3}{8}$ inches, gear F will have a speed of approximately — 286.02 revolutions per minute. The difference between the number of revolutions of gears F and F_1 will then equal 5.65 revolutions per minute, and the speed of shaft B will be one-half as great or + 2.82 revolutions per minute.

It is, of course, difficult to obtain an exact speed for shaft B, but by using gears of as fine a pitch as possible without an undue sacrifice of strength, we can obtain very nearly the exact speed desired. The center distance between the two shafts can in some cases be varied, thus increasing the range in speed reduction that may be obtained.

The possibilities of a speed reduction device of this kind is readily apparent to the designer. The speed of the shaft B with respect to shaft A is governed entirely by the speed ratio existing between the gears D and D_1. The speed of shaft A may be as high as the successful operation of the gears will allow, and still by the right combination of gears it may be possible to obtain a very low speed for shaft B. At the same time the whole gear assembly is very compact, and may be installed where other forms of speed reduction mechanisms cannot be used. In order to obtain the best results, all the running parts should, of course, be well oiled and this may be readily accomplished by running the whole assembly in an oil bath or in a case partly filled with oil. Operated under these conditions, the friction loss is very low.

Differential Mechanism for Reduction of 840 to 1. — In machine design large reductions in speed are often obtained by the use of a differential mechanism, especially when the reducing unit must occupy a comparatively small amount of space. A differential mechanism in which a reduction of 840 to 1 is secured by using only two spur gears, two worms, two worm-wheels, two bevel gears, two bevel pinions and a yoke, is shown in Fig. 5. The reduction ratio of this gearing may be altered to suit conditions by simply varying the number of teeth in the two spur gears. Dimension X is approximately 7⅛ inches on this particular lay-out.

Spur gear A is the driving member of the differential unit. It is mounted on the same shaft as worm C, and through these two parts drives worm-wheel E. Gear A also meshes with gear B, and thus drives a shaft running parallel to that on which it is mounted. On the second shaft is a worm D by means of which power is transmitted to worm-wheel F. Gears A and B revolve in opposite directions, and as the thread of both worms is right-hand, worm-wheels E and F also rotate in opposite directions. Pinned to the adjacent sides of the two worm-wheels are bevel gears G, which, together with their respective worm-wheels, are free to turn on the driven shaft. These bevel gears mesh with two idler pinions on studs at

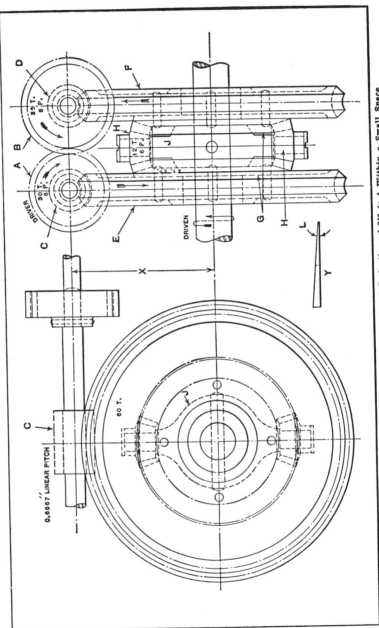

Fig. 5. Geared Mechanism for Obtaining a Speed Reduction of 840 to 1 Within a Small Space

opposite ends of driving yoke J, which is pinned to the driven shaft. The idler pinions are held on the studs by means of screws.

The large reduction obtained by this unit is due to giving spur gear B five more teeth than gear A. As a consequence, gear B makes only 30/35 revolution per revolution of gear A. The ratio of each set of worm-gearing is 60 to 1, and so for each revolution of gear A, worm-wheel E makes 1/60 revolution. The 30/35 revolution imparted to gear B at the same time causes worm-wheel F to turn 1/70 revolution in the opposite direction to that in which worm-wheel E is revolved. Because of the two worm-wheels revolving in opposite directions, the result on yoke J is the same as if worm-wheel F were held stationary and worm-wheel E were moved the difference between 1/60 and 1/70 revolution. This would mean a forward movement of worm-wheel E of 1/420 revolution.

As the center of driving yoke J is located half way between bevel gears G, the actual movement of the yoke per revolution of gear A will be only one-half the difference between the forward and backward movements of the two worm-wheels, or 1/840 revolution. This can be easily proved by constructing a triangle as shown at Y, in which the line adjacent to angle L represents the pitch diameter of pinions H, and the length of the opposite side equals the distance that a point on the pitch circle of the bevel gear on worm-wheel E would move during 1/420 revolution. Then if the adjacent side is bisected and a perpendicular is erected at that point, the length of the perpendicular will be one-half the length of the opposite side of the triangle; this perpendicular line will equal the distance a point on the center line of the yoke, located from the center of the driven shaft a distance equal to the pitch radius of bevel gears G, will move during the movement of the bevel gear on worm-wheel E. This yoke movement equals 1/840 revolution.

Compound Differential Gears for Varying Speeds.— The differential speed-changing mechanism shown in Fig. 6 has spur gears and pinions but no internal gear. This is a compound or reverted train and is intended for an automatic screw machine of the heavier class in order to provide a slow and powerful movement to the spindle for heavy thread-cutting operations, or for any other heavy work which requires a powerful drive. The gearing is contained within the spindle driving pulleys on the back shaft of the spindle head. There are three pulleys and the slow speed is obtained by shifting the belt to the center pulley A, and engaging the sliding clutch B with gear C; as this clutch slides upon a square shaft and cannot revolve, the gear C is held stationary. There are two sets of planetary pinions D and E located diametrically oppo-

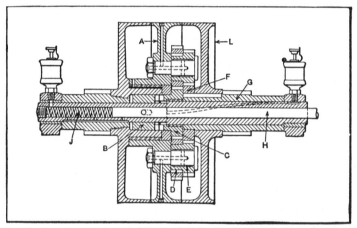

Fig. 6. Compound or Reverted Train of Epicyclic Gearing for Reducing Speed

site. The pinions on each stud are locked together but they are free to revolve about the stud. Pinions D rotate around the fixed gear C, while pinions E revolve the driven gear F at a slow speed, but with considerable power. The gear F is keyed to the extension of pinion G which meshes directly with the front spindle gear of the machine. When this slow speed is not required, the clutch B is disengaged, so that the entire train of differential gears is free upon the loose center pulley A. Two spring plungers (not shown) attached to pulley A engage the rim of pulley L and cause both pulleys to revolve together when the slow-speed attachment is not engaged, so that the planetary pinions will not revolve upon their studs at this time. The clutch B is shifted by a cam-operated rod H acting in conjunction with a spring J.

With this arrangement of gearing, the differental action and reduction of speed is the result of the difference in the diameters of pinions D and E and their mating gears. When the slow-speed attachment is operating, the larger pinions D roll around the stationary gear C and force gear F to follow slowly in the same direction. This action will be more apparent if that part of the larger pinion D which is in engagement with stationary gear C, at any time, is considered as a lever pivoted at the point where the teeth mesh with the stationary gear. As the pinion D revolves and the imaginary lever swings around its fulcrum, the teeth of the smaller pinion E in contact with gear F force the latter to move in the same direction in which the rolling pinions D and E and pulley A are moving.

Slow Starting Motion for Textile Machine. — The slow starting motion attachment shown in Fig. 7 is applied to textile machines that are used for winding the warp threads onto a loom beam, as a precaution against breakage of the threads, caused by sudden starting. Shaft A, which is a short auxiliary shaft on which the pulleys and other parts are carried, is mounted in a bearing which is bolted to the frame of the machine. Bushing B is fastened by a set-screw to A and forms the bearing for the loose pulley C. The slow-motion pulley D, which has a twelve-tooth gear cast on its inside hub, turns freely on the shaft A. The casting E, which is also fastened by a set-screw to A carries the steel pinion F, the shank of which revolves in a bearing at the end of the casting. This pinion has twelve teeth and on the opposite end of the shaft a gear of nineteen teeth is assembled, which meshes with the twelve-tooth gear on the slow-motion pulley D. The forty-tooth clutch gear G meshes with F and is loose on shaft A. This gear has a five-tooth clutch, as shown in the right-hand view at g. The driving pulley H is also loose on A and has a brake pulley I and a driving gear H_2 cast integral with it, the gear H_2 driving the main gear of the machine. On the inside of pulley H are three short studs J, each of which carries a small pawl K.

The belt-shifting mechanism is not shown in the illustration, but it is operated by a foot-treadle, which is fastened to the treadle shaft. On the end of the treadle shaft, just inside the frame, is a segment arm, which meshes with the teeth on one end of a double gear. The teeth on the opposite end of this gear operate a sliding rack which projects from the frame, just above the pulleys. The belt guide is bolted to this rack, and as the rack is run outward and the belt shifted to pulley C, the brake, which is also attached to the treadle shaft, is

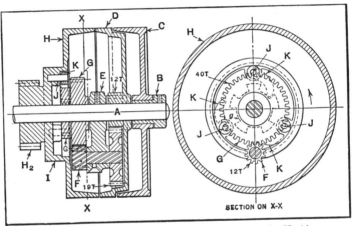

Fig. 7. Slow Starting Motion Mechanism of Textile Machine

brought into contact with I and the machine is quickly stopped. This prevents the ends of yarn that are being wound on the beam from becoming slack or entangled, which would be a troublesome and awkward condition.

When the belt is shifted from C to D, the number of revolutions per minute of gear G is reduced to about one-fifth that of pulley D as is shown in the ratio of the teeth $\dfrac{12 \times 12}{19 \times 40}$, or approximately 1/5. The relation of the three pawls K to the five teeth on the clutch g is such that one pawl is always down; that is, one will always drop into the position occupied by the

upper pawl, as shown in the right-hand view of the illustration. The pawl being thus engaged, the engaging tooth of the
clutch, which is revolving counter-clockwise and at reduced
speed, pushes against the pawl, driving pulley H, and incidentally gear H_2, slowly in the direction shown by the arrow.
When the belt is shifted onto pulley H, which is also a loose
pulley, gear H_2 will assume the full speed, pawls K will simply
drag over the teeth in the clutch, and pulley D will become
idle, as a result of the disengagement of the pawls and clutch.

Rotary Speed Varied Each Half Revolution. — An electric
switch testing machine required that shaft A (see Fig. 8)

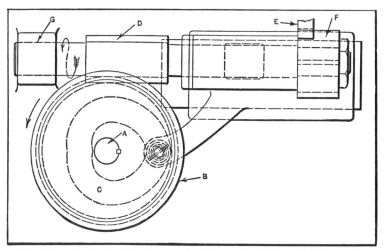

Fig. 8. Mechanism for Changing Speed of Driven Shaft Every Half Revolution

make one-half revolution in three seconds at a uniform speed,
and the following half revolution in four seconds, also at a
uniform speed. This result is obtained by means of a cam C
(on the driving worm-wheel B), which imparts a uniform
reciprocating motion to the driving worm D, causing the
worm-wheel alternately to be advanced and retarded as the
worm moves first with and then against the worm-wheel
rotation.

The mechanism is driven by gear E, which meshes with
pinion F. This pinion is attached to the worm-shaft and is

wide enough to provide for the lengthwise movement of the worm. The worm-wheel B and cam C are integral. The cam roller is carried by an arm which is part of the slide that forms a bearing for one end of the worm-shaft. The other end is supported in bearing G through which it is free to slide when the worm is moved axially.

Worm-wheel B has 56 teeth and worm D has 4 threads per inch. The speed of worm D is 8 revolutions per second. When shaft A is being turned one-half revolution in three seconds, the worm moves with or in the direction of rotation of the worm-wheel B, so that B is turned somewhat faster than it would be if worm D were not moved axially. The increased movement causes shaft A and worm-wheel B to turn one-half revolution in one second less than when worm D moves backward against the rotation of worm-wheel B.

Since worm D makes 8 revolutions per second, there are 24 revolutions in 3 seconds, and at the same time the worm advances one inch due to the action of the cam. Now the pitch of the worm thread and circular pitch of the worm-wheel is $\frac{1}{4}$ inch; hence, 1 inch axial movement of the worm is equivalent to 4 teeth of the wheel, so that the total movement equals $24 + 4 = 28$ teeth, or one-half revolution, as the wheel has 56 teeth. When worm D moves backward against the rotation of the worm-wheel, it makes 32 revolutions in four seconds, but the 1 inch axial movement, in effect, subtracts motion equivalent to 4 teeth, or $32 - 4 = 28$, or one-half turn in four seconds, as compared with three seconds for the opposite direction.

Two-gear Clock Mechanism of 12 to 1 Ratio. — In the design of mechanisms in general, eliminating useless parts and obtaining the desired result by using the most simple and direct means is often the most essential part of the problem, especially when even a single unnecessary part would greatly increase the manufacturing cost. The interesting feature of the mechanism to be described is that it contains only two gears which give the same speed ratios between the driving and driven members as are ordinarily obtained with a train of four gears. One application of the mechanism is to clock-work or time-pieces. The hour hand of an ordinary clock is driven through a compound gear train, called "dial gears," which serves to turn the hour hand one-twelfth revolution while the minute hand makes a complete revolution. The two-gear mechanism which is shown in Fig. 9, enables the

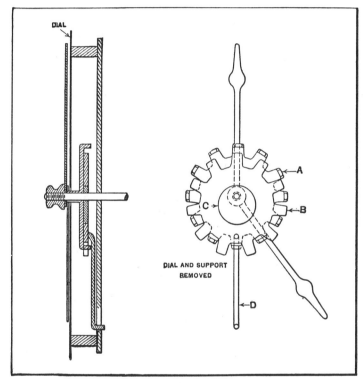

Fig. 9. Two-gear Clock Mechanism of 12 to 1 Ratio

same 12 to 1 ratio between the minute and hour hands to be obtained.

The hour hand is attached to a gear A having 12 teeth, the ends of which are bent so that they lie parallel to the axis, similar to a crown gear. Within gear A there is another gear B which has 11 straight or radial teeth. This inner gear is mounted on an eccentric C attached to the shaft for revolving the minute hand. As this shaft turns, the eccentric causes the axis of gear B to revolve, but the gear does not turn about its own axis, as there is a rod D which prevents such movement. The lower end of this rod is bent at right angles and engages a slot, thus allowing the planetary movement of gear B, but preventing it from revolving about its own axis.

Each revolution of the minute-hand shaft and eccentric, causes the teeth of gear B to withdraw from whatever tooth spaces of gear A they happen to be in mesh with and engage the next successive tooth spaces; consequently, gear A is advanced an amount equal to one tooth or one-twelfth revolution for each complete turn of the eccentric and minute-hand shaft. In this way, the desired speed ratio between the minute and hour hand of a clock is obtained. Of course, this speed ratio might be varied to suit different requirements. This mechanism is not intended only for clockwork, but can be applied wherever a simple reducing device is required in mechanisms of the type used in connection with counting and recording instruments.

The relative positions of the driving and driven parts during any fractional part of a revolution depend upon the length of the rod D attached to gear B. If this rod were of infinite length, a uniform rotation of the eccentric shaft would cause the driven part also to revolve at a uniform rate; but with rods of ordinary length, the motion of the driven part (which, in the case of a clock, is the hour hand) is accelerated and then retarded as it moves from one figure to the next on the dial. If rod D were dispensed with and gear B prevented from rotating by a pin attached to some point within the gear teeth, and engaging a fixed slot as in the case of rod D, the driven part would remain nearly stationary during a large part of the driver's revolution, and the movement from one division point to the next would be quite rapid. This rapid motion would be desirable for counting mechanisms in order to have the indicating hand opposite the numbered divisions, except when rapidly moving from one position to the next successive number on the dial.

Gearless Variable-speed Transmission.— There are many machines and mechanical units that varying circumstances make it desirable to be able to drive at a barely perceptible speed, an intermediate speed, or a high speed. A patented variable-speed unit of this type is so arranged that while its operation is entirely mechanical, any speed from zero to maximum is obtainable without the use of a single gear. The changes in speed are made without shocks or undue stresses.

Fig. 10 shows an assembly drawing of the unit. The drive is delivered to the transmission by shaft A on which tight and loose pulleys are mounted. The shaft, of course, may also be driven direct by motor or other means. Two cranks on shaft A impart motion through connecting-rods C to oscillate a shaft D on each side of the transmission. At the forward end of each of these shafts is a crank E in which there are two blocks F which are adjustable along grooves in the crank. It will be seen that the blocks in each crank are connected by means of the link G to another crank H which is bushed on the shaft B. The position of blocks F on cranks E is adjustable by turning handwheel J to raise or lower screws K, which are each connected by means of a link to the respective blocks F. The pairs of blocks on the opposite sides of the transmission are adjusted in unison by sprocket L, mounted on the same nut as handwheel J, which drives sprocket M through a chain. Instead of these sprockets and chain, screws K are sometimes connected by means of a shaft and helical gears. Then, too, instead of handwheel J being mounted directly on the nut of one of screws K, the handwheel may be placed in a convenient position for the operator, and connected to the nut by a long shaft and bevel gears.

It will be evident that at each revolution of shaft A, cranks E are rocked once forward and backward, with the result that cranks H oscillate similarly. The angular movement of cranks H becomes less as blocks F are moved from the extreme end of cranks E toward the axis of the cranks. When the center of the blocks coincides with the axis of the cranks, no movement is imparted to links G and cranks H. Integral

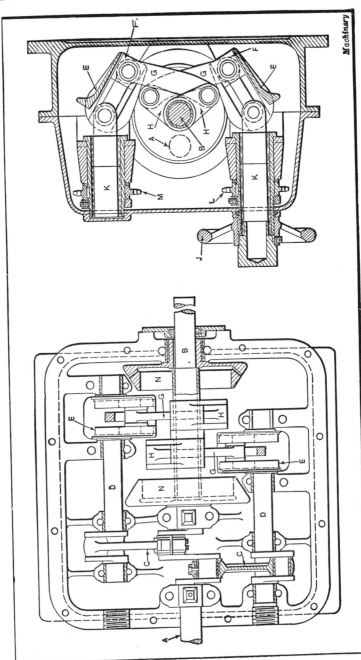

Fig. 10. Assembly Drawing Showing the Construction of the Variable-speed Transmission

with each crank *H* are two eccentrics which impart radial movements to two impeller parts. These serve as a quick-acting clutch to transmit the drive to drums *N* when they are expanded. Drums *N* are both keyed to shaft *B* to drive it when they are rotated, both drums being turned in the same direction. The two drums are engaged simultaneously so as to obtain a double drive. The impeller devices consist of an ingenious design in which rollers are employed with the double eccentrics to give the quick locking and unlocking action.

From the foregoing description it will be obvious that with a machine equipped with this transmission in the main drive, it is possible to obtain any feed or speed between zero and maximum of the tool or work. With a speed of, say, 1150 revolutions per minute of shaft *A,* the maximum speed of shaft *B* would be about 800 revolutions per minute. The housing is oil-tight, dust-proof, and filled with lubricant which is supplied to all parts by the splash obtained from the movement of the cranks, etc.

Frictional Speed-changing Devices.— Friction gearing of various forms is applied to some classes of machinery as a means of obtaining speed changes. The frictional type is simple in design and has the further advantage of providing very gradual speed changes. If a definite relation, however, must be maintained between the driving and driven members, the frictional transmission is not suitable, but, in some cases, the fact that it is not positive and tends to slip when subjected to excessive loads is a desirable feature, as it serves to protect the driven mechanism against excessive stresses.

Fig. 11 shows a type of frictional speed-changing mechanism which has been quite generally used, the details of construction being modified somewhat, owing to variations in the amount of power to be transmitted and other factors affecting the design. The particular arrangement referred to is applied to a running-balance indicating machine. The motor which drives the machine revolves the leather-faced driving disk *A* which is in contact with a steel wheel *B*. **The** vertical shaft passing through the driven wheel transmits motion to a horizontal shaft (not shown) at the top of the machine, which, in turn, revolves whatever part is to be tested for running balance. Variations in the speed of the work are obtained by changing the position of wheel *B* relative to the axis of the driving disk *A*. The adjustments of wheel *B* are controlled by a hand lever provided with a notched quadrant for holding it in a given position. This hand lever is connected with the slide of wheel *B* by link *C*. A reversal of

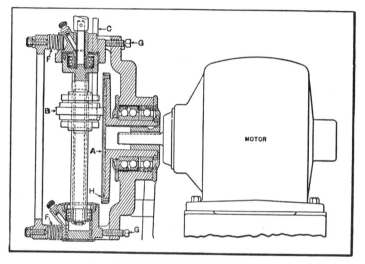

Fig. 11. Speed-changing Mechanism of Friction Disk and Wheel Type

motion is obtained by simply shifting wheel *B* to the opposite side of the axis of the driving disk. The wheel is held against the leather-faced disk with sufficient pressure by means of springs *F* which are provided with screws for varying the compression. If the leather disk becomes flattened out or thin from wear, the wheel *B* may be adjusted inward by means of stop-screws *G*. The leather disk is held in place by a retaining ring *H*. The adjustments for changing the speed should only be made when the driving disk is running.

Friction Disk and Epicyclic Gear Combination for High Velocity Ratio. — A very high velocity ratio or great reductions of speed, as well as extremely small variations of speed, may be obtained by the mechanism to be described. This mechanism (see Fig. 12) is a combination of friction disks and a train of epicyclic or differential gearing. The two disks D and E are free to revolve upon the vertical shaft C, and the hubs of these disks form the bevel gears F and G. Between these two bevel gears are the additional gears T and J mounted

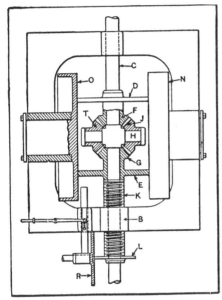

on pin H, which is attached to shaft C. The disks D and E are in frictional contact with wheels N and O, and their position is regulated by screw K, which is rotated through disks L and R. If wheel N is revolved and disks D and E are equidistant from the axes of wheels N and O (as shown in the illustration), both disks will revolve at the same speed, but in opposite directions. As gears F and G also rotate at the same speed, the inter-

Fig. 12. Combination of Friction Disks and Epicyclic Gear Train for Obtaining Great Reduction of Speed

mediate gears T and J merely revolve idly upon pin H, which remains in one position. Any change in the position of disks D and E relative to the wheels N and O will result in reducing the speed of one disk and increasing the speed of the other one; consequently, gears T and J begin to advance around whichever gear F or G has the slower motion, so that pin H and shaft C revolve in the same direction as the more rapidly revolving gear. If disks D and E are only moved a small amount from the central position, the differential action in the gearing and the motion of shaft C will be at a very slow rate. The direction of rotation may be changed by moving the disks upward or downward relative to the central position shown by the illustration.

Friction Speed-changing Mechanism of Disk, Ball and Roller Type. — A speed-changing mechanism of the friction type which was designed for use in naval nautical instruments,

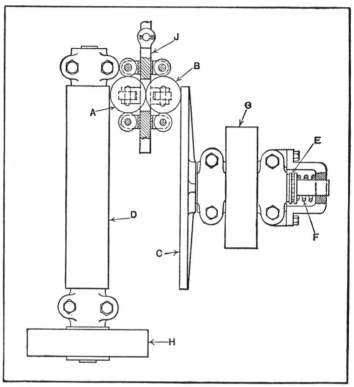

Fig. 13.   Speed-changing Mechanism of Friction-disk and Ball Type

is shown in Fig. 13. In this application of the device the power transmitted is very small, say about 0.01 horsepower. It has also been applied to mechanical fuel stokers, the power transmitted in this service being from one to two horsepower.

For transmitting up to two horsepower, excellent results may be obtained. This transmission is particularly useful when frequent changing of speed is required under conditions that would preclude the use of gearing for the purpose. A remarkable feature of this device is that the force required to effect a change from full speed in one direction to full speed in the reverse direction is so small as to be negligible, and this operation is unaccompanied by any jar or shock.

As will be noted from the illustration, the device consists essentially of two balls A and B revolving in contact with each other between a driving disk C and a driven roller D. The shaft of the driving disk is mounted in a ball thrust bearing E back of which there is a spring F. The disk may be driven by a gear or pulley G or it may be connected directly to an electric motor or gasoline engine. The power is transmitted from the driven roller by pulley H. The two balls employed to transmit the power from disk C to roller D are mounted in a carrier J in which they are loosely held by small rolls mounted in brackets. The carrier is so mounted as to be capable of movement across the face of the disk.

The balls A and B are capable of two movements; that is, they may rotate on their own centers and also roll across the disk and roller between the two latter members. The disk, the balls, and the roller are all made of hardened steel. When the balls are near the edge of the disk the roller runs at its highest speed. Upon moving the ball carrier nearer the center of the disk, the roller gradually loses speed, until at the center it will cease to rotate. As the carrier continues to move, passing the center, the roller rotates in the opposite direction.

The pressure of the spring F is not excessive, and while the tractive force between the balls would appear at first to be small, there is actually no slippage in practice. This is in accordance with the law of friction which states that the amount of friction is independent of the amount of surface in contact but depends entirely upon the pressure. An illustration of this law may be observed in the case of locomotive driving wheels, where the tractive force of, say, six points of contact with the rails, is sufficient to draw a heavy train.

Friction Roller Between Cones. — Many speed-changing mechanisms of the friction type have opposing cones which are connected by some intermediate member that may be adjusted to vary the speed. The use of an ordinary belt has already been referred to. Fig. 14 shows an arrangement for regulating the speed of a driven shaft, by changing the position of a wheel *A* placed between the driving cone *B* and the driven cone *C*. These two cones are made of cast iron and the bearing surface of the intermediate wheel is formed of leather disks held in place between two flanges or collars. This

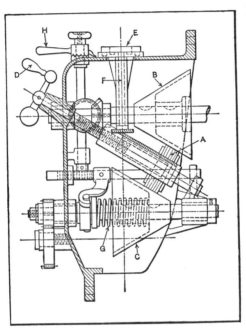

particular mechanism is used for varying the feeding movement of a cold-metal saw. The handle *D* connecting with a screw is used for controlling the position of the intermediate wheel and the rate of speed. A dial at *E* shows the rate of feed per minute, this dial being connected through shaft *F* and a gear at the lower end with a rack on the adjustable member, so that any change in the position of the wheel is

Fig. 14. Friction Cones and Intermediate Wheel for Varying Speeds

indicated by the dial. The lower friction cone is held in contact with the wheel by means of a spring *G*, the tension of which may be regulated by lever *H*. This lever is provided with graduations so that the same tension as well as the rate of feed per minute may be duplicated.

Band or Ring Between Cones.—Another method of transmitting motion from a driving to a driven cone is shown in Fig. 15, which illustrates the Evans friction cones. The two cone-pulleys are not directly in contact with each other, but bear against a band or ring of leather which serves to transmit the motion. The speed of the driven cone is varied by simply shifting this leather ring so that it bears against a larger or smaller part of the cones. If cone *A* is the driver, the speed of cone *B* would be gradually increased if belt *C* were shifted toward the right, since the practical effect of this shifting movement is to increase the diameter of the driving pulley. This mechanism is used ordinarily as a variable-speed counter-shaft. There are two general methods of

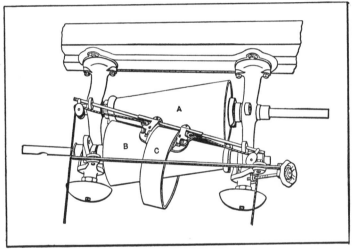

Fig. 15. Friction Cones Which Transmit Motion Through Adjustable Leather Ring or Belt.

starting or stopping the driven members. Some friction cones are so arranged that the leather ring is shifted to a parallel part of the cones for disengaging the drive, and others are so designed that one cone is raised and lowered by the shifting lever, thus starting and stopping at the same speed.

Spherical Rollers Between Disks. — The necessity for accurately adjusting or controlling the speeds of driven shafts has resulted in the development of a variety of variable-speed mechanisms. Among these is the friction-driven type of variable-speed mechanism. The design shown in Fig. 16 has a

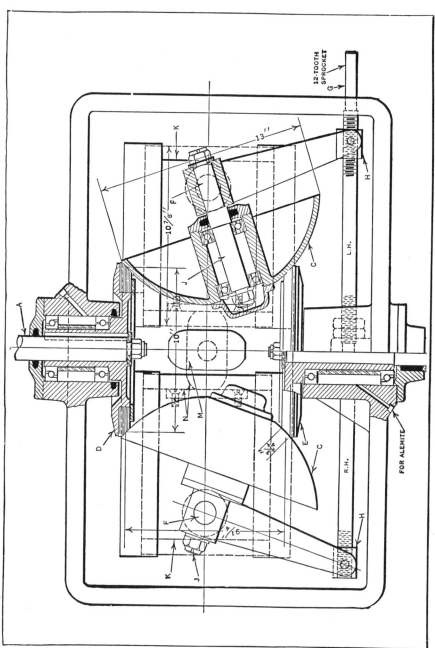

12-TOOTH
SPROCKET
G

H

L.H.

K

13″

10⁷/₈″ F

C

J

A

-10

D

M

N

E

FOR ALEMITE

C

R.H.

F

6¼″

K

J

H

Fig. 16. Variable-speed Friction-Drive Speed-changing

maximum driving ratio of 4 to 1, and provides an accurate means of adjusting the speed of the driven disk E and its shaft within the required range of from 2 to 0.5 revolutions per minute when the driving shaft A revolves at a constant speed of 1 revolution per minute. Like most friction drives, this design is employed where the load is light.

The drive consists primarily of two spherical members C which engage the friction disk D of the driving shaft and the disk E secured to the driven shaft. The variation in speed is obtained by rotating the spherical members C about the points F. When the spherical members are in the positions shown, the driven shaft will be rotated at approximately its slowest speed, as the driving disk D makes contact with the spherical members at their outer edges, while the driven disk E makes contact near the centers of the members C, where the surface speed is comparatively low.

The operator adjusts the speed by rotating the shaft G, a sprocket and chain arrangement (not shown) being employed for this purpose. The right- and left-hand threads on shaft G, together with the nuts H attached to the arms of the bearing shafts J, provide means for accurately positioning the spherical members C to give the desired speed.

The members C are mounted on slides K. These slides are connected by springs which serve to keep the friction members in driving contact with the disks D and E. The friction drive can be disengaged at the will of the operator by means of a lever outside the housing, which controls the cam M. When the cam is turned, the slides K are forced apart, thus disengaging the friction driving members and causing the driven shaft to stop. The dotted lines at N show the position of the cam and slides when the drive is stopped.

Concave Friction Disks and Inclined Wheel. — The frictional variable-speed transmission shown in Fig. 17 is an example of the type having annular concave frictional surfaces engaged by an intermediate wheel the inclination of which is varied for changing the speed. The principle upon which the device operates is illustrated by the diagram at the

left. The two disks having annular concave surfaces are
rotated from some source of power and run loose on shaft *A*
which is driven at a variable speed. The intermediate wheel
D is pivoted at *O* to arm *B*, so that it can be inclined as indicated by the dotted lines. The drive to shaft *A* is transmitted
through arm *B*. When wheel *D* is parallel to shaft *A*, as
shown in the illustration, and the two disks *G* and *C* are revolving in opposite directions at the same speed, wheel *D*
will simple revolve about pivot *O*, and arm *B* and shaft *A*
will remain stationary. If wheel *D* is inclined, however, as

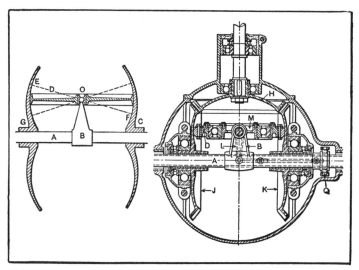

**Fig. 17. Variable-speed Transmission Having Annular Concave Surfaces and
Inclinable Friction Wheel**

indicated by the dotted line *EF*, the contact surface at *E* will
be revolving at a higher circumferential speed than the surface *F* on disk *C;* consequently, pivot *O*, arm *B*, and shaft *A*
will be given a rotary motion, the rate of which depends upon
the angularity of wheel *D*. The greater the angularity, the
greater will be the difference in the diameter of the contact
surfaces of disks *G* and *C* and the higher the speed of shaft *A*.
By inclining wheel *D* in the opposite direction, the rotation of
shaft *A* can be reversed.

A variable-speed mechanism designed on this principle is shown at the right of the diagram in Fig. 17. A bevel gear *H* mounted on the end of the driving shaft revolves the two bevel gears *J* and *K* mounted on shaft *A*, which is the driven member. These bevel gears *J* and *K* have annular concave surfaces which engage the cork surface of wheel *D*. This wheel revolves on an annular ball bearing, the inner race of which is attached to ring *M* pivoted on a stud carried by arm *B*. The angular position of wheel *D* is controlled by a lever *L* integral with the pivoted ring *M*. This lever is connected with ring *Q* which is engaged by a forked lever similar to the form used for shifting clutches.

An objection to variable-speed mechanisms based on this principle is that the variation of speed does not change the torque, so that, even though there is considerable speed reduction, the torque will not be proportionally greater, because the limiting factor for the torque is the frictional adherence between the driving and driven contact surfaces, and this frictional resistance is independent of the speed at which the shaft *A* is running; consequently, while variable-speed devices in general are of such construction that the torque increases when the speed decreases, in the present case the speed is variable, while the torque remains constant. As the main feature of variable-speed devices is often not the variation of speed as much as the increased torque obtained by a decrease in speed, the objection referred to is one of great importance.

Materials for Friction Gearing. —In the selection of material for the driving member of friction gears, good frictional qualities combined with a reasonable degree of durability are essential. In order to determine the relative merits of different kinds of friction materials, tests were conducted by Professor Goss of the University of Illinois. The driving and driven wheels used for these tests were each 16 inches in diameter and were mounted on parallel shafts and run with the peripheries or edges in contact. The driving wheels were made of the fibrous materials referred to later and were 1¾ inches wide; the driven wheels were ½ inch wide and made of cast iron, aluminum and type metal. It was found that leather fiber is exceptionally strong and has a high coefficient of friction when in contact with cast iron or an aluminum alloy — materials commonly used for the driven member. Straw fiber has a somewhat lower frictional value, and is not so durable as leather fiber, but nevertheless is satisfactory and has the advantage of being obtained readily. Tarred fiber is exceptionally strong, but its frictional coefficient is comparatively low. Sulphite fiber has the highest coefficient of any of the materials listed, but it is the weakest. Leather is inferior to plain straw fiber as to both frictional qualities and strength. A summary of other results of these tests follows.

Power Transmitted by Friction Gears. — The power transmitting capacity of friction wheels of given size and running at a given speed depends upon the pressure of contact and the coefficient of friction. Since the life of the fibrous or leather driving wheels depends upon the contact pressure (the diameter decreasing due to a yielding of the material as the pressure becomes excessive) the allowable working pressure is determined with reference to durability. Allowable pressures per inch of face width should be approximately as given in

Allowable Working Pressures and Coefficients of Friction

Material	Allowable Working Pressure per Inch	Coefficient of Friction when Driver is in Contact with	
		Cast Iron	Aluminum
Leather fiber.......	240	0.31	0.30
Straw fiber.........	150	0.26	0.27
Tarred fiber........	240	0.15	0.18
Sulphite fiber......	140	0.33	0.32
Leather............	150	0.14	0.22

the accompanying table, which applies to the 16-inch wheels previously mentioned. This table also includes working values for the coefficients of friction. The frictional coefficient for the wheels tested approaches its maximum value when the slip between the driving and driven wheel amounts to 2 per cent; the coefficient diminishes when the slip exceeds 3 per cent. As a general rule, the percentage of slip, according to the results of the tests previously referred to, should not be less than 2 per cent nor more than about 4 per cent.

The number of horsepower transmitted by friction gearing may be determined by the following general formula in which H = the number of horsepower; D = diameter of driving wheel in inches; N = revolutions per minute; P = allowable working pressure in pounds per inch of face width (see table); W = face width in inches; f = coefficient of friction (see table).

$$H = \frac{\pi DPWNf}{33,000 \times 12}$$

For a given coefficient and contact pressure, this formula may be simplified by determining the value of the expression $\frac{\pi Pf}{33,000 \times 12}$ and inserting this value X in the formula, $H = DNWX$. The fibrous wheel should always be the driver to avoid wearing a flat spot on it in case the driven wheel stalls, and as rigid a support as possible is essential.

Driving Disk Engaging Side of Driven Wheel. — When the driving disk engages the side of the driven wheel (in order to provide for speed changes and possibly a reversal of rotation) pure rolling action is not obtained because the driver makes contact with the driven disk at various diameters; consequently the velocity of the driven disk at one side of the driver differs from the velocity at the other side where contact is at a smaller or larger radius, depending upon the side. In order to avoid an excessive amount of slippage between the driving and driven disks, the ordinary running positions of the driver should be such that the minimum distance from the center of the driver face to the center of the driven wheel, will not be less than twelve times the width of the driver face. For example, if the driver face is ½ inch, the minimum distance to the center of the driven disk should preferably be 6 inches. When the driver is closer than this minimum distance, the coefficient of friction is reduced and also the power-transmitting capacity.

Obtaining Contact Pressure. — The method of applying contact pressure is adapted to conditions, but, in general, the lever-operated eccentric box or thrust box is commonly used; it is a simple method for giving hand or power control. In some cases, more elaborate devices are used. The pressure may be positively applied and it may be made to vary automatically as the load increases or decreases. As friction is essential to the operation of this type of gearing, care should be taken to prevent any great reduction of the driving power by the accumulation of grease or other foreign matter on the friction surfaces. Rigid support for the friction wheels and the maintenance of a good contact between the working surfaces are also of importance. Friction gearing is not a suitable form of transmission where it is essential to maintain a prescribed relation between driving and driven parts of a mechanism throughout an entire cycle of operations. In some cases, however, a transmission which is not positive is preferable in that it constitutes a safety device and prevents the transmission of shocks or an excessive amount of power to parts of a mechanism which might thereby be injured. Friction gearing is also very simple in design, and operates smoothly and quietly.

Multiple-disk Type of Speed-changing Mechanism. — The variable-speed mechanism shown in Fig. 18 is an ingenious design used on certain cylindrical grinding machines for changing the rotary speed of the part being ground and also the rate of the table traverse. Three levers grouped around a dial at the front of the machine are used for controlling the mechanism. The position of lever A governs the rotary speed of the work, and another lever in front of the circular dial (not shown in the illustration) serves to change the rate of the table traversing movement. These changes of work speed and table traverse are entirely independent. The long lever R is used for starting and stopping the rotation of the work and the traversing movement of the table simultaneously. The mechanism is driven from a driving shaft which runs at a constant speed and connects with coupling B. The sprocket

C is connected to the reversing mechanism and drives the table traverse. Another sprocket (not shown) is connected by a pair of silent chains and a splined shaft, with a driving member for the headstock.

The mechanism operates as follows: The shaft *F* carrying coupling *B* drives shafts *G* and *H* at a constant speed through spur gearing. The shafts *G* and *H* carry a series of hardened steel disks mounted on square portions of the shafts. These

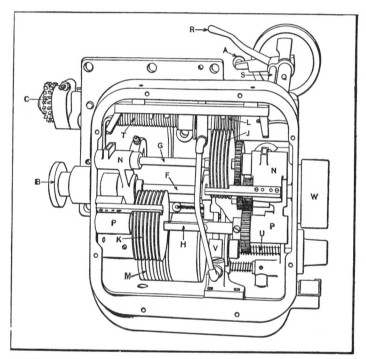

Fig. 18. Multiple Friction Disk Type of Speed-changing Mechanism

disks *J* and *K* are ground slightly convex and each group of disks intermeshes with another group or series of hardened steel disks *L* and *M*. Each of these driven disks has a rim at the periphery so that the point of contact with the driving disk is always at the outer edge. The shafts *G* and *H* are mounted in swinging brackets *N* and *P*, both of which pivot on shaft *F*, thus allowing the position of disks *J* and *K* to be varied relative to the disks L and M. If the convex disks J are swung towards the recessed disks L, the surfaces of disks J, which actually do the driving, decrease in radius and, consequently, the speed of disks L and their shaft also decreases. The lever A controls the position of bracket P and the speed of the headstock, whereas the lever at the front of the dial (not shown) controls bracket N and the feeding movements of the table. Motion is transmitted to these brackets through bevel pinions meshing with segment gears on the brackets.

With this mechanism, slight variations in speed may be obtained while the machine is in motion. When lever R is shifted for stopping the machine, a cam at the end of shaft S operates a lever which relieves the pressure applied to disks L and M by the springs shown at T and U. This lever also applies brakes which quickly stop the table and headstock. When the lever is raised for starting the mechanism, the disks L and M grip the intermeshing disks J and K, and the driven members are started without shock, the action being very similar to the well-known multiple-disk friction clutch. A plunger pump at V pumps oil from the bottom of the case to a distributor at the top which lubricates the entire mechanism.

Governors for Speed Regulation. — When the regulation of speed is automatically controlled, some form of governing mechanism of the centrifugal type is commonly employed. Many of the governors used on steam engines depend for their action upon the effect of centrifugal force on a rotating element. In the case of a "fly-ball" governor, weights or balls attached to pivoted levers are revolved by the engine and if the speed increases above normal, the balls or weighted levers move outward from the axis of rotation, owing to the increase in centrifugal force. This change in the position of the revolving balls may be transmitted through suitable connecting levers and rods to a valve which partly closes, thus reducing the steam supply. When a governor of this type is applied to a Corliss engine, the release of the steam valves and the point of cut-off is controlled directly by the governor. Most governors of the fly-ball type have one or more springs which tend to resist the outward movement of the revolving balls.

The inertia or centrifugal-inertia governor, which is used so extensively, is attached to the fly-wheel and regulates the speed by varying the position of the eccentric or crankpin that operates the valve. The general principle upon which this type of governor operates is illustrated by the design shown in Fig. 19. This particular governor has an inertia bar A with enlarged ends to increase the weight at the ends. This bar is pivoted at B where there is a roller bearing to reduce the frictional resistance. The eccentric C is attached

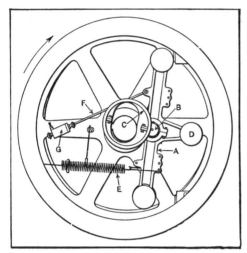

to the inertia bar and it has an elongated hole or opening to permit movements relative to the crank-shaft. Directly opposite the eccentric is a third weight D, which balances the effect of gravity on the eccentric. A heavy coil spring E is attached to the inertia bar. A rod F is pivoted to the bar

Fig. 19. Centrifugal-inertia Type of Engine Governor

on the opposite side of bearing B and is connected to a loose-fitting piston in the oil dashpot G.

The flywheel revolves in the direction shown by the arrow and speed variations cause a slight movement of the inertia bar about its bearing in one direction or another, thus changing the position of the eccentric, which changes the point of cut-off. If the speed increases, the inertia bar lags behind momentarily and the steam is cut off earlier during the stroke because the eccentric swings inward and shortens the travel of the valve. If a sudden increase of load should cause the engine to run slower, lever A, as a result of its inertia, would tend to continue running at the faster speed, which would swing the lever forward about bearing B in the direction of rotation, thus increasing the valve travel and admitting more steam to the cylinder by delaying the point of cut-off. The spring end of the inertia bar is the heavier and the speed of rotation depends entirely upon the equilibrium between the centrifugal force acting upon the inertia bar and the tension of the spring, while the actual movement of the governor parts is effected by the inertia of the weighted end of the bar. The sensitiveness of the governor may be varied by adjusting a by-pass valve upon cylinder G. Other governors of this general type vary in regard to the form of the weighted lever and the arrangement of springs or other details. The inertia type is preferable to the purely centrifugal design for engines subjected to sudden and decided load changes.

CHAPTER XII

DIFFERENTIAL MOTIONS

WHEN a motion is the resultant of or difference between two original motions, it is often referred to as a differential motion. The differential screw is a simple example of a motion of this kind. This is a compound screw from which a movement is derived that is equal to the difference between the movements obtained from each screw. The diagram A, Fig. 1, illustrates the principle. A shaft has two screw threads on it at e and f, respectively, which wind in the same direction but differ in pitch. Screw f passes through a fixed nut and screw e through a nut that is free to move. The motion of the movable nut for each revolution of the screw equals the difference between the pitches of the threads at e and f.

This combination makes it possible to obtain a very slight motion without using a screw having an exceptionally fine pitch and a weak thread. Another form of differential screw is shown at B, which illustrates a stop that enables fine adjustments to be obtained readily. The screw bushing g is threaded externally through some stationary part and is also threaded internally to receive screw h which is free to move axially but cannot turn. Both screws in this case are right-hand, but they vary as to pitch. If bushing g has a pitch of $\frac{1}{32}$ inch or 0.03125 inch and screw h a pitch of $\frac{1}{36}$ inch or 0.02777 inch, one complete turn of g will advance screws h only 0.00348 inch (0.03125 — 0.02777 = 0.00348), because, as bushing g advances $\frac{1}{32}$ inch, it moves screw h back a distance equal to the difference between the pitches of the two threads. By turning the bushing only a fractional part of a turn very small adjustments may be obtained.

Differential Motion of Chinese Windlass. — The Chinese windlass shown by the diagram C, Fig. 1, is another simple example of a differential motion. The hoisting rope is arranged to unwind from one part of a drum or pulley onto another part differing somewhat in diameter. The distance that the load or hook moves for one revolution of the compound hoisting drum is equal to half the difference between the circumferences of the two drum sections.

The well-known differential chain hoist illustrated at *D* operates on the same general principle as the Chinese wind-

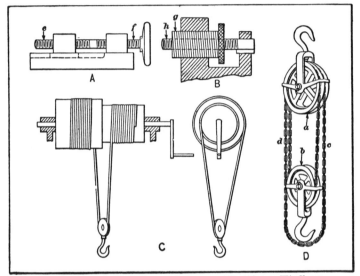

Fig. 1. (A and B) Differential Screws; (C) Chinese Windlass;
(D) Differential Hoist

lass. The double sheave *a* has two chain grooves differing slightly in diameter, and an endless chain passes over these grooves and around a single pulley *b*. This pulley *b* and the hook attached to it is raised or lowered, because, for a given movement, a greater length of chain passes over the larger part of sheave *a* than over the smaller part. If the upper sheave is revolved by pulling down on the side *d* of the chain that leads to the groove of smaller diameter, the loop of chain passing around pulley *b* will be lengthened, thus lowering the pulley; the opposite result will be obtained by pulling down on chain *c* which leads up to the larger diameter of the sheave.

Differential Motions from Gearing. — Most differential motions are derived from combinations of bevel or spur gearing. The epicyclic bevel gear train illustrated by diagram *A*, Fig. 2, is applied to many mechanisms of the differential type, and its action under different conditions should be thoroughly understood. The shaft *a* has mounted on it two bevel gears *b* and *c* and an arm *d*. The arm is attached to the shaft and carries a pinion *e* which meshes with each gear and is free to revolve upon the arm. There are several conditions that can exist with a gear train of this kind.

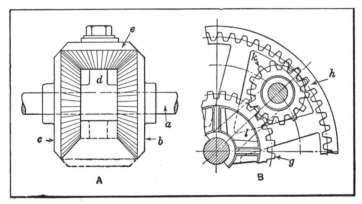

Fig. 2. Epicyclic Trains of Bevel and Spur Gearing

First, assume that gear *b* is stationary and *c* loose on the shaft. If the shaft and arm *d* is revolved, motion will be transmitted from arm *d* to gear *c*, through pinion *e*, and gear *c* will make two turns for every one of arm *d* and in the same direction as the arm. If gear *b* should rotate instead of being stationary, this motion, combined with that of the arm, would modify the motion of gear *c* and it would also make a difference whether gear *b* turned in the same direction as the arm or in an opposite direction.

Second, suppose the preceding conditions are reversed and one of the bevel gears *b* or *c* is revolved while the other gear remains stationary, and that arm *d* carrying the bevel pinion constitutes the driven element. With only one gear revolving, the arm will turn in a direction corresponding to that of the gear and at half its speed. If both gears rotate in the same direction at different speeds, the arm will follow in that direction and with a speed intermediate between the two. If the gears are driven in opposite directions at different speeds, the arm will follow the more rapidly moving gear, and if the speeds are equal, pinion e will revolve upon the arm, but the latter will remain stationary.

Third, assume that arm d remains stationary and gears b and c are loose on the shaft. If gear b is the driver ,the pinion e will simply transmit motion to gear c in the opposite direction, the three gears in this case forming a simple train with pinion e acting as the idler. The force tending to rotate arm d will be twice the force transmitted from gear b to gear c. A practical application of this last principle is found in the Webber differential dynamometer. The arm of this dynamometer which supports the scale pan and weights corresponds to arm d and is pivoted on a shaft carrying two bevel gears. On the arm and meshing with these two bevel gears are bevel pinions and the amount of power transmitted through this train of gearing is measured by the weights in the scale pan. The combination of gearing illustrated by diagram A usually has two or more pinions meshing with the bevel gears. In many cases, there are two pinions located diametrically opposite, as indicated by the full and dotted lines. The addition of other pinions, however, does not affect the action of the gearing.

Differential Spur Gearing. — The diagram B, Fig. 2, shows an arrangement of spur gearing which gives a differential motion. This combination consists of ordinary spur gear g, an internal gear h, and a pinion k. This pinion is free to turn on a stud that is attached to arm l. In the application of this gearing, there are three possible conditions. In the first place, the internal gear h may be stationary, and the gears g and k may revolve. Second, the arm l may be stationary, in which case either the internal gear h or gear g may be the driver. Third, gear g may be stationary and the motion be transmitted

in either direction between gear h and arm l. Fig. 3 shows
a practical application of this gear combination. In this design, there are two intermediate pinions (corresponding to
k in diagram B, Fig. 2) which are mounted on an arm and
located diametrically opposite. This arm is keyed to the end
of a shaft. The large internal gear is stationary and forms
part of a casing enclosing the gears. The central gear is keyed
to another shaft which is in line with the shaft carrying the
pinion arm. This arrangement is simply used to obtain a

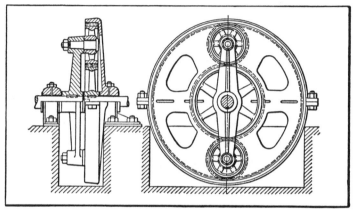

Fig. 3. Epicyclic Gearing for Obtaining Speed Reduction by Differential Motion

reduction of speed. The design is compact, although dif-
erential or epicyclic gearing, in general, is inefficient as a
transmitter of power. Such gear combinations, however, have
certain mechanical advantages, and they are often utilized by
designers for a variety of purposes as indicated by the different mechanisms to be described.

**Differential Motion between Screw and Nut Rotating at
Different Speeds.** — Variations of movement are sometimes
obtained by the differential motion between a revolving screw
and a nut which is rotating about the screw at a different
speed. One application of this principle is illustrated by the
variable-speed mechanism of a milling machine shown in Fig.
4. This mechanism is designed to increase the efficiency of
a machine by accelerating the speed of the table when the

cutters are not at work. The machine table moves rapidly up to the cutting point, then the speed is reduced while milling and, after the operation is completed, the table is quickly returned to the loading position so that the idle or non-cutting period is reduced.

This mechanism is located beneath the machine table C, which is traversed by a screw D, that passes through the plain bearings E, F, and G, mounted upon the base of the machine. The pinion H is confined longitudinally between bearings E and F, and it is splined to screw D, so that the latter must

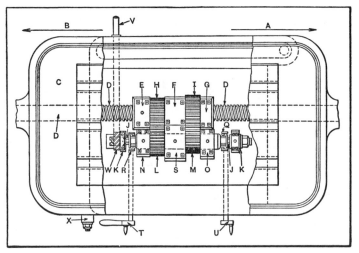

Fig. 4. Variable Feeding Mechanism Which is Partly Controlled by the Differential Movement Between a Revolving Screw and Nut

turn with the pinion but is free to slide in a lengthwise direction. The hole through gear I is threaded to fit screw D so that it is practically a nut and gear combined. The auxiliary shaft J supported in bearings K carries two pinions, L and M, which are loosely mounted upon the shaft. This shaft J is rotated continuously in one direction through spiral gears W from the driving shaft V. Within the housings N and O are clutch sleeves which encircle the shaft J. The sleeves are splined to the shaft, but are free to slide upon it, and they may be locked with teeth formed on pinions L and M. These clutches are controlled by levers T and U at the front of the machine which are connected by the shafts shown, with the clutch shifting devices at R and Q. The action of the clutches is controlled automatically by adjustable stops located on the front of the machine table.

The clutch connecting with gear L is first engaged by hand lever T. The table then moves forward rapidly (in the direction indicated by arrow A) as gear H revolves screw D and causes it to turn through the gear nut I which is held stationary at this time. Just before the milling cutter begins to act upon the work, lever U strikes a stop, thus engaging the clutch with gear M. The gear nut I is then revolved in the same direction as gear H but at a slower speed, so that the forward movement of screw D is reduced, because of the differential action between the screw and nut. Both sets of gears continue to operate while the cut is being taken; when the milling operation is completed, another stop engages lever T, thus stopping the rotation of gears L and H. As the gear nut I continues to revolve about the screw, the movement of the machine table is reversed, since screw D is not rotating. The motion continues in the direction indicated by arrow B until a third stop to the right of lever U trips the latter, thereby stopping gear I and the table movement. The table is now in position for removing the finished parts and replacing them with others that require milling.

Differential Feeding Mechanism for Revolving Spindle. — The spindle of a horizontal boring, drilling, tapping, and milling machine is given a lengthwise feeding movement by the differential action between the revolving spindle and a revolving nut which engages a helical groove in the spindle. The spindle is driven by a large gear A (see Fig. 5) which connects with the back gearing of the machine. The hub of this gear has two keys which engage the splined spindle. The sleeve B on which gear A is mounted has gear teeth cut in one end which mesh with three planetary pinions D that engage one side of the double internal gear E. The other side of this internal gear meshes with pinions N. These pinions, in turn, mesh with gear teeth formed on the rotary nut L which engages directly with a spiral or helical groove cut in the spindle. A flange on this nut rotates between large ball thrust bearings, as shown, in order to take the end thrust in either direction.

When nut L rotates at the same speed as the spindle, the latter does not move in a lengthwise direction, but, by revolving nut L either faster or slower than the spindle, a feeding movement in one direction or the other is obtained. The rotation of nut L is regulated by the gearing at G. When the feeding movement is stopped, gear F, which carries the

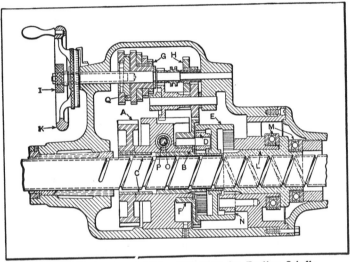

Fig. 5. Mechanism of Differential Type for Feeding Spindle
in Lengthwise Direction

planetary pinions D, does not revolve and nut L rotates with the spindle, which, therefore, remains in the same longitudinal position. When gear F which is connected indirectly with the feed change-gears G is revolved by these gears, the nut L is revolved independently of the spindle and at a different rate of speed.

Application of Floating Lever Principle.— What are known as "floating" or "differential" levers are utilized in some forms of mechanisms to control, by the application of a small amount of power or force, a much greater force such as would be required for moving or shifting heavy parts. Floating levers are commonly applied to mechanisms controlling the action of parts that require adjustment or changes of position at intervals varying according to the function of the apparatus subject to control. The initial movement or force may be derived from a hand-operated lever or wheel, and the purpose of the floating lever is to so control the source of power that whatever part is to be shifted or adjusted will

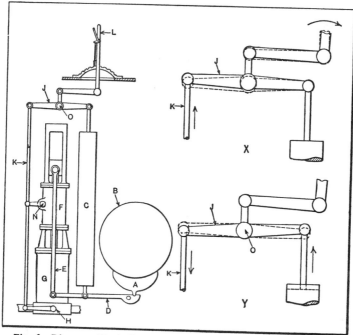

Fig. 6. Diagrams Illustrating Application and Action of Floating Lever

follow the hand-controlled movements practically the same as though there were a direct mechanical connection. A floating lever is so called because it is not attached to fixed pivots and does not have a stationary fulcrum, but is free to move bodily, or to "float" within certain limits and in accordance with the relative forces acting upon the different connections.

Fig. 6 illustrates one application of the floating lever. The diagram at the left represents an auxiliary braking apparatus for a large hoist. The brake shoe A is applied to the brake drum B whenever the dead weight C rests upon the lever D. This lever is connected by rod E with a cross-head attached to the upper end of a piston rod extending through the oil cylinder F and into the steam cylinder G. When steam is admitted beneath the piston in cylinder G by opening a valve at H, the weight is raised and the brake released, and, if for any reason the steam pressure should be suddenly reduced, weight C would fall and the brake be applied automatically. The movements of the piston in cylinder G and, consequently, of weight C are controlled by hand lever L through floating lever J, in such a manner that the weight rises and falls, as the lever is shifted, practically the same as though the force for moving the weight were derived directly from the lever by means of a rigid mechanical connection. The action of the mechanism is as follows: If the weight is down and the brake applied, and lever L is moved from its central position to the right, the left-hand end of lever J will be raised (as shown on an exaggerated scale by diagram X), thus lifting rod K and opening valve H; this valve has no lap, so that any movement of the lever admits steam to the cylinder. As soon as the piston begins to rise, the right-hand end of lever J also rises (see diagram Y) and turning about pivot O immediately begins to close the steam valve. If the lever L is moved through a small arc, the valve is closed quickly and the weight only rises a short distance; on the contrary, if the lever is thrown over to the extreme position, the piston and weight must move upward a proportionately greater distance before the valve is closed. If the lever, after being thrown to the right, is moved towards the left, valve H opens the exhaust port and the weight descends; as soon as it begins to move downward, the left-hand end of the floating lever is raised, which tends to close the exhaust port and prevent further downward motion.

An apparatus of this kind responds so quickly to adjustment that the weight follows the motion of the hand lever almost instantaeously and the end of the floating lever connected to rod K has very little actual movement. The oil cylinder F is used to stabilize the action of the weight and prevent overtravel which would occur if there were only the cushioning effect of steam. The by-pass valve N controls the flow of oil from one end of the cylinder to the other as the piston moves up or down, so that the motion of the weight ceases as soon as the steam and oil valves are closed.

Controlling Mechanism of Steering Gear. — The practical effect of the floating lever previously described for controlling the movements of power-driven apparatus may be obtained by other mechanical devices, examples of which are found on steamships for controlling the action of the steering engines. Engines used for this purpose are commonly equipped with a control valve which distributes steam to the engine valves. The latter are generally of the hollow piston type and are arranged to receive steam either at the ends or in the center, the exhaust varying accordingly. The admission of steam either to the ends or in the center is governed by the position of the control valve. For instance, if the control valve is moved in one direction, steam may be admitted to the ends of the engine valves and be exhausted in the center. If the control valve were moved in the opposite direction, this order would be reversed and also the direction in which the engine rotates; therefore, each engine valve requires but one eccentric, the control valve acting as a reversing gear. The mechanism which operates this control valve is so designed that, when the engine is set in motion to move the rudder either to port or starboard, this same motion is utilized to shift the control valve in such a way that the movement of the rudder coincides with the motion of the steering wheel. While the floating lever has been used in connection with this controlling mechanism, the common form of control depends upon the action (which is often differential) either of gearing or of a screw and nut.

With the arrangement illustrated at A, Fig. 7, the control valve of a steering engine is governed by the action of a screw that is operated by the steering wheel, and a nut that is revolved by the engine. The shaft *a* is connected with the steering wheel and transmits rotary motion to screw *b* which is splined to, and free to slide through, gear *c*. The rod *d* serves to operate the control valve of the steering engine. Any rotary motion of shaft *a* moves screw *b* in a lengthwise direction in or out of the nut on worm-wheel *e*, unless this nut is revolving at the same speed as the screw. The action of the mechanism is as follows: If worm-wheel *e*, which meshes with a worm on the steering engine crankshaft, is stationary, the rotation of shaft *a* will turn screw *b* in or out of the nut and shift the control valve, thus starting the engine in one direction or the other, depending upon which way the control

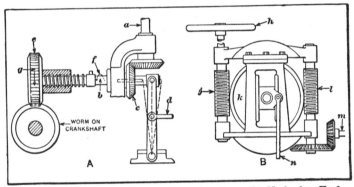

Fig. 7. (A) Controlling Device for Steering Gear; (B) Mechanism Used as Substitute for a Floating Lever

valve was moved. As soon as the engine starts, worm-wheel *e* and the nut begin to revolve, which tends to move the screw and control lever in the opposite direction. Suppose screw *b* were revolved in the direction shown by the arrow *f*, thus moving the screw and control lever to the right; then, as the engine starts, worm-wheel *e* and the nut revolve as shown by the arrow *g*. Now as soon as the rotation of shaft *a* and screw *b* is stopped or is reduced until the speed of rotation is less than that of worm-wheel *e*, the screw is drawn back into the nut and the control valve is closed. If the steering wheel and screw *b* were turned slightly and then stopped entirely, the rudder would only be moved a corresponding amount.

because the control valve would soon be shifted, by the action of worm-wheel e, to the closed position. Steering engines, in general, are equipped with some form of stopping device which automatically limits the movement of the rudder.

Rolling Worm-wheel Type of Controlling Mechanism. — The ingenious substitute for the floating lever illustrated at B in Fig. 7 depends for its action upon a worm-wheel which is interposed between two worms. The handwheel h controls the rotation of worm j, which meshes with the worm-wheel k. The worm l on the opposite side of the worm-wheel is rotated by whatever apparatus is to be controlled. The shaft of the worm-wheel is journaled in boxes which are free to slide up and down the vertical slides in the framework shown. Any vertical displacement of the worm-wheel is transmitted to rod n which operates the valve, clutch, or other mechanical device used for starting, stopping, and reversing the driving machinery. Assume that the mechanism is at rest with the worm-wheel midway between its upper and lower positions in the vertical slides of the housing. When the handwheel h is revolved in a direction corresponding to the motion desired, worm j revolves, and worm l is stationary, since the mechanism is not yet in motion; therefore, the rotation of the handwheel has the effect of rolling the worm-wheel k between the two worms either up or down, depending upon the direction in which the handwheel is rotated. Any vertical displacement of the worm-wheel will, through the medium of controlling rod n, start the power-driven machinery. This motion is immediately transmitted to shaft m and worm l which acts to move worm-wheel k in the opposite direction vertically, provided worm j is stationary or is revolving slower than l. The result is that the power-driven member is moved or adjusted proportionately to the rotation of the handwheel h. The handwheel, for instance, might be turned to a position corresponding to a certain required adjustment, which would then be made automatically.

Control Mechanism having Differential Bevel Gearing. — The steering gear controlling mechanism illustrated in Fig. 8

operates on the same general principle as the design previously described, although the construction is quite different. The control valve, in this case, operates with a rotary motion, instead of moving in a lengthwise direction. Shaft *A* is revolved by the steering wheel and transmits rotary motion to shaft *B* through the gearing shown. The differential action for regulating the position of the control valve is obtained by means of three gears *C, D,* and *E.* Gear *C* is keyed to shaft *B,* and gear *E* on the extended hub of worm-wheel *F* is free to revolve about shaft *B.* Gear *D* interposed between gears

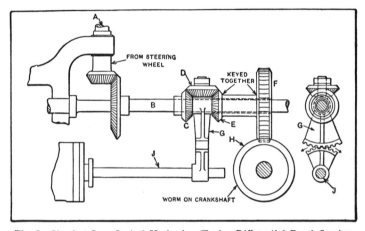

Fig. 8. Steering Gear Control Mechanism Having Differential Bevel Gearing

C and *E* is mounted upon a segment gear *G* which engages another segment gear on the control valve spindle *J.* If shaft *B* is revolved while gear *E* and the worm-wheel are stationary, gear *D* rolls around between the gears and, through the segment gear, turns the control valve, thus starting the steering engine and with it the worm *H* on the crankshaft which drives worm-wheel *F* and gear *E.* As soon as the rotation of shaft *B* is stopped, gear *E* which has been revolving in the opposite direction to that of *C* rolls gear *D* back to the top position, thus closing the control valve and stopping the engine. If gears *C* and *E* are revolved at the same speed, gear *D* simply rotates between them and the control valve remains open. If the speed of gear E exceeds that of C, the valve begins to close, and if C revolves faster than E, the valve is opened wider and the engine continues to operate. This general principle has been applied to various classes of mechanisms.

Differential Governors for Water Turbines. — Many of the automatic governing devices used for controlling the speed of water turbines have a differential action. A simple form of governor is illustrated in principle by the diagram A, Fig. 9. An auxiliary water motor drives the bevel gear a by belt d, and bevel gear c is driven by belt e from a shaft operated by

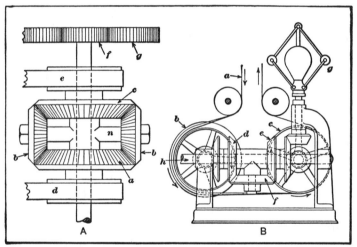

Fig. 9. Differential Governing Devices for Water Turbines

the turbine to be governed. Both gears a and c are loose on their shaft, but the arm n which carries the bevel pinions b is fast to the shaft. On one end of the shaft there is a pinion f which meshes with a rack g that operates the turbine gate, and thus controls the flow of water to the turbine. As the auxiliary motor has no work to do except to drive part of the governing mechanism, it runs at practically a constant speed; the variations due to the rise or fall of the water level are so small a percentage of the total head of water that the speed of this motor is little affected. It will be assumed then that the speed of gear a is practically uniform. The speed of gear c, however, changes with an increase or decrease of the load upon the turbine, and, as gear c runs faster or slower than gear a, the arm n follows it around one way or the other and thus opens or closes the turbine gate.

The governor shown at B also has a differential action, but it is controlled by centrifugal force acting on a fly-ball governing device. The governor is operated by a belt a connected with the turbine. This belt passes around idler pulleys and over the wide-faced pulleys b and c. These pulleys, through bevel gearing, drive the differential gearing composed of gears d, e, and f. Gears d and e are loose from their shafts and pinion f is pivoted on an arm that is keyed to the shaft. Gear e is connected by the gearing shown with a centrifugal governing device at g. The belt pulley b is conical and the diameter at the center is the same as that at pulley c. When the turbine is operating at normal speed, the belt is at the center of the conical pulley b and, consequently, gears d and e revolve at the same rate of speed in opposite directions. The result is that the arm carrying pinion f remains stationary. If the turbine begins to run too fast, the balls at g move outward under the action of centrifugal force, and belt a is shifted by a mechanism not shown to a smaller part of the conical pulley b. The resulting increase in the speed of gear d causes the arm carrying pinion f and the shaft h to which it is attached to revolve in the same direction as gear d. As a result of this movement, the turbine gate is lowered by means of gearing not shown, and the speed of the turbine wheel is reduced. If the turbine should begin to run more slowly than the normal speed, the shifting of belt a by governor g would cause gear d also to revolve slower, thus turning shaft h in the opposite direction and raising the gate.

Another modification of the differential governor is shown by the diagram, Fig. 10. This type of governor is equipped with two sets of epicyclic gearing. The gears A and B are free to turn on the shaft, but may be retarded by brake bands at E and F. The inner gears C and D are driven by belts connected in some way with the turbine. One of these belts is open and the other crossed, so that the gears revolve in opposite directions. The brake bands are so arranged that, when one tightens, the other loosens its grip on the brake drum. Both of these bands are operated by a shaft *G* and the tightening of the bands is effected by a double ratchet mechanism (not shown) having two pawls. One pawl rotates shaft *G* in one direction and the other in the opposite direction. When the speed increases or decreases, one pawl or the other is operated by a fly-ball governor driven from the turbine. As the result of this motion of the pawl, one band is tightened and the other released, so that one of the gears

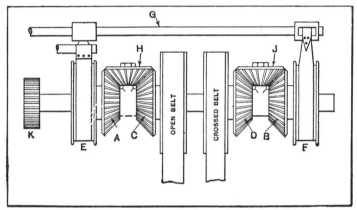

Fig. 10. Differential Governing Mechanism Controlled by
Ratchet-operated Brakes

A or *B* is held with a greater or less degree of friction or is prevented from turning altogether, while the other one runs free. If gear *A* is held by the brake, the arm carrying pinion *H* will begin to turn in the same direction in which gear *C* turns, whereas, if gear *B* remains stationary, the arm carrying pinion *J* will follow gear *D*; consequently, the pinion *K* on the end of the shaft will by means of a rack raise or lower the turbine gate. This governor depends for its sensitiveness upon the fly-ball governing device, and for its power upon the transmitting capacity of the open and cross-belts.

Differential Gearing of Automobiles. — One of the important applications of differential gearing is found on automobiles. The object of transmitting motion from the engine to the rear axle through differential gearing is to give an equal tractive force to each of the two wheels and, at the same time, permit either of them to run ahead or lag behind the

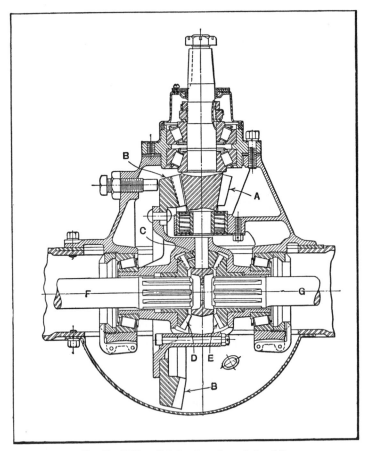

Fig. 11. Differential Gearing of an Automobile

other as may be required in rounding curves or riding over obstructions. The axle is not formed of one solid piece, but motion is transmitted to the right- and left-hand wheels by means of separate sections, the inner ends of which are attached to different members of the differential mechanism.

The principle of this mechanical movement will be understood by referring to Fig. 11. The propeller shaft extends from the transmission case where speed changes are obtained, and revolves the bevel pinion A which drives the large bevel gear B. Gear B is attached to a casing which contains the differential gearing. As this casing revolves it carries around with it a "spider" upon which is mounted either three or four equally-spaced pinions C. These pinions are free to turn about the bearings formed on the arms of the spider, and they are located between and mesh with the side gears D and E. These side gears are mounted upon the splined ends of the right- and left-hand axles F and G and the side gears rotate with these axle sections.

Under ordinary conditions, the rotation of gear B causes gears D and E to both revolve at the same rate of speed, since the connecting pinions C are moved around with the casing, but do not revolve. To illustrate the action, assume that the wheels are jacked up and are simply revolving in one position; then, if one wheel is held from turning so that, say, gear E is stationary, the rotation of bevel gear B will roll pinions C around on gear E with the result that gear D will revolve twice as fast as when gear E is revolving with it and at the same speed. On the other hand, if the opposite wheel and gear D were held stationary, the gear E would run at twice its normal speed; moreover, if the speed of either of the gears is reduced, the other side is speeded up a corresponding amount. The differential gearing is ordinarily incorporated in the rear axle, except when power is transmitted to the wheel by means of side chains, in which case the differential is in the countershaft. Gears A and B usually are either the "spiral bevel" or the "hypoid" type.

Speed Regulation through Differential Gearing. — When the speed of a driven part is governed by drives from two different sources, differential gearing may be used to combine these drives and allow any variations in speed that may be required. An application of this kind is found on the fly frames used in cotton spinning for drawing out or attenuating the untwisted fiber or roving, by passing it between different pairs of rolls which move, successively, at increased speeds. After the fiber is attenuated, it is wound on bobbins and at the same time given a slight twist. The diagram, Fig. 12, represents a mechanism for controlling the speed of the bobbins, one of which is indicated at B. This bobbin receives its motion through a train of gearing connecting with the main shaft of the machine and also through another combina-

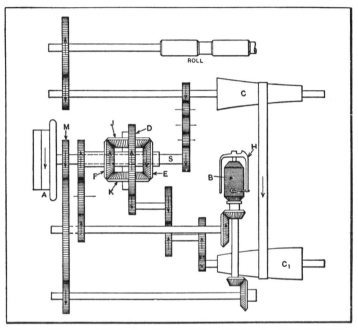

Fig. 12. Diagram of Mechanism Having Differential Gearing Through Which Speed Changes are Transmitted

tion of gearing which is driven by a pair of cone-pulleys for decreasing the speed of the bobbin as the roving is wound upon it and the diameter increases.

The main shaft is driven by pulley A and motion is transmitted through shaft S and the gearing shown to the cone C and the rolls, one of which is indicated at the upper part of the diagram. The cone C and the rolls move at a constant speed, and the roving is delivered by the rolls at a uniform rate of speed. On the shaft S there is a bevel gear E, which is one of the gears of an epicyclic train that is commonly known as the "differential motion." The large gear D corresponds to the arm of the gear train, since it carries the two intermediate bevel pinions J and K. This gear D is driven from the lower cone C_1 which is connected by belt with the upper cone. Bevel gear F which meshes with the pinions carried by gear D is loose on shaft S and is connected through gearing with the bobbin B. With this arrangement, the speed of the bobbin depends first upon the speed of bevel gear E, which is constant, and also upon the speed of gear D, which may be varied by shifting the position of the belt on the cones. Any variations in the relative speeds of gears D and E will produce twice the variations in the speed of the bobbin.

The roving is wound on the bobbin in successive helical layers by means of the flyer H driven at a constant speed by gear M on shaft S. The roving passes from the rolls to the flyer, and entering the top of its hollow spindle, is threaded down through one arm of the flyer and then wound on the bobbin. The flyer and bobbin revolve in the same direction, but the bobbin has a higher velocity and, for that reason, draws the roving from the flyer and winds it in successive layers as the bobbin travels up and down, so as to cover its entire surface. As each successive layer is added, the bobbin increases in diameter, and its speed relative to that of the flyer must be decreased in order to prevent breaking the roving. This change of speed is transmitted to the bobbin through the differential gearing referred to by shifting the belt on the cone-pulleys.

Differential Gear and Cam Combination. — The differential gear and cam combination described in the following is used on fly-frames in conjunction with the same general class of mechanism illustrated by the diagram, Fig. 12. This mechanism differs from the differential ordinarily used in that it has no epicyclic train of gearing. As previously explained, a differential motion is employed in connection with a shifting belt and cone-pulleys for changing the speed of the bobbins.   The differential action is obtained, in this case, by means of a crown gear A (Fig. 13) which is attached to the main driving shaft B; the crown gear C secured to sleeve E, which carries the bobbin driving gear F, and the double crown gear D, which is mounted on a spherical seat and engages gears A and C at points diametrically opposite.   This double crown gear operates in an oblique position, so that a small part of the gear meshes with gear A on one side and a small part on the other side meshes with gear C.   The spherical bearing allows the intermediate crown gear D to swivel in any direction, and it is held in position by a cam surface on the edge of sleeve G.   The gear C has the same number of teeth

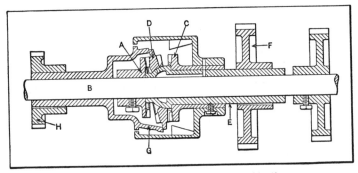

Fig. 13. Differential Gear and Cam Combination

as the intermediate gear D, but gear A has a somewhat smaller number of teeth.

The differential action is obtained by the relative motions between gear A and cam G.   This cam is driven from the lower belt-cone of the machine which is connected with gear H.   If cam G were revolving at the same speed as gear A, the same teeth on gears A and D would remain in contact and the entire gear combination would act practically the same as a clutch.   As soon as the speed of the cam differs from that of gear A, the position of intermediate gear D is changed so that different teeth are successively engaged.   As the result of this differential action, the speed transmitted to gear C is either increased or decreased.   The extent of the differential motion depends upon the difference between the speeds of gear A and cam G. As this difference diminishes, the speed of gears D and C increases; inversely, as the speed of cam G is reduced, the speed of gear C is also reduced, since the motion from gear A is lost as the result of differential action. The advantages claimed for this mechanism are quiet operation and reduction of friction.

Differential Hoisting Mechanism. — An ingenious method of utilizing differential action to vary the speed of a hoisting mechanism is illustrated by the diagram, Fig. 14, which represents the crane to which this mechanism is applied. There are two chains attached to the crane hook. One of these chains

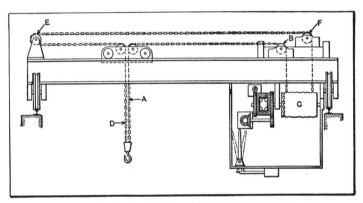

Fig. 14. Crane Equipped with Differential Hoisting Mechanism Shown Diagrammatically in Fig. 15

A passes over a pulley on the trolley and over pulley B to the winding drum C. The other chain D passes upward over its trolley pulley to the left, and over pulley E to pulley F, and then down to a drum located back of drum C. These chains may be wound upon their respective drums either in opposite directions or in the same direction, and at varying rates of speed. If both drums are rotated in opposite directions at the same speed the effect will be to raise or lower the hoisting hook, whereas, if the drums rotate in the same direction and at equal speed, the chain will be taken in by one and given off by the other, thus causing the hook and its load to be carried

horizontally without raising or lowering it. Any difference
in the speed of the two drums when moving either in the same
or opposite directions will evidently cause the hook to move
both vertically and horizontally at the same time.

The mechanism for operating the two hoisting drums is
illustrated diagrammatically in Fig. 15. There are two electric motors J and K. Motor J drives the worm-wheels L in
opposite directions and also the attached bevel gears. The
other motor K drives the spur gears M and the upper bevel
gears. The intermediate pinions N between the bevel gears
revolve on arms Q which are keyed to the shafts of their re-

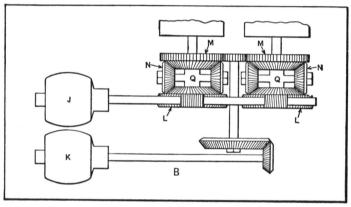

Fig. 15. Differential Hoisting Mechanism

spective drums. The bevel gears with which the pinions mesh
are loose on their shafts. With this arrangement, if motor
K is stationary, motor J will drive the drums in opposite directions and raise or lower the hook as previously explained.
On the other hand, with motor J stationary, motor K will
operate the drums in the same direction and move the crane
hook horizontally. As these motors may be reversed or operated together at varying speeds, any desired combination of
movements and speeds for the hook and its load may be
obtained.

Variable and Reversing Rotation for Feed-Rolls. — The
requirement of the mechanism here described, which is used

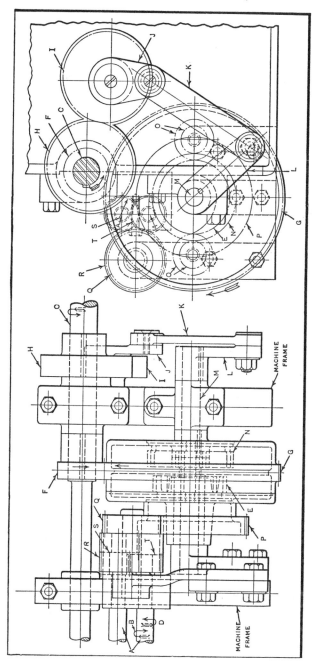

Fig. 16. Special Design of Differential Mechanism for Transmitting to Feed-rolls a Variable and Reversing Rotation

on a cotton combing machine, is to rotate feed-rolls A (see Fig. 16) with their respective top rolls (not shown) 1.4 revolutions in the direction of arrow B while driving shaft C makes 0.6 revolution, and, during the remaining 0.4 revolution of shaft C, to rotate feed-rolls A 0.7 revolution in the direction of arrow D.

The object is to feed approximately 4 inches of cotton

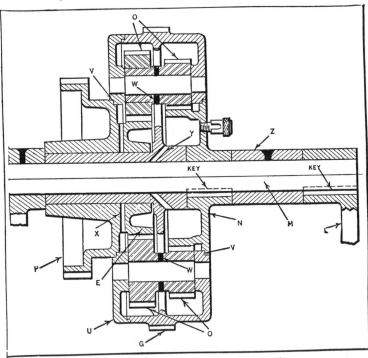

Fig. 17. Cross-section of Differential Gearing

forward, as indicated by arrow B, and then reverse and feed approximately 2 inches of material backward, as indicated by arrow D, and repeat for each revolution of driving shaft C, thus performing a doubling process in conjunction with other elements of the machine.

The required result was obtained by combining a constant motion and a variable oscillating motion, the two motions co-acting on a common gear E (see also Fig. 17) through an epicyclic or differential gearing combination. Referring to Fig. 16, it will be seen that the constant motion is effected by driving pinion F and housing gear G; also that the variable oscillating motion is produced through eccentric gears H and I, crank-arm J, connecting link K, rocker arm L, rocker shaft M, and rocker shaft gear N. Through the planetary gears O these two motions are permitted to combine and drive feed-rolls A forward and backward as mentioned, through gears E, P, Q, R, S, and T.

The timing of the forward and backward rotation of rolls A is controlled by eccentric gears H and I, which transmit a quick-return motion to rocker arm L for the forward rotation of rolls A, as indicated by arrow B, and next a slower motion in the opposite direction to rocker arm L for the backward rotation of the rolls A, as indicated by arrow D. Eccentric gear H is keyed to driving shaft C, while I is keyed to the hub of crank arm J, which revolves on a suitable stud fastened to the frame of the machine.

Housing gear G, which runs loose on rocker shaft M, and all other gears and bearings inside of it, are splash-lubricated through oil-holes X, Y, and W, Fig. 17. By pouring one-half pint of oil into the housing, three weeks' supply of lubricant is provided, which is well distributed to all bearings by planetary gears O revolving through it. Oil-tight joints are provided for gears N and P, as shown at V. Gears E and P are integral and revolve on the extended hub of housing gear G. Rocker arm L and rocker shaft gear N are keyed to rocker shaft M. This mechanism replaced a cam and clutch arrangement which was too noisy and did not wear well.

Differential Speed Indicator. — A sensitive speed-indicating device which shows variations of speed between two rotating parts is shown, partly in section, in Fig. 18. This indicator operates on the differential principle. It is equipped with two cylindrical rollers; one roller is shown at A and the other is located in a similar position on the opposite side of the vertical center line. The axes of the roller shafts are in the same vertical plane, and on the ends of these shafts are mounted belt pulleys C. These pulleys are connected with the shafts the relative speeds of which are to be compared. Each roller A is in contact with a spherical steel ball B three inches in diameter. The ball is held in position by a small stop D at the rear and by a small roller E at the front. This roller

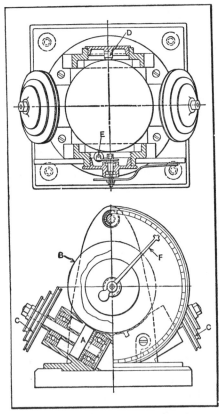

is mounted on an arm fixed to a spindle which is free to rotate and to the outer end of which is attached the pointer F. When both the supporting rollers A are driven at the same speed and in the same direction, the spherical ball will rotate about a transverse horizontal axis and will carry the wheel E vertically up or down, as the case may be. The direction of movement will be indicated by the pointer F.

If either of the supporting rollers runs faster than the other, the ball will rotate about some inclined axis and wheel E will naturally turn so that its axis is parallel to that about which the sphere rotates.

Fig. 18.   Differential Speed Indicator

This instrument is said to be very sensitive as an indicator of speed variations. For instance, it is claimed that a difference in the speed of the rollers due to a variation of 0.001 inch in the diameter of driving pulleys having a nominal diameter of $2\frac{1}{2}$ inches can be detected.

CHAPTER XIII

STRAIGHT-LINE MOTIONS

A COMBINATION of links arranged to impart a rectilinear motion to a rod or other part independently of guides or ways is known either as a *straight-line* motion or a *parallel* motion, the former term being more appropriate. Mechanisms of this type were used on steam engines and pumps of early designs to guide the piston-rods, because machine tools had not been developed for planing accurate guides. One application

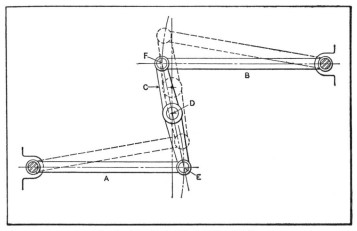

Fig. 1. The Watt Straight-line Motion

of straight-line motions at the present time is on steam engine indicators for imparting a rectilinear movement to the pencil or tracing point. The principle of the well-known parallel motion, invented by James Watt in 1784, is illustrated by the diagram, Fig. 1. Links *A* and *B* are free to oscillate about fixed pins at their outer ends, and are connected by link *C*. A point *D* may be located on the center-line of link *C,* which follows approximately a straight line when links A and B are given an oscillating movement, because, when A moves from its central position, the center of pin E moves to the left along its circular path while the center of pin F moves to the right. As the motion of point D is affected by both links A and B, it moves very nearly in a straight line, provided D is correctly located and the angular motion of the links does not exceed about 20 degrees. Very few straight-line mechanisms produce a motion which is absolutely straight, and the general practice is to so design them that the guided part will be on the line when at the center and extreme ends of the stroke.

Scott Russell Straight-line Motion. — The mechanism illustrated in Fig. 2 will give an exact straight-line motion, but

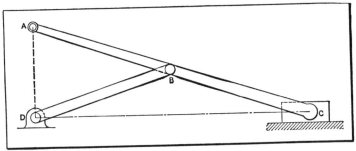

Fig. 2. The Scott Russell Straight-line Motion

it is necessary to have an accurate plane surface upon which block C can slide. In addition to this sliding block, there are two links AC and DB. The link DB is one-half the length of AC and the shorter link is connected at a point B midway between A and C. The shorter link oscillates about a stationary pivot at D as end A is moved up or down along the straight line AD. Since AB, DB, and BC are equal, a circle with B as the center will intersect points A, D, C for any angle DCA; consequently, the line AD, traced by point A is perpendicular to DC, since ADC is always a right angle.

Instead of having guides or a plane surface for the sliding block C, the mechanism is sometimes modified by attaching the block end of link AC to another link which is free to oscillate about a fixed pivot so located that the link will be perpendicular to the line CD, when in its mid-position. The longer this link and the greater the radius of the arc described by the connecting point at C, the more nearly will C move in a straight line; hence, the longer this link, the less point A deviates from a straight line. This modification of the Scott Russell straight-line motion is sometimes called the *grasshopper motion*.

Straight-line Motions for Engine Indicators.— Some form of straight-line motion is necessary on a steam engine indi-

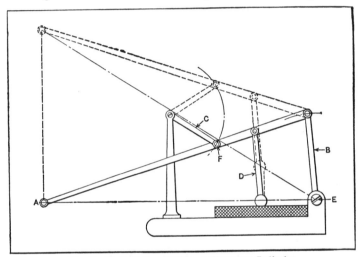

Fig. 3. Straight-line Motion of Thompson Indicator

cator in order that the motion of the indicator piston will produce a parallel movement of the tracer point or pencil, which draws a diagram on the paper or indicator card. The cylinder of the indicator is open at the bottom and is connected by suitable pipes with each end of the steam engine cylinder, so that the under side of the indicator piston is subjected to the varying pressure acting upon the engine piston. The upward movement of the indicator piston resulting from the steam pressure is resisted by a spiral spring of known resilience, and a rod extending above the piston connects with some form of link work designed to give a straight-line motion to the tracer point. When the engine is running and the indicator is in communication with the steam cylinder, variations of pressure will be recorded by the vertical movement of the pencil or tracer which is brought into contact with paper wound about a cylindrical drum that is rotated by the reciprocating motion of the engine cross-head.

The straight-line or parallel motion of one indicator is shown in Fig. 3. The arm A which carries the pencil at its outer end is pivoted to link B which, in turn, is pivoted to the top of the indicator. As arm A moves upward, the outer end is guided along a straight line by link C, which oscillates about

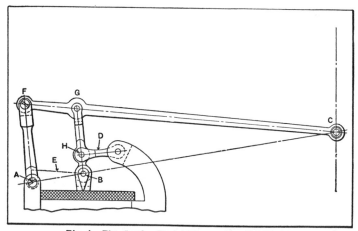

Fig. 4. The Crosby Straight-line or Parallel Motion

a fixed pivot and is connected to arm A at F. This mechanism is so proportioned that a line from A to E intersects the point at which link D is attached to the piston.

The straight-line motion of another steam engine indicator is shown in Fig. 4. This mechanism, like the one previously described, is so arranged that the fulcrum A of the entire mechanism, the connection B, and the pencil point C are always in a straight line. The fundamental principle of this mechanism is that of the pantograph. If link D were removed and replaced by another link at E, both parallel and equal in length to FG, this would result in a well-known form of pantograph mechanism. The length of link D to replace E may be determined as follows: The procedure is to first ascertain, by trial, a convenient location for the point at which link D is to connect with link BG. The path followed by point H as end C is moved along a straight line is plotted on a large scale for all positions on the linkage within the required range of movement. This path will be approximately the arc of some

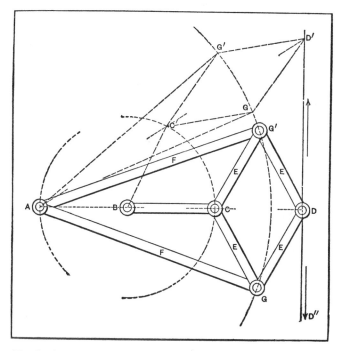

Fig. 5. Arrangement of Peaucellier Linkage for Straight-line Motion

circle, and the fixed pivot for link D is located at the center of this circle. If a link at E were actually used in place of link D, a straight-line motion at C could be obtained, providing the pivot B had a straight-line motion. Any form of guide intended to insure a straight movement at B would be objectionable, since it is desirable to reduce the friction of mechanisms of this type to a minimum. It is also essential to have the parts as light as possible in order to minimize the inertia and the effect of momentum, which is especially troublesome when taking cards from engines operating at high speed.

With the parallel motion of another indicator, a pin on the pencil arm corresponding to the one shown at F in Fig. 4 engages a curved slot in a stationary plate which is secured to the indicator in a vertical position. This curved slot takes the place of a link, and its curvature is such as to compensate for the tendency of the pencil to move in an arc.

Peaucellier Straight-line Motion. — The link mechanism shown in Fig. 5 will give an exact straight-line motion. This

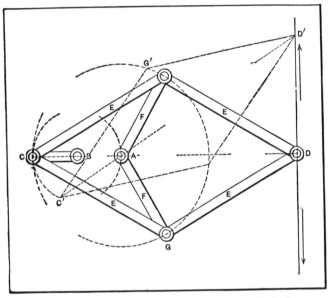

Fig. 6. Modification of Peaucellier Straight-line Mechanism

mechanism was invented by Peaucellier, a French army officer. It is composed of seven links moving about two fixed centers of motion, A and B. The four equal links E form a rhombus; the links F are equal, and the center B is midway between A and C. If the point D be moved in the direction of the arrows, it will be constrained to move in the straight path $D'D''$, which is perpendicular to the line of centers $ABCD$. This may be tested experimentally. The path of the point C is the circumference $AC'C$; and the path of GG' is the arc described with the radius F. If the center-line of the links E and F be assumed in any position such as $AC'D'$, it will be found that the rhombus the sides of which represent the length of the links E takes the position shown in the drawing.

In Fig. 5, the centers A and B are external to the links E. A variation of the linkage is shown in Fig. 6, in which the centers A and B are within the rhombus. The links F are equal, and center B is midway between A and C, as in Fig. 5. The corresponding links and points in the figures are labeled with the same letters; it may be shown experimentally that the point D is compelled to move in a straight line perpendicular to the line of centers $CBAD$.

CHAPTER XIV

MISCELLANEOUS MECHANICAL MOVEMENTS

THE mechanisms described in this chapter are of such a miscellaneous character that they cannot be placed in any of the general groups or classifications covered by preceding chapters. They are included in this treatise, however, to add to the variety of the mechanisms described.

Mechanism to Insure Full-stroke Movement of Operating Lever. — Mechanisms are sometimes so arranged that hand-operated movements are, to some extent, controlled mechanically, to prevent motion in the wrong direction or incomplete action. The full-stroke ratchet mechanism shown in Fig. 1 is used on an adding typewriter to prevent the operator from starting handle A and not completing the required movement. For instance, if handle A is in the upper position, as shown at the left, any downward movement must be continued until the handle has made a complete stroke before it can be reversed for returning it to the original or upper position. Similarly, if the lever is at the lower end of its stroke, as shown by the view to the right, any upward movement must be completed before the direction of motion can be reversed. This positive control of the action of handle A is obtained in a very simple manner. As the handle is moved downward or upward, pawl B is carried with it. This pawl is pivoted to part D and normally held in a vertical position by a spring. When handle A is at the upper end of its stroke, as shown at the left, and a downward movement is started, pawl B engages sector C and its upper end swings to the right; as the downward movement of handle A continues, pawl B engages successive notches in sector C, and locks into one of these notches if an attempt is made to return handle A before the downward stroke is completed. When handle A has been pushed all the

398

way down (as shown to the right), pawl B drops into the enlarged notch E of sector C where there is enough room to permit the pawl to swing around to the vertical position; consequently, as soon as handle A is moved upward, the top of pawl B swings to the left and again engages successive notches in sector C, thus preventing any return of handle A to the lower position until the pawl has cleared the upper end of the sector and again swings to a vertical position.

Lock to Prevent Reversal of Rotation. — Some shafts must be free to rotate in one direction but be locked instantly against

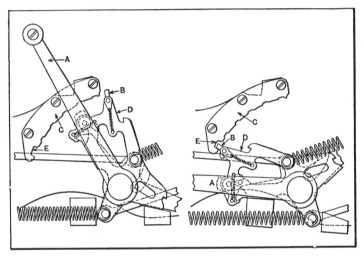

Fig. 1. Full-stroke Mechanism to Prevent Starting the Operating Lever and not Completing its Movement

a reversal of rotation. A ratchet mechanism may be objectionable because of its noise and backlash. Under these conditions, the arrangement shown in Fig. 2 was found to be satisfactory. Shaft A is free to rotate clockwise, but a reversal is not allowed, although when the shaft is not running clockwise, there is always a tendency toward reversal because of torque exerted on the shaft. Keyed to the shaft is a ring B into which are cut three wedge-shaped recesses containing rollers C. Ring B rotates within bracket D, which may be bolted to the wall or some immovable structure. Ring B

and bracket D are kept in alignment by retaining plates E, which are bolted to the ring. The tool-steel rollers C are kept in contact with ring B and bracket D by means of light springs F which are riveted to their keepers.

When shaft A and ring B rotate clockwise, the rollers tend to move relatively in the opposite direction, thus compressing springs F. This movement releases any wedging action between the roller and members B and D, although the rollers always remain in contact with these members. Any back-

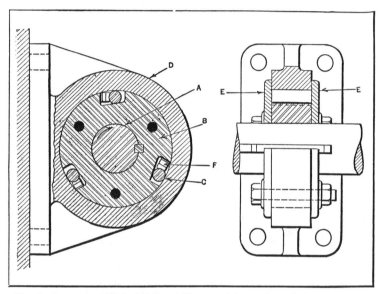

Fig. 2. Reversal of Shaft Rotation is Prevented by Wedging Action of Rollers

ward or counter-clockwise rotation is stopped instantly, because the rollers become wedged between parts B and D, thus locking them together. It is evident that the greater the torque counter-clockwise, the greater will be the locking effect within the limits of the strength of materials used. This simple contrivance proved to be very effective for preventing a reversal of shaft rotation.

Device to Rotate Shafts Synchronously in Opposite Directions. — Two shafts which are in alignment are rotated synchronously and in opposite directions by the simple arrangement shown by the diagram Fig. 3. The two shafts S and S_1 have cranks of equal throw. On the outer ends of these cranks are universal joints. Balls K are shown, but

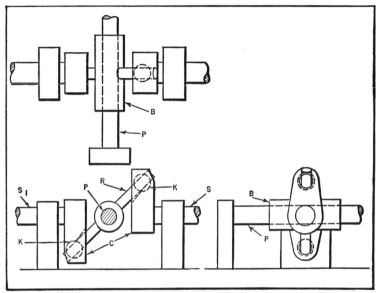

Fig. 3. Diagram Showing Device for Rotating Shafts in Opposite Directions, the Shafts Being in Alignment

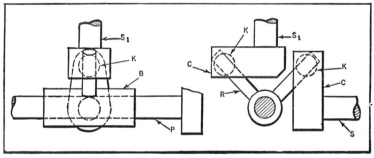

Fig. 4. Transmission Shown in Fig. 3 Applied to Shafts Located at Right Angles

any equivalent joint is satisfactory. The arms R of rocking beam B are free to slide through holes in balls K. This rocking beam is free to move axially and rock upon fixed shaft P, which is located at right angles to shafts S and S_1 and in their planes. This fixed shaft is also midway between the planes of rotation of the ball centers.

This is a flexible arrangement, as it may be applied to shafts located at any angle. Fig. 4 illustrates shafts at right angles. The connecting member is the same in principle. The arms of the rocking beam are at right angles, and the axis of the fixed shaft in this case is at the vertex of the angle made by the driving and driven shafts. If the driving and driven shafts are not in alignment, the rocker arms must be offset the same amount as the shafts. One driver using one rocking beam and having arms suitably arranged can operate a number of shafts parallel to the driving shaft or making an angle to it, provided the driving and driven shafts have a common plane; or in case the driven members are angularly disposed, provided the planes of the driven members have a common vertex through which the axis of the driver passes.

This mechanism is adapted for complete enclosure and oil-immersed operation. This form of transmission was applied where gears were not desired, although if gears had been used, five would have been necessary.

Disengagement of Worm-gearing for Rapid Adjustment. — A hand-operated winding drum used for adjusting the height of airplane models in a "wind tunnel" is so arranged that the drum may be rotated either through worm-gearing for precise adjustments, or directly, by disengaging the worm-gear, when rapid adjustments are desired. The airplane model is supported by wires (not shown) which pass up over pulleys and then to the winding drum B (Fig. 5), one wire extending forward and the other backward relative to drum B. The rotation of this drum winds or unwinds both wires simultaneously. The knurled handwheel C is used for fine adjustments, motion being transmitted through a single V-thread worm and a straight-faced worm-wheel D located at one end of the drum. If a rapid adjustment is required, the worm-wheel is disengaged from the worm merely by pulling the worm-wheel and drum axially on bolt F; the drum is then

turned directly by hand, the flange *G* being knurled to provide
a better grip.

When the outward pull is released, spring *H* immediately
forces the worm-wheel back into mesh, thus relocking the setting. The handwheel *C* is graduated on top so that the amount
of adjustment can be determined. Although this device is
fitted to laboratory apparatus, the idea of sliding a worm-
wheel axially out of mesh with the worm might be utilized
for other purposes; however, it is evident that only straight-

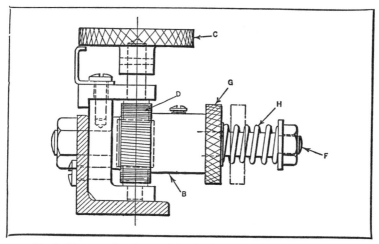

Fig. 5. Worm-gearing Arranged for Quick Disengagement When Rapid
Adjustment is Required

faced worm-wheels could be used. For some applications, it
might be preferable to replace the knurled portion of the
winding drum with a spur gear arranged to slide into engagement with driving gears when the worm-wheel was disengaged. Another variation for possible application to small
lifting blocks would include a friction brake for rapid lowering by gravity.

Centrifugal Chuck-closing Mechanism. — The automatic
chuck-closing mechanism illustrated in Fig. 6 is operated by
the action of centrifugal force upon balls or spherical weights
which move outward when chuck rotation begins, thus auto-

matically closing the collet chuck upon the work. This device is used on a plain screw machine.

The aluminum body *A,* which contains the chuck-closing mechanism, is mounted at the rear of spindle *C.* Sixteen equally spaced steel balls *B* are located in slots formed around the edge of ball-holder *D.* This ball-holder is free to slide forward or backward for opening and closing the chuck, and it is centered on three supports *E* that form part of body *A.*

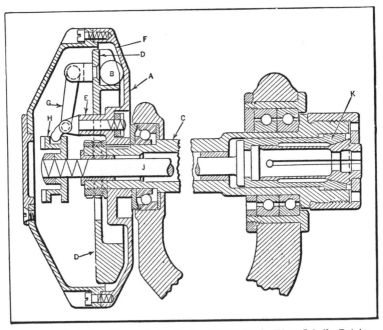

Fig. 6. Mechanism Which Automatically Closes Chuck When Spindle Rotates

The front or chuck end of the spindle is a standard type, and has a collet chuck, as shown. When the spindle and chuck-closing mechanism begin to revolve, balls *B* move outward, due to centrifugal force, and as they engage the inclined surface *F,* ball-holder *D* is pushed backward with a force which increases as the rotary speed increases. This backward movement of *D* causes levers *G,* acting through collar *H,* to push rod *J* and chuck sleeve *K* forward, thus closing the collet

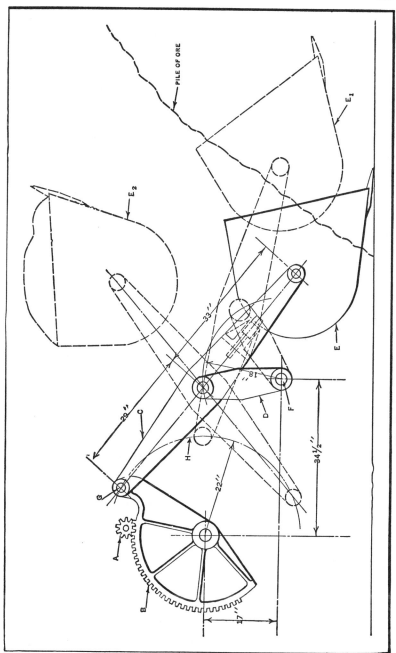

Fig. 7. Scooping Mechanism for Shovel Truck

chuck about the work. When the machine is stopped, the balls return to their inner positions and the spring collet chuck opens. This machine has a friction clutch on the spindle with a foot-treadle which controls starting and stopping.

Mechanical Scooping Motion. — A mechanically reproduced scooping motion incorporated in a truck designed principally for handling lead ore is illustrated by the diagram Fig. 7, which merely shows those parts that are essential to the motion required. Pinion A, which is driven by a motor, meshes with the gear sector B. Arm C is pivoted to one end of gear B; the other end is pivoted to bucket E and the middle part of arm C is pivoted to rocker arm D, which swings about a fixed pivot F.

This combination, when proportioned according to the dimensions given, provides the required scooping action. The truck is driven forward to locate the bucket close to a pile of ore. Then as pinion A turns gear B, pin G moves downward along arc H, and the bucket E is forced into the pile, as at E_1, at the same time being forced upward with an efficient scooping action similar to that obtained with the large steam shovels. The dotted lines at E_2 indicate the elevated position of the bucket when loaded.

This mechanism has proved to be a great time- and labor-saver, as the truck loads itself and at the same time lifts the ore high enough so that it can readily be charged into the furnace. The truck is also very compact, so that it can be run into a freight car for unloading. The complete mechanism permits sluing the loaded bucket 90 degrees each way from the central position, so that the truck can be run up a narrow aisle for charging furnaces with the ore.

Power Press Stock Gage. — This stock gage is made part of the press equipment instead of part of the die, as is customary. The mechanical gage finger that has been developed embodies a very simple tripping device. In the diagram, Fig. 8, the device is shown in its operating position on the back of a straight-side blanking press. A hook-block A slides in gibs in the cross-arm L, and carries on its lower end the stop-finger M secured by the set-screw N. The threaded stem Q of the slide extends through the stop-lug H and carries stop-nuts above and below this lug. The ram of the press K carries the trunnion block J in which swings the hook B on the

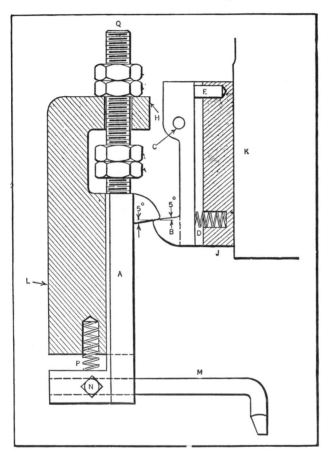

Fig. 8. Stock Gage Which Moves with the Punch After Part Has Been Blanked but Holds Stock During Blanking Operation

cross-pin C. Hook B is pressed outward by the spring D until the upper end strikes the stop-pin E. The radius of the tips of the hooks is approximately 1/16 inch.

In operation, the action is as follows: With the sliding hook down, the stop-nuts are against the top of lug H, and the tip of the finger M rests on the die. The ram K descends carrying J and B with it. The large radii of A and B engage pressing B back against the spring D until B passes A. On the up stroke, the sloping surfaces of A and B engage and B lifts A until the stop-nuts strike the under side of the lug H, which prevents further movement of A. The reaction between the sloping surfaces of B and A forces B and A apart, so that they unhook, allowing A to drop through the force of

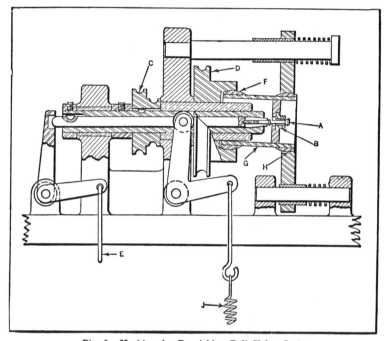

Fig. 9. Machine for Burnishing Ball Valve Seats

gravity and the action of spring P. On the next stroke of the press, the action is repeated. It will be noted that the finger M moves with the punch after the stock has been blanked, but holds the stock stationary during the blanking. This one feature largely reduced the number of spoiled blanks caused by movement of the stock during blanking.

Eccentric Motion for Burnishing Ball Valve Seats. — This mechanism for burnishing a ball valve seat B (see Fig. 9) causes the valve A to rotate about its own axis, and at the same time this axis has a planetary conical motion, the apex of the cone being at the center of the spherical surfaces of the valve and seat; hence, local irregularities are eliminated and the density of the metal increased.

The valve A is held by a spindle collet which is closed by a foot-pedal attached at E. The valve seat B fits into a hexagon socket which prevents it from turning. The spindle pulley C turns 1500 revolutions per minute, and pulley D revolves 100 revolutions per minute. This pulley D has a ball socket F bored off center, which gives the axis of sleeve G a conical

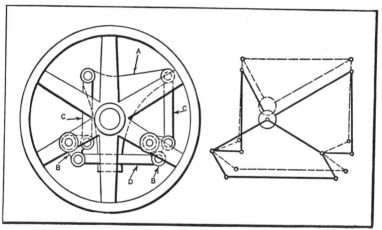

Fig. 10. Shaft Coupling Designed for Lateral or Angular Displacement

motion, the apex of the cone being at point A which is also the center of the ball socket H. Sleeve G does not revolve. A foot-pedal attached to spring J is used to hold the valve seat B against valve A with sufficient pressure to burnish the valve seat properly. Both valves and seats are of brass, and oil is used in burnishing.

Shaft Coupling Rigid in Torsion Only.—The linkage shown in Fig. 10 was devised in order to secure a flexible coupling that would permit a comparatively large and constantly changing amount of misalignment. With this arrangement, misalignment does not cause the coupling to heat or become noisy. Mounted on the driven member so they can rotate are two bellcranks *B*, which are joined by a link *D* as shown, so that they rock in unison. These bellcranks are also connected with the driving spider *A* by links *C*, which should be provided with ball brasses if the angular misalignment is excessive. A line diagram of the linkage is shown at the right in the illustration. The broken lines indicate the position taken by the links and the bellcranks when vertical displacement takes place.

Angular Transmission for Shafts. — If a designer is seeking a means for transmitting power between two shaft ends

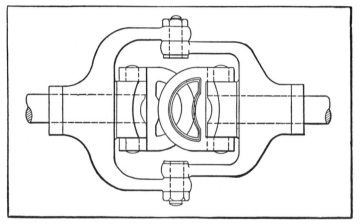

Fig. 11. Transmission with Swiveled Ends

at right angles where the four requirements are positive motion, flexibility, compactness, and quietness of action, what form of mechanical motion will he adopt? A survey of the small group of mechanisms available for angular drives will reveal a rather meager list from which to choose. There are angular transmissions that are positive and compact, but are neither flexible nor quiet. There are transmissions that are flexible and quiet, but neither positive nor compact. It may reasonably be assumed, then, that a description of any new mechanical movement that meets these requirements will be of interest to designers and power transmission engineers.

The use of the Bartlett angular transmission to be described is not confined to right-angle transmissions, but is applicable to any shaft angle from 0 to about 120 degrees, although at the latter angle the contact between the rubbing surfaces would be considerably reduced. If the shaft ends are swiveled, as in Fig. 11, a complete angular sweep of 180 degrees, or 90

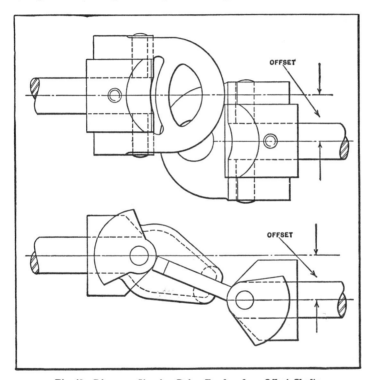

Fig. 12. Diagrams Showing Drive Employed on Offset Shafts

degrees in either direction, is possible; and by adding a ring and a second swivel-pin, a universal joint of a hitherto unattainable magnitude of angular sweep may be produced. Such a universal joint has the unusual property of maintaining a uniform angular velocity ratio of 1:1 between the driving and the driven shafts. It is well known that the common type of universal joint produces an increasing departure from a uniform ratio as the shaft angle increases; for example, with

a shaft angle of 30 degrees, Hooke's joint causes a total variation of about 8¼ degrees between the two shafts twice in
each revolution, and the actual variation from uniform angular
velocity is 28.87 per cent.

Where the shafts are parallel, but considerably offset, this
drive can perform the same function as the well-known Old-
ham coupling, the angular velocity ratio still being uniform.
Two positions of the drive when used in this way are shown

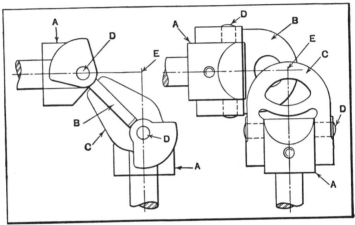

Fig. 13. Transmission Used for Right-angle Drive

in Fig. 12, where the positions in the two cases differ by a
quarter of a revolution of the shaft.

When used as a right-angle drive for lineshafting, the drive
is connected direct to the shaft ends, which are preferably
mounted in self-aligning bearings secured to a special hanger.
This hanger provides a rigid connection between the bearings
and allows for adjustment in all directions. It also provides
a convenient surface on which an oil-case of simple design
may be attached.

From Fig. 13 it is possible to obtain a good idea of the
construction and action of the parts when the transmission
is used for a right-angle drive. The two hubs A are keyed
to the shaft ends. Each hub carries two hardened steel pins
D over which the driving members B and C are free to turn

through an angle of somewhat more than 45 degrees on each side of the shaft axis. The member C is of cast iron, semi-steel, or bronze, and is slotted as shown to provide a sliding fit for the member B, which is of steel with the working surfaces hardened and ground. The openings in the semicircular parts are for the purpose of reducing the weight, and the extra metal on the opposite side of the pins acts as a counterbalance, but is not needed except for high speeds.

For a perfectly uniform angular velocity ratio, the following conditions must exist:

1. The center line of each pin must intersect the axis of the shaft.

2. The center lines of the pins must lie in the mid-plane between the sliding surfaces of the tongue and slot.

3. The axes of the two shafts must intersect in a point E.

4. The center lines of the two pins must be equally distant from the intersection E of the shaft axes.

If these four conditions hold, it must follow that for all positions of the sliding members, the center lines of the two pins lie in the same plane, which is the mid-plane of the tongue and slot. Also this mid-plane is always inclined at equal angles with the axes of the two shafts. Hence any angular motion of one shaft must be accompanied by an equal angular motion of the other shaft.

While deviation from the four conditions mentioned will affect the uniformity of the velocity ratio to some extent, it is not necessary to hold to the same degree of precision in mounting these units as would be required for gears.

This drive is actually a flexible shaft coupling applied to shafts whose axes are set at angles varying from 0 to 90 degrees. When the shafts are not in perfect alignment, the bearings are relieved of undue stress, and the action is smooth and quiet. Shaft misalignment is not possible with gears which mesh properly, nor, in general, with most other types of angular drives.

Water-operated Automatic Switching Mechanism. — In making steel wire, a billet of the proper size is heated and reduced in cross-section in a series of mill stands. The steel emerges from the last or finishing stand in the form of a small rod which is several thousand feet long and moves at a speed of 25 miles per hour or more. This rod is guided through pipes to a horizontal reel, the speed of which is automatically regulated according to the speed of the finishing stand, and this forms the rod into a coil. The coiled rod is removed from the reel by a suitable mechanism and then reduced to a wire by cold-drawing through a series of dies.

In order to obtain high efficiency, the rods follow one another very closely; in fact, there is only a space of a few feet

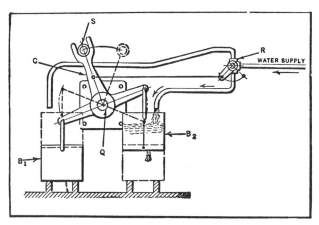

Fig. 14. Mechanism of Automatic Switching Device

between the rear end of one rod and the forward end of the next rod. Since it would be impossible to clear the reel in these short time intervals, it is necessary to employ two reels and coil the rods alternately on them. This requires a switching device which directs the rods to the two reels in alternation.

Before describing the switching device, it might be well to summarize the conditions: (1) The rods are white-hot; (2) they move at a high rate of speed; (3) the time available for the switching device to act is very short, and the action must be instantaneous; (4) the rods are of different lengths, so that switching must occur at irregular intervals; (5) the device must possess a maximum degree of reliability, because its failure would cause considerable danger and a decrease in production. The device to be described meets all of these conditions in a very simple manner.

Pipe S (see Fig. 14) can swing around a fulcrum in such a manner that while one end always registers with a pipe leading to the finishing stand, the other end will register with either a pipe leading to the right-hand reel or with another pipe leading to the left-hand reel. Bellcrank C can swing around stationary pin Q, moving pipe S into either of the two necessary positions. A three-way valve R is operated by the bellcrank. Parts B_1 and B_2 are buckets, each having a hole in the bottom. Water flows into each bucket alternately, in a volume exceeding that of the outflow at the bottom, so that the water level in the bucket gradually rises.

In the position shown in full lines, bucket B_2 is receiving water and becoming slowly heavier, while bucket B_1 is emptying itself and becoming lighter; hence the bellcrank tends to swing into the dotted position, thereby reversing the three-way valve, which diverts the stream of incoming water to bucket B_1 and allows bucket B_2 to become empty.

Now the only obstacle to such a motion of the bellcrank is furnished by the rod which, running from pipe S to whichever reel pipe it is in line with, bridges the gap between the two pipes and acts, so to speak, as a splice. But at the instant that the tail end of the rod leaves pipe S, the motion of the bellcrank takes place. This action is, of course, reversed when the tail end of the next rod leaves pipe S. The amount of water is so regulated that the scraping action on the side of the rod is slight and therefore not injurious. This device has provisions for manual control in case of emergency.

Mechanical Lapping Mechanism for Plane Surfaces. — The mechanical movements required in accurately lapping the parallel surfaces of small brass rings are obtained from the mechanism shown in Fig. 15. This mechanical lapping process enables one man with one machine easily to lap 2500 rings a day to within 0.0002 inch of parallelism.

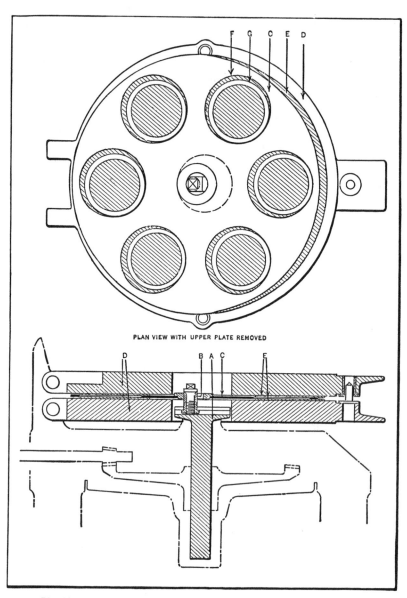

Fig. 15. Mechanical Lapping Mechanism for Finishing Small Brass Rings

Referring to the illustration, a vertical shaft A is rotated at 150 revolutions per minute. An adjustable eccentrically located roller B on the upper end of the shaft engages a hole in a driving disk C, the hole being about half again as large as the roller. Thus the disk is oscillated in a horizontal plane 150 times a minute. As the roller is smaller than the hole in the disk, and therefore assumes a position eccentric to the disk, and as the point of eccentricity is continuously progressing around the circle, due to the inertia of the unequally divided mass of the disk, the disk itself assumes a slow motion of rotation.

The driving disk oscillates and rotates between two plates D, which are nominally stationary, but which have a certain amount of freedom in all directions. These plates are covered, on adjacent faces, with disks of abrasive cloth E. The driving disk has openings F cut in it to receive the rings G, these openings being on a circle which is of the same diameter as the mean diameter of the abrasive disks.

When the roller is adjusted to the proper eccentricity for the size of the rings, the oscillation of the driving disk will pass the rings over the outside and inside edges of the abrasive disks an equal amount. The openings in the driving disk are a little larger than the rings, so that the latter are free to rotate around their own centers. They do this because the path around the plates, of any points on the outside edges of the rings, is greater than that of any points on their inside edges; therefore the lapping action is equalized as the rings twist around their own centers.

Thus, by an ingenious adaptation, a simple crank translates to a number of work-pieces, at one time, an oscillatory motion, as well as motions of rotation around both their individual and common centers, and a complication of mechanical movements is obtained, without precise construction or skilled attention, which produces precision results—a good example of simplification in machine design.

Pantograph Mechanisms for Reproducing Motion on a Different Scale. — A pantograph is a combination of links which are so connected and proportioned as to length that any motion of one point in a plane parallel to that of the link mechanism will cause another point to follow a similar path either on an enlarged or a reduced scale. Such a mechanism may be used as a reducing motion for operating a steam engine indicator, or to control the movements of a metal cutting tool. For instance, most engraving machines have a pantograph mechanism interposed between the tool and a tracing point which is guided along lines or grooves of a model or pattern. As the tracing point moves, the tool follows a similar path

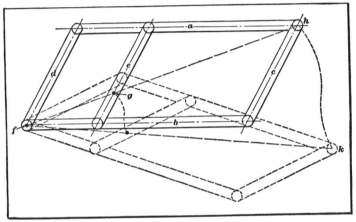

Fig. 16. Pantograph for Reproducing Motion on a Reduced or Enlarged Scale

but to a reduced scale, and cuts the required pattern or design on the work.

A simple form of pantograph is shown by the diagram, Fig. 16. There are four links, *a*, *b*, *c*, and *d*. Links *a* and *b* are equal in length, as are links *c* and *d*, thus forming a parallelogram. A fifth connecting link *e* is parallel to links *c* and *d*. This mechanism is free to swivel about a fixed center *f*. Any movement of *h* about *f* will cause a point *g* (which coincides with a straight line passing through *f* and *h*) to describe a path similar to that followed by *h*, but on a reduced scale. For instance, if *h* were moved to *k* following the path indicated by the dotted line, point *g* would also trace a similar path.

Another form of pantograph mechanism is shown at *A* in Fig. 17. This pantograph, which is sometimes called "lazy tongs," is used to some extent for obtaining the reduction of motion between an engine cross-head and the indicator drum when taking indicator cards. The pantograph is pivoted at *b* by a stud which may be secured to a block of wood or angle iron attached to a post or in any convenient place. The end

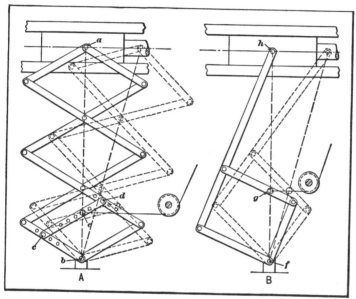

Fig. 17. Pantograph Mechanisms Applied to Engine Cross-head to Reduce Motion When Taking Indicator Cards

a has a pin which is connected to the cross-head of the engine. The cord which transmits motion to the indicator drum is attached to the cord-pin *e* on a cross-bar. This cross-bar may be placed in different positions relative to the pivot *b*, by changing screws at *c* and *d*; the cord-pin *e*, however, must always be in line with the fixed pivot *b* and pin *a*. The position of the cross-bar in relation to pivot *b* determines the length of the travel of cord-pin *e* and, consequently, the rotary movement of the indicator drum and the length of the diagram which the pencil traces upon the indicator card. The objection to this reducing mechanism is the liability of lost motion resulting from wear in the numerous joints.

The pantograph reducing mechanism shown at B in Fig. 17, has four links joined together in the form of a parallelogram, and one of the links is extended and pivoted to the engine cross-head. The swiveling movement of the pantograph is about the fixed pivot f, and the cord which operates the indicator drum is attached at g. As the illustration indicates, this point of attachment g coincides with a line passing through the pivots f and h, the same as for the pantograph shown at A. If $F =$ the length of the engine stroke and L, the length required for the indicator diagram,

$$F : L = fh : fg, \quad \text{or} \quad \frac{F}{L} = \frac{fh}{fg}.$$

Action of an Adding Mechanism.— The adding mechanism to be described is applied to a machine which is a typewriter and adding machine combined. With this machine debit and credit accounts may be written down indiscriminately; each set of items added, and the total amount printed beneath each vertical column. The writing is done on the typewriter in the regular way, and the figures are set up and printed with the adding mechanism at the same time that the reading matter is written. Two adding mechanisms or "accumulators" are required, one being for the debit and the other for the credit column. This machine may also be used in various other ways. For instance, a list of items may be printed in a series of vertical columns and these columns added to obtain the total amount in both horizontal and vertical directions; finally, these totals, both horizontal and vertical, may be added together to obtain the grand total. Discounts may also be reckoned, amounts may be subtracted from each other, and many other operations performed in connection with commercial work.

The adding keyboard is composed of nine vertical rows of nine keys each. The lower key of each row is numbered one, the next two, and so on up to nine. Of the nine vertical rows, the first on the right is for units, the next, for tens, etc., or, since the reckoning is usually in dollars and cents, the first row is for cents, the next, for dimes, and the succeeding rows, for dollars. Figs. 18 and 19 show diagrammatical sections through the machine along the line of any one of the vertical rows of adding keys, which are shown at *G*. Other important parts of the mechanism are the rack *A*, the type sector *F*, by means of which the numbers are printed on paper carried by roller *K*, and the accumulator wheels *B*, by which the addition is performed. These parts, as well as the other moving mechanism shown, are duplicated for each one of the nine rows of keys, there being nine racks, nine type bars, nine sets of accumulator wheels, etc., in all.

The adding mechanism is operated by the movement of rack *A*. This movement takes place under the influence of spring *O* whenever stop *N* is swung back as shown in Fig. 19 by the operation of the handle of the machine. The length of the movement which spring *O* thus gives to rack *A* is determined by keys *G*. If the figure $476.34 is set up on the keyboard, for instance, key "4" will be depressed in the cents column, and, when the movement of the rack takes place, the rack teeth beneath the accumulator wheel will have moved four spaces.

This is more clearly seen in Fig. 18 where each of the keys *G* is shown to be mounted on a stem which carries a stop *H* at the lower end. When the keys are depressed, these stops come into line with corresponding steps formed at the left-hand end of the rack *A*. These steps are so proportioned that when key 1 is depressed, for instance, the rack is allowed to move one tooth before striking its abutment. When key 2 is depressed, it moves two teeth and so on, as shown by the numbered arrows at the lower part of the illustration. When no key is depressed, indicating zero, then a stop is interposed which prevents any movement of the rack. When key 9 is depressed, the rack takes the full movement of nine teeth allowed by the striking of the projections on the under side of the rack against supporting bar *J*.

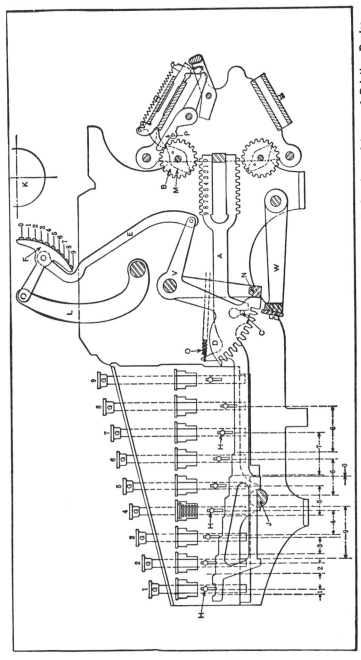

Fig. 18.　Diagrammatical Cross-section of Adding Mechanism Showing Relation of Keys, Racks, Accumulators, and Printing Device

Each rack A has cut in it a slot engaging pin C in sector D. Each sector is, in turn, connected by link E with the type bar F having numbers from 0 to 9. Whenever a key (key 4, for instance) is depressed as shown in Fig. 19, and the rack is allowed to move four teeth backward under the influence of spring O, the type bar F is thereby set at the corresponding figure. The throwing forward of lever L to which the type bar is pivoted then prints this figure "4" on paper wrapped about roll K. It is important to remember that rack A and type bar F are positively connected under all conditions. It should, perhaps, be mentioned that the teeth in sector D simply provide for more accurate alignment of the type in printing than would otherwise be possible. Just before the printing stroke takes place, arm W swings up, carrying a plate which enters the corresponding tooth space in each one of the nine sectors D, aligning all the figures on type bars F and giving a good, evenly printed number on paper.

The Accumulator Mechanism. — The accumulator mechanism, by means of which the adding is done on the machine previously referred to, will now be described. There are ordinarily nine accumulator wheels for each of the nine racks. This particular machine, however, has two sets of nine wheels each, one set being above rack A (see Fig. 19), and the other below it. The upper one is the debit accumulator for addition in the debit column and the other is the credit accumulator for the credit column. Only the upper or debit accumulator will now be considered. This set of nine accumulator wheels, of which only one is shown at B, may be swung into and out of engagement with the teeth of racks A, at will. These accumulator wheels have 20 teeth each; they could have ten, except for the fact that it would make them inconveniently small. Each wheel is provided with a two-tooth ratchet M positively pinned to it. This ratchet spans ten of the wheel teeth between its points. Pawl P is adapted to engage the teeth of ratchet M, and is connected with the mechanism by means of which the tens are carried from one column to another (that is, from one accumulator wheel to another) as

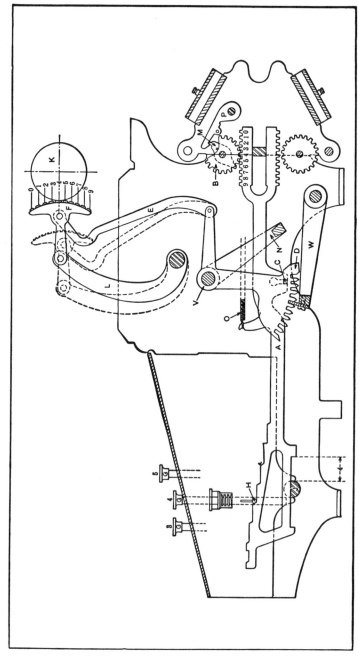

Fig. 19. Adding Mechanism Arranged for Printing the Number 4 and Adding it into the Accumulator

will be described in connection with diagrams Figs. 20 and 21.

Order of Operations for Adding. — Figs. 20 and 21 show, in diagrammatical form, the method of procedure followed in the simple problem of adding 4 to 9, and obtaining the sum 13. At *A,* Fig. 20, the machine is shown "clear," that is, with the accumulator wheels at zero, which means that one tooth

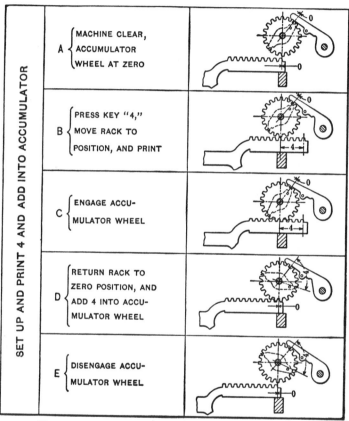

Fig. 20. Diagrams Illustrating Action of Adding Mechanism

of the two-tooth ratchet is up against the hook of the pawl. Key 4, corresponding to the number to be added, in this case, into the accumulator wheel, is now depressed and the operating handle of the machine is pulled over. The first thing that takes place is that the rack is allowed to move four teeth

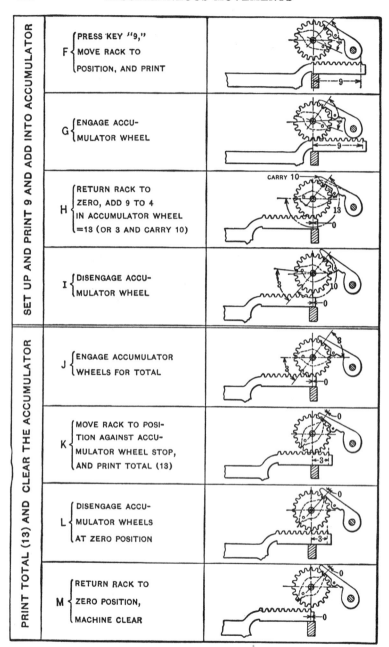

Fig. 21. Continuation of Diagrams Illustrating Operation of Adding Mechanism

to the right, as shown at *B* (see also Fig. 19). In this position, the number "4" is printed. Next (as shown at *C* in Fig. 20) the mechanism automatically throws the accumulator wheel down into engagement with the rack. Then as the operator allows the handle to return, the rack moves back to the zero position again as shown at *D,* carrying the accumulator wheel with it a space of four teeth from its zero position. The mechanism then disengages the accumulator wheel, leaving the machine ready for the next operation with the 4 added into the accumulator, as shown at *E.*

To add 9 to the 4, key 9 is depressed and the operator pulls the handle. This results in a movement of nine teeth of the rack as shown at *F* in Fig. 21. The figure 9 is then printed. The accumulator wheel is next engaged, as at *G.* Then the rack is returned to the zero position as at *H,* and the accumulator wheel is disengaged as at *I.* This evidently moves the accumulator wheel $9 + 4 = 13$ teeth as shown at *H.* In doing this, one of the teeth of the two-tooth ratchet lifts the pawl as it passes under it. This raising of the pawl operates a spring-loaded mechanism, which shifts the next accumulator wheel (that for the tens column) one tooth, when the wheels are returned from engagement in operation *I.* This operation corresponds to that of "carrying" when adding with pencil and paper, except that it is done automatically. This carrying mechanism will not be described in detail as the parts are small and rather complicated, although the action is simple. The mechanism may be understood more clearly by considering the actions of the wheels when every one of them in the accumulator, from cents up to the millions of dollars, is set at 9 — that is, when they are set up for 9,999,999.99. Now suppose that one cent is added, so that the first wheel is moved beyond 9 — that is, to 0. The tooth of the ratchet *M* will then pass under the first pawl, raising it. When the accumulator wheels return from engagement, this raising of the first pawl releases a spring-loaded mechanism which moves the next wheel from 9 to 0. This, in turn, moves the next wheel from 9 to 0 and so on until each one of the row has been advanced one tooth, setting the whole row at 0,000,000.00. This operation is done so rapidly that one cannot distinguish between the successive operations, but each one is dependent upon the preceding one. The operations required for finding a total are shown at J, K, L, and M, Fig. 21. The first thing the operator does is to depress the "debit total" key at the left of the keyboard, the sum having been added into the upper or debit accumulator. He then pulls the operating handle, and the accumulator wheels are engaged with the racks as shown at J. The next operation is the release of the racks so that the springs move them toward the right. There are, in this case, no keys depressed in the keyboard, so that the racks would move the full distance of nine teeth, were it not for the fact that they have to carry the accumulator wheels with them, and the ratchets on these wheels come in contact with the pawls, thus arresting their movement and stopping the movement of the racks.

The previous operation of adding 9 to the 4 in the wheel set the "units wheel" three teeth beyond the point of the ratchet, and the "tens wheel," one tooth beyond the point of the ratchet. It is evident, then, that in operation K the units rack will be allowed to move three teeth and the tens rack one tooth. This will evidently set up the unit type bar at "3" and the tens type bar at "1." On the return of the handle, the printing mechanism is operated, transferring the total "13" to the paper. The accumulator wheel will then be released, and the rack will be allowed to return to the zero position as shown at M. This leaves all the accumulator wheels back in the zero position, with the teeth of the ratchets back against the pawls, leaving the machine "clear" and ready for the next operation.

It might have been desired to print a sub-total instead of a total; that is, a total for the addition as far as it had proceeded, but not to clear the machine, thus permitting more figures to be set up and printed and added into the same sum. Sub-totals can be printed at any point in the adding up of a line of figures, as required, by a simple change in the operation shown at *J, K, L,* and *M* in Fig. 21. This consists simply in allowing the wheels to remain in engagement at *L,* so that the racks, when they return in operation *M,* will bring the wheels to the same position as they had in *J,* thus leaving the totals still set up in the accumulator. Since there are two independent accumulators, it is evident that a number can be added into either one or both of them; or a total or sub-total can be taken from one of them and added into the other — all depending upon the manipulation of the keys and the time of throwing the accumulator wheels into and out of action.

This adding machine has what are known as "controlling keys." These are named "non-add," "debit add," "debit sub-total," "debit total," "credit add," "credit sub-total," "credit total," "repeat," and "error." The pressing down of the non-adding key permits the printing of a number without adding. In other words, this keeps the accumulators permanently out of engagement with the racks. The debit and credit add keys permit a number to be printed and added into the corresponding accumulator, even though the carriage is not set in the proper position for that accumulator. The use of these keys, therefore, gives a flexibility to the machine which is necessary for special operations such as horizontal adding. The debit and credit sub-total keys take and print a total from either the debit or credit accumulators without clearing the accumulators. The debit and credit total keys, on the other hand, take the total from either the debit or credit accumulators, as the case may be, and clear the accumulator after the total is printed. The pressing down of the repeat key holds in the downward position whichever of the number keys have been depressed, allowing the same number to be repeatedly printed and added as many times as the operating handle is pulled. This is useful in multiplying by repeated additions and for other similar uses. The pressing of the error key will release every other key on the keyboard, both of the number keys and of the operating keys as well.

The keyboard is provided with an interlocking mechanism

connected with the controlling keys of the machine and with the operating lever. This mechanism, among other things, prevents the keys from being pressed down or changed after the operating lever movement is started. The keyboard also has a connection with an error key, the pressing of which releases all the keys that may be depressed at the time. Means are also provided for automatically releasing and returning the keys after each operation.

Accumulator Controlling Mechanism. — The engagement of the accumulators with the racks, and their release, in the

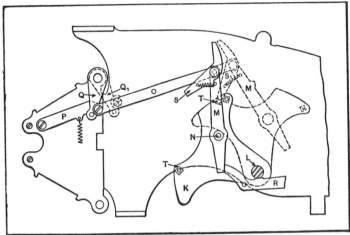

Fig. 22. Flying Lever Connection Between Operating Shaft and Accumulators of Adding Mechanism

operation of the adding mechanism previously described, is effected as follows: The sector K (see Fig. 22) is directly connected with the operating shaft L controlled by the operating handle. It is provided with connections with both accumulators, although this illustration only shows the connections with the debit accumulator. Flying lever M is connected with the debit accumulator by means of links O_1 and bellcrank Q. Member P is simply a spring detent to locate Q for either the engaged or disengaged position of the accumulator wheels.

As sector K starts on its stroke toward the dotted position, flying lever M is carried with it, owing to the resistance which the end of the latter meets with against abutment R. When K has gone far enough so that the end of the lever has dropped off R, the lever M becomes free. The movement has been sufficient, however, to move accumulator lever Q to position Q_1 which throws the wheels into engagement. If it had been desired to throw the wheels into engagement at the end of the stroke instead of at the beginning, detent R would have been withdrawn from the position shown, leaving flying lever M free. Near the end of the stroke of K, however, the end of the pawl S would have struck stud T, making M and K solid, for all practical purposes, and moving Q to the position Q_1

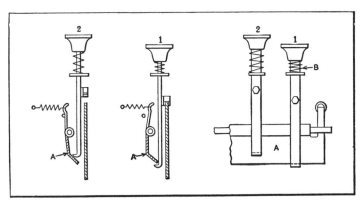

Fig. 23. Simple Arrangement for Holding in the Downward Position Only One Key at a Time in a Row of Adding Machine Keys

at the end of the stroke. If it had been desired to keep the accumulator wheels out of engagement altogether, R would have been lowered out of the position shown, and S would have been moved to a position clear of stud T. Then flying lever M would have been entirely free of K, and no movement of Q would have taken place. The provisions for throwing the accumulator out of engagement at either the commencement or end of the return stroke are similar to those just described.

Adding Machine Key Control. — The keyboard of an adding machine is said to be "flexible" when so arranged that, if a key has been depressed, it will stay down, but the pressing

down of another key in the same vertical column will release
the first key.   With this arrangement, if an attempt were
made to depress two keys successively, the releasing of one
by the downward action of the other would eliminate a possible error.   As a further advantage, if the wrong key were
pressed, the depression of the right one restores the wrong
one to its normal position.   The simple, but ingenious, device
for controlling the action of the keys on one of the commercial adding machines is illustrated in Fig. 23.   If key No. 1
is depressed, the lower hooked end of the stem on which it
is mounted springs past the end of a long pivoted strip *A*
that extends throughout the entire length of the vertical row
of keys.   The result is that the key is held in the downward
position by this hooked end until some other key is depressed.
For instance, if the operator presses down on key No. 2, this
will swing the strip *A* about its pivot to allow the hooked end
of the stem to pass, and this movement of strip *A* releases the
hooked end of key No. 1 which immediately is forced upward
to its normal position by a spring *B*.   In the same manner, any
key which may be pressed down will throw back the strip and
release any other key which may at the time be depressed.

CHAPTER XV

HYDRAULIC TRANSMISSIONS FOR MACHINE TOOLS

WITHIN recent years many standard machine tools have been designed with hydraulic feed mechanisms built in, as a part of the machine. Among the first tools to be so equipped were broaching machines. The production capacity, flexibility of control, and low maintenance cost of these hydraulically operated tools attracted the attention of many machine tool builders and users. Later, grinding machines and drilling machines were successfully equipped with hydraulic feeds. Following this, several lathe and chucking machine manufacturers and builders of milling machines began the development of hydraulically equipped machines.

In practically all new applications of hydraulic transmission to machine tools, no accumulators are employed. Thus, the new system of feeding or driving consists essentially of an oil-pump and a cylinder having a piston driven by oil circulated by the pump and controlled by piping and valve equipment, to give the piston any movement required for feeding or driving. Where hydraulic rotary drives have been applied, a motor similar in construction to the rotary oil circulating or driving pump takes the place of the cylinder and piston arrangement.

Although the basic principle of hydraulic operation of feeds and drives appears simple, the actual development of a practical system involves considerable engineering. The requirements of one machine may be met by comparatively simple equipment, whereas the hydraulic operation of another machine may require a system of piping, specially designed control valves, two or more driving pumps and hydraulically driven devices of various kinds.

Advantages of Hydraulic Operation. — The following is a brief resumé of the principal advantages claimed for hydraulic, as compared with mechanical, operation of machine tool feeds and drives.

1. Straight line or rotary transmission of power at any desired point of application.

2. Higher cutting speeds.

3. Longer life of cutting tools.

4. Greater flexibility of speed control.

5. Quick reversal of feed, with practically no shock.

6. Simple and efficient control, both hand and automatic, of all rapid traversing, feeding, and reversing movements.

7. Quiet operation.

8. Low power consumption, which varies automatically to meet resistance offered to cutting tool or driven member.

9. Safety insured by relief valves, which can be set to stop the feeding movement at any predetermined pressure.

10. Ability to "stall" against obstruction, thus protecting parts against breakage and providing an ideal method of cutting shoulders to exact positions and facing to length, by using positive stops.

11. "Slip," which permits movement to slow up when tool is overloaded without "windup" of mechanical feed gear.

12. Fewer moving parts.

13. Provision for checking and comparing action or condition of cutting tools by pressure gage, which indicates cutting force.

14. Adaptability for operating auxiliary devices, such as work-holding clamps, clutches, diamond dressing tool, indexing pins, etc.

15. Comparatively simple centralized control over one or any number of hydraulically operated units arranged or located according to requirements.

16. Greater flexibility of design which permits feed to be adjusted to most efficient rate after machine is assembled.

17. Reliability and low up-keep cost due to comparatively few wearing parts.

Two Fundamental Circuits Employed.— The designs of practically all hydraulic feeds and drives recently applied to machine tools have been based on one of the two fundamental hydraulic circuits shown diagrammatically in Figs. 1 and 2, or a combination of these two circuits. In both circuits, the pistons *A,* which impart the required feeding movements, are driven by a liquid delivered from their respective pumps *B.* The pressure is low if the piston encounters no resistance, and the distance it is moved corresponds to the discharge rate of the pump, whether the resistance be high or low.

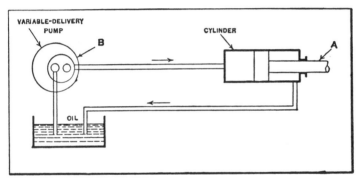

Fig. 1. Hydraulic Feed with Closed Circuit

In the case of the circuit shown in Fig. 1, the rate of feed is changed by varying the volume of liquid delivered by the pump *B.* With the circuit shown in Fig. 2, the feed is controlled by opening or closing the choke valve *C,* causing more or less of the liquid from the constant-volume pump *B* to be by-passed by valve *D,* but admitting liquid to cylinder *E* at the rate necessary to give the required feed.

Pressures Used for Feeds and Drives.— Both high- and low-pressure systems are used, depending on the type of machine and its requirements. For instance, the reciprocating tables of light weight internal grinders may be driven by constant-displacement low-pressure gear pumps, as comparatively little pressure is required to move the tables of such machines. On the other hand, a certain car-wheel boring machine is equipped with four variable-delivery pumps, each having a capacity of 3060 cubic inches per minute at a pressure of 1000 pounds per square inch. Two of these pumps are used for feeding the two boring heads, which have 9-inch feeding cylinders. This equipment gives an available feeding force of 60,000 pounds on each carriage. The other two pumps are used for operating the mechanism for chucking the car wheels.

Variable-delivery Pumps Arranged with Closed Hydraulic Circuits. — For simplicity, Fig. 1 shows the pump pushing

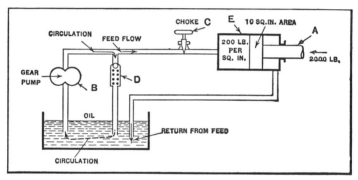

Fig. 2. Hydraulic Feed with By-passed Circuit

the piston outward, a valve which must be used to reverse the flow in the discharge and return pipes on the in stroke being omitted in the diagram. The complete arrangement gives a definite rate of feed proportional to the metered discharge of the pump less leakage from the closed pressure side of the circuit through the pump pistons and the feed piston. The rate at which the piston moves will never exceed the rate corresponding to the fixed discharge of the pump, and greater or less resistance to the movement of the piston will only raise or lower the pressure in the cylinder and connecting pipe to the pump without materially changing the speed at which the piston moves.

Perhaps the most outstanding advantage claimed for this system of hydraulic transmission of power is the sensitive and easy control of the speed of the driven mechanism. If we pass the entire output of the pumping unit to the driven unit, the speed of the latter may be regulated by varying either the speed or displacement of the pumping unit. This constitutes a very efficient means of control, and one that is limited only by the mechanism involved.

The by-passed circuit illustrated diagrammatically in Fig. 2 shows the gear pump B pushing the piston A at a rate corresponding to only a fraction of the displacement of the gear pump. The excess displacement escapes through a relief valve D into the oil-pot, and is again taken up by the gear-pump suction pipe and continuously circulated.

Types of Pumps Used in Feeding Machine Tools. — At present there are three types of pumps in general use for feeding machine tools:

1. An accurately made gear pump capable of delivering a constant volume of oil at a constant pressure. These pumps are usually arranged to deliver oil at pressures up to 250 pounds per square inch. A relief valve is used in connection with this type of pump for maintaining an even pressure.

2. A multiple-piston pump with variable stroke. This type of pump is built to deliver a variable amount of oil at pressures up to 1000 pounds per square inch.

3. A pump which combines the first and second types and is arranged to deliver a large volume of oil at about 250 pounds pressure from a gear pump, and a smaller volume at a higher pressure from a variable-delivery piston pump. Both pumps are built into one housing and interlocked as to control.

There is nothing unusual about the gear pumps employed, which are simply required to deliver the necessary volume of oil at a constant pressure. The variable-stroke pumps or units, such as the Oilgear automatic variable-delivery pump are necessarily more complicated. This pump was designed for use in equipping the smaller sizes of milling, boring, drilling, and similar machines with hydraulic feeds. The following specifications for this pump may be of interest to designers:

Forward and reverse feeds and forward and reverse rapid traverse are provided. Either hand or automatic control may be employed. Pipe connections with the gear pump are provided for operating fixtures or other auxiliary equipment. Different positions of the control valve give full speed (rapid approach), feed forward, neutral, feed reverse and full speed reverse (rapid return). When used with a 3⅞-inch cylinder, this pump has a feeding range of from 1.66 to 23 inches per minute and a reverse feed of from 3.32 to 46 inches per minute. The rapid traverse speed in either direction is 93 inches per minute. The maximum working pressure is 1000 pounds per square inch and the power consumption at maximum capacity is 2 horsepower. The drive shaft speed is 860 revolutions per minute or lower. The pump is about 19½ inches high.

Rotary Drives for Long Strokes and Rotary Tables. — When very long table strokes are required, as in the case of some types of grinding machines, or when the table has a rotary movement, as in some milling machines, it is desirable to use a rotary motor in place of the feed cylinder. The working parts of a motor of this type are identical with those of the corresponding variable-stroke pump. The displacement of one such motor is 4.6 cubic inches per revolution; and the maximum torque, 690 inch-pounds at 1000 pounds per square inch. The maximum speed is 860 revolutions per minute, and the output at this speed, 9.4 horsepower.

Operating Multiple Feeds. — To have complete individual speed control of two or more hydraulic cylinders or motors, each cylinder must be driven by its own pump and the entire flow from the pump must go through that cylinder. However, drilling machines of the multiple-spindle type having several feed cylinders operated simultaneously by oil supplied by a single gear pump have proved practical. If sufficient oil is pumped at all times, so that under the worst condition of usage there is still a slight amount being by-passed through the relief valve, the rate of movement of the piston of any one cylinder can be changed without any material change in the rate of travel of the pistons in the other cylinders.

If the volume of oil delivered is not sufficient to maintain the pressure adjusted by the by-pass valve in the case of two or more motors or cylinders operated in parallel, the motor or cylinder encountering the least resistance may take the entire flow until its stroke or work is finished or stopped by closing a valve. The remaining units will operate successively according to the order of their resistance values. The total time required for all the cylinders to perform this work under this condition will be the same as though they operated simultaneously, assuming, of course, that the volume of oil delivered by the driving pump is the same. By employing a variable-delivery pump for changing the volume of oil flow in a system of this kind, the operator can obtain any desired feed for each cylinder as it comes into operation.

Cylinders Operated in Series. — If the speed control required on two cylinders is simultaneous and proportional, the two cylinders may be placed in series in a closed circuit with a single pump. Such a circuit with the three pumps in series is shown diagrammatically in Fig. 3. This circuit is applicable to a drilling machine equipped with three drilling heads.

The movement of such heads may be coordinated mechanically by racks and pinions or by linkage, but the hydraulic method is more flexible, less liable to damage through breakage, and in some cases, cheaper. The three cylinders must be graduated to the speed requirements of the heads. If one head is to move twice as fast as another, its cylinder must have one-half the volume. Also, each cylinder must be so designed that the volume displaced in its piston-rod end is equal to the volume displaced in the head end of the next succeeding cylinder. This is evident from the diagram, as the oil supplied to the head end of each cylinder after the first one, comes from the rod end of the preceding cylinder.

In order to keep the movement of such a set of pistons properly coordinated, the pistons must run against their cylinder heads at the termination of each cycle, so that they always start the next cycle in the same relation. The relief valves

A, B, and *C,* permit oil to pass around any piston that has stalled against its cylinder head during the back stroke, thus bringing all the pistons successively back against their cylinder heads.

Use of Multiple Transmitter. — The series circuit of Fig. 3 divides the total working pressure into as many parts as there are cylinders. This reduces the maximum working pressure available in each cylinder, and tends toward large cylinder diameters. For this and other reasons, it is sometimes better to use a multiple transmitter, consisting of one double-acting

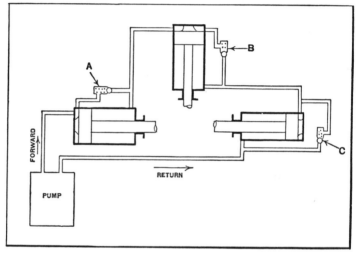

Fig. 3. Diagram of Cylinders Operated by Series Circuit

cylinder reciprocated by the pump and operating several cylinders whose piston-rods are attached to a single cross-head. Each of these secondary cylinders acts as a pump or impeller for its own individual driven or feeding cylinder. This system establishes several separate closed hydraulic circuits, each of which operates its feeding cylinder at definite speeds, the pressure in each separate circuit depending on the resistance against the piston-rods of the respective feeding cylinders and on the corresponding piston areas.

The strokes of all the impelling cylinders are the same, but their diameters and the diameters of the feeding cylinders may be varied to give any desired feeding forces and strokes to the respective feeding cylinders, provided the totals are within the power capacity of the pump. All speed variations and distances traveled by the pistons of the respective feeding cylinders are proportional to the speeds and distances traveled by the piston of the main cylinder connected to the pump.

In this case also, it is necessary to provide means similar to those shown in Fig. 3 for bringing each of the feeding cylinders against this cylinder head at the end of every cycle

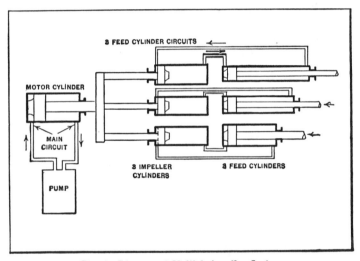

Fig. 4. Diagram of Multiple-Impeller System

to keep the pistons in coordination. A system of this kind is indicated in Fig. 4, only the principal circuits being shown. In practice, the circuits of both Figs. 3 and 4 require low-pressure make-up lines from the pump to each circuit, and other details, which are omitted for the sake of clearness.

Slip and its Effect. — A tool-holder fed by an oil-pressure piston and a volumetric pump cannot be used to chase a thread, because its rate of feed is not absolutely constant. There is always a certain amount of leakage (across the bridges and through the plunger fits) in the pump, and this leakage is greater as the cut becomes heavier and the oil pressure rises. If the feed is low, in inches per minute, this leakage may be a considerable percentage of the pump delivery. This is the "slip," and in the early designs of hydraulically fed tools, was generally assumed to be a defect. Geared feeds do not slip, and the tool must cut the given thickness of chip, whether the material is hard or soft, the cut deep or shallow. If the tool manages to back off slightly due to "windup" in the rods and gearing, the lost travel must be made up, and the average thickness of chip throughout the cut must be equal to the geared feed rate.

If the tool is fed by oil from a volumetric pump, it will never move faster than the nominal feed rate, but higher cutting pressures will cause the feed to slow up. The travel lost by this increased slippage is never made up. Hence the tool is not damaged by being forced to maintain the given feed rate, as may be the case when overload causes winding up of the mechanical feed gear.

The rate of leakage in a hydraulic feeding system depends upon the pressure, and the pressure is directly caused by the resistance encountered by the tool. Consequently the flow of oil actually delivered by the pump into the feeding cylinder is less as the resistance increases. In actual practice, this reduction of feed with increasing pressure may be a very significant fraction of the theoretical feed rate, especially with heavy cuts at slow feeds.

For instance, a standard 3⅞-inch diameter feed cylinder has a piston area of 11.8 square inches and can deliver a net feeding force of 11,800 pounds to a cutting tool. When working at this maximum pressure of 1000 pounds per square inch, the slip of the entire apparatus would quite likely amount to 15 cubic inches of oil per minute. In ordinary cuts, such a feeding cylinder usually operates at pressures of 250 or 300 pounds per square inch, and the slip is, say, 5 cubic inches per minute.

Therefore, if the feed were adjusted to give 4 inches per minute under 250 pounds pressure, the additional slip of 10 cubic inches of oil as the pressure rises to nearly 1000 pounds would reduce the rate of feed by nearly ⅞ inch per minute, leaving a net feed of about 3⅛ inches per minute during the excessively heavy cut. This amounts to a 20 per cent reduction in the 4-inch per minute rate of feed. If the feed rate were set at 16 inches per minute, the reduction would be 5 per cent, as the amount of slip is practically constant for given pressures. These rates are based on a type of pump having relatively large leakage through a distributing valve. Other types would show about one-half as much slippage.

Speed-changing Hydraulic Transmission.— The hydraulic transmission illustrated by Figs. 5 and 6 is so designed that the speed of the driven pulley may be varied from zero up to the full speed of the driving pulley, so that this mechanism may be utilized as a clutch or for changing speeds. This transmission is intended for general application.

The driving pulley A on shaft B (Fig. 5) revolves gear C and two idler gears D and E (see Fig. 6), These idler gears are housed in case F to which the driven pulley G is attached. The gears referred to act as a pump. and circulate oil through ports H, J, K, and L (as indicated by the arrows), provided the ports in the cylindrical or plug valves M and N are open.

If valves M and N are fully open, the gears will rotate freely, because the oil can circulate through the passageways without resistance; consequently, the driven member and its pulley will remain stationary. If, however, the valves M and N are closed gradually, there will be a corresponding increase in resistance to the rotation of the gearing, and as a result, the driven member will rotate at a rate of speed depending upon the amount of resistance. When valves M and N are completely closed, all rotation of the gears is prevented, and the driving and driven members rotate at the same speed. The transmission then acts like a clutch in engagement, whereas when valves M and N are fully open and the driven member is stationary, the action is similar to a clutch that has been disengaged. Thus it will be seen that the gears revolve as a unit only when the valves are fully closed, and they rotate

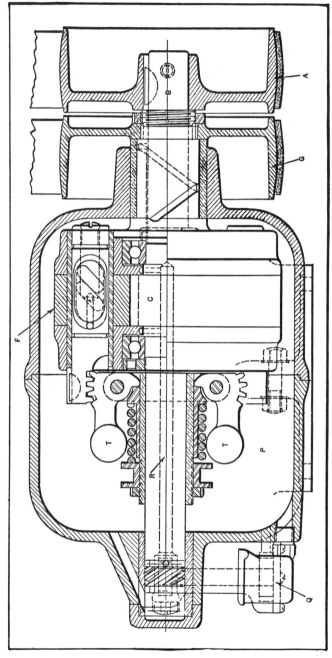

Fig. 5. Hydraulically Controlled Transmission Which May be Used Either as a Clutch or for Varying Speeds

about their axes when the valves are partially or entirely open for the purpose either of varying the speed or discontinuing the drive entirely.

The main supply of oil is in the main casing at P (Fig. 5)

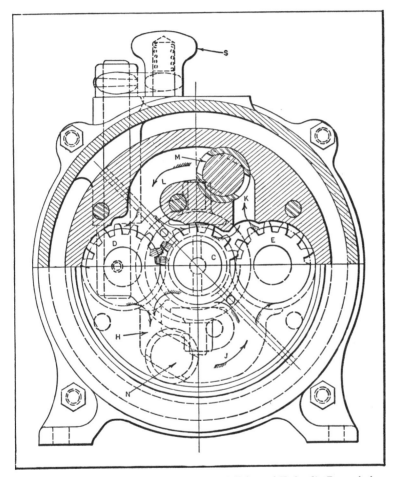

Fig. 6. Cross-section, Showing Gears and Control Valves of Hydraulic Transmission

and a small pump at Q, driven through spiral gearing from the main shaft, forces the oil through a central opening R in this shaft. Lever S (Fig. 6) serves to control the positions of valves M and N and the speed variations. Any variation in the speed for which the mechanism is set, caused by changes in load, is regulated by the centrifugal governor T (Fig. 5).

In order to relieve the oil pressure at the points where the teeth of gears C, D, and E intermesh, small radial holes are drilled through these teeth and connect with diagonal holes leading to the spaces between the teeth, thus relieving the oil pressure and lessening friction. This transmission is also designed to provide reversal by special arrangement of gearing connection with the driven member. The hydraulic feature of the transmission, however, is the same as described.

CHAPTER XVI

AUTOMATIC FEEDING MECHANISMS

MACHINES which operate on large numbers of duplicate parts which are separate or in the form of individual pieces are often equipped with a mechanism for automatically transferring the parts from a magazine or other retaining device, to the tools that perform the necessary operations. The magazine used in conjunction with mechanisms of this kind is arranged for holding enough parts to supply the machine for a certain period, and it is equipped with a mechanical device for removing the parts separately from the magazine and placing them in the correct position wherever the operations are to be performed. The magazine may be in the form of a hopper, or the supply of parts to be operated upon by the machine may be held in some other way. The transfer of the parts from the hopper or main source of supply to the operating tools may be through a chute or passageway leading directly to the tools, or it may be necessary to convey the parts to the tools by an auxiliary transferring mechanism which acts in unison with the magazine feeding attachment. These automatic feeding mechanisms are usually designed especially for handling a certain product, although some types are capable of application to a limited range of work. The feeding mechanisms described in the following include designs which differ considerably, and illustrate, in a general way, the possibilities of automatic devices of this kind.

Attachments having Inclined Chutes.— One of the important applications of magazine feeding attachments is in connection with the automatic screw machine. Most of the parts made on these machines are produced directly from bars of stock, but secondary operations on separate pieces are some-

447

times necessary, and then an automatic or semi-automatic attachment may be employed to transfer the parts successively to the machine chuck where the tools can operate upon them. Many of these attachments have magazines which are in the form of an inclined chute that holds the parts in the

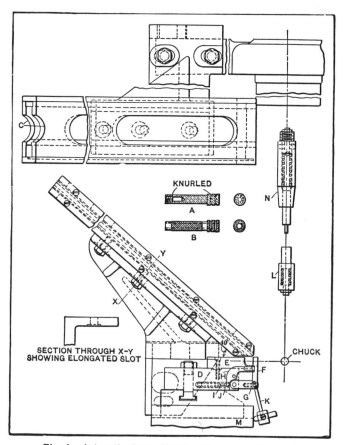

Fig. 1. Automatic Screw Machine Magazine Attachment

correct position and from which they are removed, one at a time, by a transferring device. An example of this type of magazine attachment is shown in Fig. 1. This attachment was designed for feeding the handles of safety razors. The preliminary screw machine operations involve turning, forming, knurling, drilling, tapping, and cutting off the handle, thus producing a piece of the form shown at A in the illustration. These partly finished handles are then placed in the chute or slide of the feeding attachment, from which they are transferred to the chuck, so that a hole can be drilled clear through the handle as indicated at B, and one end of the hole be slightly enlarged. The upper and lower plates C of the chute have grooves milled in them to correspond to the enlarged parts of the handle. As each successive handle reaches the lower end of the chute and drops into the small pocket shown, a spring plunger L attached to the turret advances and pushes the work out into the chuck of the machine.

As the ends of the handles have shoulders, the pocket at the bottom is automatically enlarged to permit the passage of this shoulder. The work-carrier consists principally of two blocks D and E and a finger F. Block D is held in the cross-slide and block E is attached to the top of block D. The forward end of block E is cut out to fit the work, which is held in place by finger F. This finger is fastened to lever G, pivoted on block D, and normally held in position by a pawl H engaged by plunger I and pin J. When a piece of work drops into the pocket in block E and the front cross-slide has advanced far enough to bring the work in line with the hole in the chuck, the enlarged part of the plunger L trips the finger F after the work has been partly inserted in the chuck. This action is caused by the contact of plunger L with a beveled edge on pawl H which disengages the V-shaped end of the pawl from a groove in lever G and, at the same time, pushes back spring plunger I, thus allowing finger F to drop away from block E. The pawl H serves as a locater for the work and, when disconnected from lever G, it swings down and the work is pushed into the chuck by plunger L which is held in the advancing turret. After a piece has been inserted in the chuck, the cross-slide, as it moves outward, brings trip K against casting M which, through the combined action of lever G, pawl H, and spring plunger I, closes the work-carrier. The piece in the chuck is forced in against a spring plunger

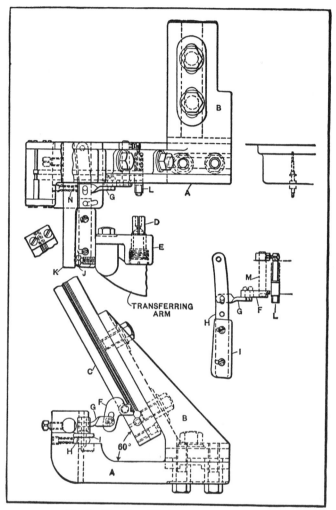

Fig. 2. Magazine Attachment for Pinion Staffs

held by feed finger N. This spring plunger ejects the work when the machining operation has been finished and the chuck is opened.

Feeding Attachment for Pinion Staffs. — The magazine feeding attachment shown in Fig. 2 was designed for handling pinion staffs of the form illustrated by the dotted lines in the upper right-hand corner of the illustration. The chute C is supported by a bracket B which is attached to a boss provided on the automatic screw machine for holding special attachments. The bracket A is attached to B and carries the mechanism for feeding the pinion staffs successively to the place where they can be removed by the transferring arm. The two main parts of the chute are grooved to fit the pinion staffs, so that the latter are held in the correct position. The operation of this attachment is as follows: The chute is filled with pinion staffs and the lower one is held back temporarily by trip F. This trip is connected to link G, which carries a pin that engages a slot cut in lever H (see detailed view). Lever H has fastened to its upper side a trip-lever plate I the inclination of which may be varied. When the transferring arm swings upward, it is stopped in the correct position by set-screw J, which engages stop K, the arm itself bearing against plate I and forcing it back, together with lever H. This action, through connecting link G, operates trip F and allows one piece to drop into the pocket formed at the end of this trip. The transferring arm carrying a split bushing D then advances and pushing back the nest L passes over the end of a pinion staff and grips it. The transferring arm then recedes and swings down to the chuck in which the pinion staff is placed. When the transferring arm descends, the spring N returns trip-lever plate I and lever H to their former position. Trip-lever F also swings back in order to catch another piece, the pinion staff in the trip being deposited in the nest L ready for transferring to the split bushing D the next time the transferring arm ascends.

Magazine Attachment for Narrow Bushings.— The narrow bushings shown at A, Fig. 3, are blanked out and drawn in a die to the shape shown; they are then turned, faced, and threaded (as indicated at B) in an automatic screw machine. Two separate operations are required, but the magazine attachment shown in this illustration is used for both. The bushings are placed in the inclined slide or chute, and the lower one is retained temporarily by a finger i, which is held upward by spring k, the exact position of the finger depending upon the adjustments of set-screw j which engages a projecting end. The transferring arm, which removes the work from the lower end of the chute and conveys it to the chuck, has a swinging or circular movement, as indicated by the dotted line. The work is gripped as the holder (shown in detail at C) advances, and then, as the transfer arm starts to swing downward toward the chuck, the finger i is depressed,

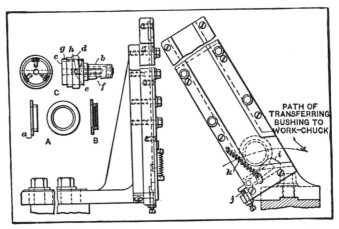

Fig. 3. Magazine Attachment for Handling Parts Shown at A and B

thus allowing the bushing to slide out of the chute. The work-holder has a taper shank b which fits into the main body c. On this body is held a ring d through which a pin is driven. The pin h in this ring d fits into an elongated hole in body c and enters spring plunger e. A slot in body c receives a flat spring g, which is provided to grip the work securely. This spring also compensates for slight variations of diameter.

The degree of inclination for chutes of magazine attachments varies from 20 to 60 degrees and depends upon the size and shape of the work. The chute should incline at a greater angle for small work than for large work. The chutes of attachments used for handling flat pieces, such, for example, as might be cut out in a blanking die, are usually held in a vertical chute instead of one that is inclined.

Hopper Feeding Mechanism for Screw Blanks.— The automatic feeding mechanism to be described is used on a thread rolling machine of the type having straight dies between which the blanks are rolled to form the threads. The faces of the dies are in a vertical position and one die is given a reciprocating motion in a direction at right angles to the axis of the screw blank. The automatic feeding mechanism

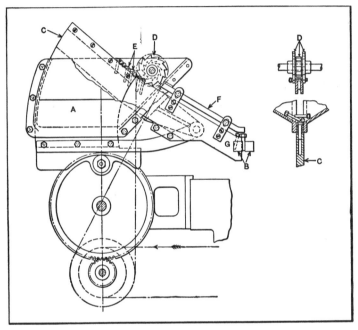

Fig. 4. Hopper-feeding Mechanism for Screw Blanks

shown in Fig. 4 is arranged to transfer the screw blanks from the hopper A to the dies at B in such a way that each successive blank is in a vertical position when caught between the dies. The hopper A, which is at the top of the machine, is equipped with a plate or center-board C which passes through a slot in the bottom of the hopper and is given a reciprocating motion by a gear-driven cam. This center-board has a vertical slot extending along the upper edge (see detail sectional view) which is a little wider than the diameter of the screw blank bodies. As the center-board moves up through the mass of screw blanks, one or more of these blanks are liable to drop into the slot and hang suspended by their heads. If a blank does not happen to be caught for any one stroke of the center-board, the mass of blanks is disturbed and it is likely that one or more blanks will fall into the slot on the next successive stroke of the center-board.

As some blanks are picked up while in a crosswise or other incorrect position, an auxiliary device is employed to dislodge such blanks. This device consists of three revolving wheels at D which have teeth like ratchet wheels. The arrangement of these wheels is shown by the detailed view. The center wheel, which is the smallest, revolves above the heads of the blanks which are moving down the slot of the center-board in the proper position, as indicated at E. The two outer wheels, which are larger than the central one, revolve close to the outer edges of the center-board. If a blank is not in the correct position, it will be caught by these wheels and be thrown back into the hopper, but all blanks that hang in the slot pass between the outer wheels and beneath the central one without being disturbed. After the blanks leave the center-board, they pass down the inclined chute G, which is provided with a guide F that holds them in position. As each successive blank reaches the lower end of the chute, it swings around to a vertical position and is caught between the dies which roll screw threads on the ends.

Feeding Shells with Closed Ends Foremost. — The possibilities of mechanical motion and control are almost boundless, if there is no limit to the number of parts that may be incorporated in a mechanism, but as complication means higher manufacturing cost, and usually greater liability of derangement, the skillful designer tries to accomplish the desired results by the simplest means possible; it is this simplifying process that often requires a high degree of mechanical ingenuity. The feed-chute shown in Fig. 5 illustrates how a very simple device may sometimes be employed to accomplish what might appear at first to be difficult. This is an attachment used in conjunction with an automatic feeding mechanism for drawing shells in a punch-press. These shells are fed from a hopper, and it is essential to have them enter the die with the closed ends down. If a shell descends from the hopper with the open end foremost, it is automatically turned around by the simple device shown. The view to the left illustrates the movements of a shell which comes down in the proper position or with the closed end foremost. In this case, the bottom of the shell simply strikes pin *B* and, after rebounding, drops down through tube *C*. If the open

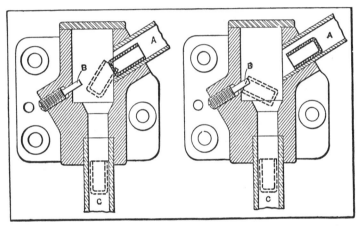

Fig. 5. Simple Attachment of an Automatic Feeding Mechanism for Turning Shells Which Enter Open End Foremost

end of a shell is foremost, as illustrated at the right, it catches on pin *B* and is turned around as the illustration indicates. If a shell enters the die with the closed end upward, the drawing punch will probably be broken.

Feeding Bullets with Pointed Ends Foremost.— An attachment for feeding lead bullets or slugs to press tools with the pointed ends foremost, regardless of the position in which the bullets are received from the magazine or hopper, is illustrated in Fig. 6. This attachment is applied to a press having a 4½-inch stroke. The bullets enter the tube *A* which connects with a hopper located above the press. An "agitator tube" moves up and down through the mass of bullets in the hopper and the bullets which enter the agitator tube drop into tube A. As each bullet reaches the lower end of this tube, it is transferred by slide C (operated by cam D attached to the cross-

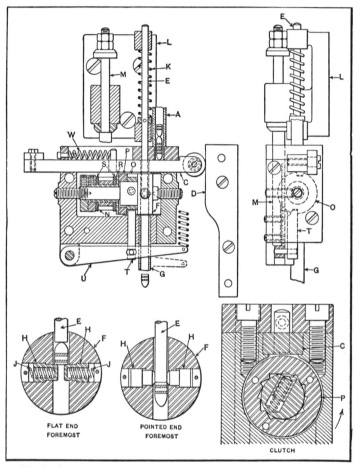

Fig. 6. Attachment for Hopper Feeding Mechanism Which Delivers All Bullets to a Dial Feed Plate with Pointed Ends Foremost

head) to a position under the rod E. The rod-holder L is also carried by the cross-head. Whenever a bullet enters tube A with the rounded or pointed end downward, it is simply pushed through a hole in dial F and into feed-pipe G leading to the dial feed-plate of the press. This feed-plate, in turn, conveys the bullets to the press tools where such operations as swaging or sizing are performed.

The arrangement of dial F is shown by the detailed sectional views at the lower part of the illustration. Whenever a bullet enters the dial with the pointed end foremost, the plungers H are pushed back against the tension of springs J and the bullet drops into the tube beneath. If the blunt or flat end is foremost, the plungers are not forced back, and as rod E is prevented from descending further, it simply moves upward against the tension of spring K as the cross-head continues its downward motion. A mechanism is provided for turning dial F one-half revolution so that every bullet that is not pushed through the dial will be turned around with the pointed end foremost before it drops into the feed-tube G. This rotary motion of the dial is derived from a rack M attached to bracket L, and a pinion N with which the rack meshes. The location of the dial is governed by an index plate O and a plunger T which enters one of the notches in the index plate; the latter is attached to dial F. A clutch P (see also detailed sectional view) is fastened to sleeve R. Fiber friction washers S are used to prevent breakage in case anything unusual should happen.

When the cross-head descends, the rack M revolves the clutch in the direction shown by the arrow. When within' one-quarter inch of the lower end of the stroke (this position is shown in the illustration), the rack M strikes lever U and disengages the index plunger T. The rack descends far enough to give it time on the return stroke to move dial F sufficiently to prevent the returning index plunger from re-entering the hole it just occupied. On the return stroke, the lost motion of the rack in its bracket provides time for the withdrawal of rod E before dial F is revolved. This lost motion can be adjusted so that the highest point of the upward stroke is reached just as dial F has turned 180 degrees, thus bringing the other index slot in line with plunger T. If the rack should move too high, the friction washers S will allow for this excess movement by slipping. This half revolution of dial F turns a bullet that is not pushed through it end for end, so that it drops down in the pipe G with the pointed end foremost. The slide C is returned for receiving another bullet from tube A by the action of spring W which holds the slide roller firmly against the cam-plate D.

Feeding Shells Successively and in Any Position.— A feeding mechanism designed to feed shells or cartridge cases one at a time and in any position is shown in Fig. 7. Owing to the weight of the heads of cartridge cases, they may readily be arranged upon a table heads downward, and the particular mechanism to be described is arranged for changing the shells from a vertical to a horizontal position before dropping them into a trough by means of which they are conveyed to the operating tools. The table A upon which the shells are placed is slightly inclined so that the shells readily slide towards a horizontal disk B which is rotated constantly by a belt and pulley. As the disk revolves, the shells are carried towards the funnel-shaped mouth of a guideway C where there is a wheel D having teeth of irregular form. This wheel is revolved in the same direction as disk B so that it continually pushes back some of the shells and prevents jamming. The shells which move too near the center of disk B to enter the mouth of the guide-way are carried around until they meet the edge of an inclined fence E, which is just above the disk near the center, but is arched near the periphery so that shells can pass under it. This fence causes the shells to move out towards the circumference of disk B, so that they may enter the guide-way as they again come around.

Just beyond the wheel D there is a feed-wheel F which has teeth of regular form that fit between the cartridge cases. This wheel is rotated in the direction shown by the arrow, so as to feed the shells forward at a definite rate along the guide-way C. This guide-way, excepting at the mouth, is only slightly wider than the shell diameter, so that all the shells in it form a continuous and orderly row. The guide-way may be curved gradually in any direction, so that the shells which

enter it with their axes vertical may be turned to any desired
position as they pass along. As previously mentioned, the
guide-way, in this case, changes from a vertical to a horizontal
position. At the end of the guide-way there is a pair of
stops that act alternately to allow one shell to issue at a time
from the guide-way. The first stop consists of a pair of
fingers G which rise up through the floor of the guide, and the

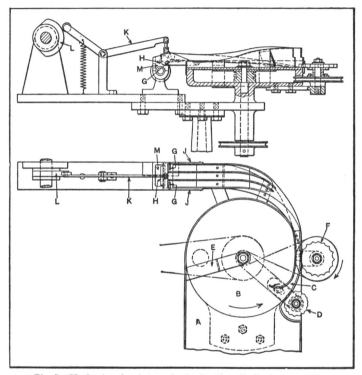

Fig. 7. Mechanism for Automatically Feeding Shells One at a Time

second stop is in the form of a gate H which moves down in
front of the foremost shell of the row. These two stops are
carried on a pivoted frame J so arranged that, as the gate rises
to allow the foremost shell to pass from the mouth of the
tube, the fingers G rise in front of the second shell to hold
back the whole row. The frame J is connected with a lever
K which is intermittently rocked by the cam L. The succes-

sive shells drop into the trough M as they are discharged from
the guide-way.

Feeding Shells Successively and Gaging the Diameters. —
The mechanism described in the following is part of a cart-
ridge-making machine, and its function is to feed cartridge
cases or shells from a tube, one at a time, and provide means
of detecting shells having heads that are over the standard
diameter. The shells are placed heads downward onto a fixed

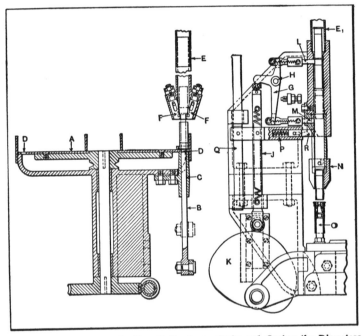

Fig. 8. Mechanism for Feeding Shells Successively and Gaging the Diameters

table from which they are pushed by hand onto a revolving
disk A, Fig. 8. This feed disk operates on the same general
principle as the one illustrated in Fig. 7. As each successive
shell passes from the guide-way of the revolving disk, it is
placed directly over a push-rod B. This push-rod is pivoted
to the end of a lever which is oscillated by a cam, thus causing
the push-rod to move vertically through a guide C and through
one of the slots D formed in the periphery of the feed disk A.

Each time the push-rod B moves upward, it pushes a shell into the end of tube E. This tube has two gravity fingers F and, as the shell rises, its rim lifts these fingers and separates them far enough to allow the rim to pass; the fingers then drop back behind the rim and prevent the shell from falling when the push-rod recedes. When this push-rod makes the next successive stroke, the shell lifted by it pushes the first shell up into tube E which is bent over to form an arch and terminates at E_1.

When the vertical section of tube E is filled and the shells passed over the top of the arch, they fall open end first down into the vertical section E_1. Just below the end of tube E_1, there is a device for releasing the shells one at a time. This consists of a three-armed lever G, which is pivoted at H and is given an oscillating or rocking movement by vertical rod J having a roller in contact with cam K, against which the rod is held by a spring. As lever G oscillates, it withdraws, alternately, two fingers L and M which project into the passageway for the shells. These fingers are withdrawn against the tension of suitable springs and the upper one catches the cartridge shells by the rim, whereas the other one extends beneath the open end. When the upper finger is withdrawn, a shell drops against the lower finger and, when the latter is withdrawn, this shell is released and, at the same time, the upper finger moves in and prevents the next successive shell from dropping out until it is released by the backward motion of finger L. As each successive shell drops, it passes through a gage N and then falls over one of the vertical pins O, which are equally spaced around the periphery of the machine table. This table is revolved intermittently in order to locate the shells beneath a series of tools carried by a tool-holder having a vertical reciprocating motion.

Attached to the rod J, there is a bar P the movements of which are steadied by a bar Q mounted in suitable guides. The bar P carries a spring plunger R having a beveled end which engages a beveled surface as shown; consequently, as rod J and bar P are lifted by cam K, plunger R is pushed back far enough to clear the rim of the descending cartridge. When rod J descends, however, plunger R moves inward and bears downward on the head of the cartridge beneath it, thus pushing it through the gage N and onto one of the series of pins O. If the rim of a cartridge should be so large that it would not readily pass through the gage, the resistance overcomes the tension of the spring that holds J into contact with the cam, and the cartridge remains in the gage until the next stroke of the machine. As the table moves around, the attendant will notice that there is a pin without a shell upon it and, therefore, he will remove the next successive shell, because, ordinarily, the shells are not so large as to resist being forced through the gage by a second stroke of the push-down bar P. If an exceptionally large head will not pass through the gage, the machine must be stopped and the shell removed by hand.

Feeding Mechanism for Taper Rolls.— The device here described was designed for taking taper rolls, of the kind used in roller bearings, from a hopper, selecting these rolls, and feeding them small end first into a centerless grinding machine. The hopper used is of the type generally applied to thread-rolling machines. The center board is arranged with a V-groove in place of the usual slot, and the rolls are picked up and allowed to slide down into the selecting mechanism lengthwise.

The mechanism is shown depositing a roll R (see Fig. 9) into the fixture ready for the feed-bar K to come back and pick it up. The feed-bar is actuated by a cam on the opposite end of the machine (not shown in the illustration). The body A of the fixture is fastened to the hopper by a bracket. Sliding on body A are two plates B and C, retained by gibs. These plates are moved inward by pawls D and outward by compression springs E. Pawls D are oscillated about pivot pins F by the action of the cam surfaces G on slides H which are dovetailed into the body.

Recessed into slides H is a dog J. Another dog L is attached to feed-bar K, and is held in adjustment by clamp W. Pivoted in the body is a bellcrank M, the forked end of which

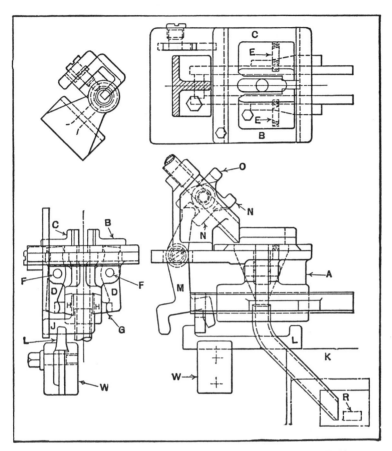

Fig. 9. Device for Feeding Taper Rolls Into Centerless Grinding Machine

straddles dog J, while the ball end meshes with a fork on the escapement pawl N. The function of the escapement pawls N and O is to cut off the feed of the rolls, as they come down from the hopper, and allow them to slide down one at a time on plates B and C.

In operation, the fixture works as follows: Feed-bar K is shown in its maximum "in position," and a roll R has been deposited ready to be picked up by the feed-bar on its return. Plates B and C are in their open positions, and escapement pawl N is shown holding back the rolls in the feed-tube. As the feed-bar returns, one of the projections on dog L engages dog J and carries slides H outward, forcing plates B and C inward through the action of pawls D which ride on the cam surfaces on slides H.

As dog J continues its outward movement it engages a prong on bellcrank M, causing it to pivot in the body and oscillate the escapement pawl enough to allow one roll to slide out on plates B and C. The remaining rolls in the line are retained in the feed-chute by pawl O, which bears on the top of the following roll. As feed-bar K completes its stroke, the roll at R is fed down by a finger (not shown), and the feed carries it between the grinding wheels on its return stroke. While the feed-bar is on its return stroke, the outer projection on dog L engages dog J, carrying the slides H inward.

Pawls D ride down the cam surfaces on the slides, and the compression springs E force plates B and C outward. During this outward movement of the plates the small end of the roll tilts downward, and as the plates continue to move apart, the small end drops through, leaving the roll suspended by its large end between the plates. The plates continue outward, allowing the roll to drop small end first through the feed-chute into the position R. Meantime bellcrank M is engaged by the dog J and oscillates pawl N downward. Pawl O is carried up and the rolls slide forward against pawl N, thus completing one cycle.

Revolving Magazine on Feeding Attachment.— The automatic feeding attachment shown in Fig. 10 has a revolving carrier of magazine B for holding the blanks to be operated on. This attachment is used for feeding the blanks from which the barrels for watch springs are made. The shape of these barrels, which are about ¾ inch in diameter, is indicated at M. The magazine wheel B is recessed, as shown by the side view, to form a pocket for the blanks, and it is provided with slots around the edge in which the blanks fit, as indicated at N. The blanks are inserted in the attachment or magazine wheel through slot C which connects with pocket D. The wheel B is rotated by a belt which transmits motion from a pulley on the front camshaft to a pulley located on shaft S. As these two pulleys are of the same diameter, the magazine wheel rotates at the same speed as the front camshaft. The blanks, as they are carried around by the wheel, drop into slide H and from there into a pocket in a bushing held by a carrier. The block I of this carrier (see enlarged detail view) is counterbored to receive a bushing O which contains plunger P, and the bushing is cut out to receive the spring fingers E. These fingers are attached to plugs F which are held in drilled holes in block I. The bushing O is free to

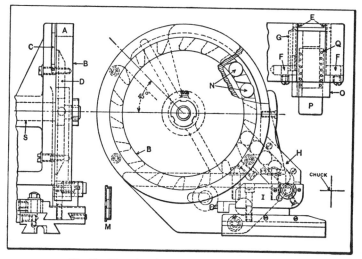

Fig. 10. Magazine Attachment of Revolving Type

slide in block I and is held back by spring G, which bears against a pin driven into the bushing. As a blank rolls down the slide H, it is deposited in bushing O. The cross-slide upon which the attachment is mounted then advances to locate the blank in line with the hole in the chuck. When in this position, the turret advances and a stop on it pushes plunger P forward, thus forcing the blank from the fingers E and depositing it in the chuck. The spring Q which returns plunger P is made much heavier than the spring G used for holding back the bushing O. The object of this arrange-

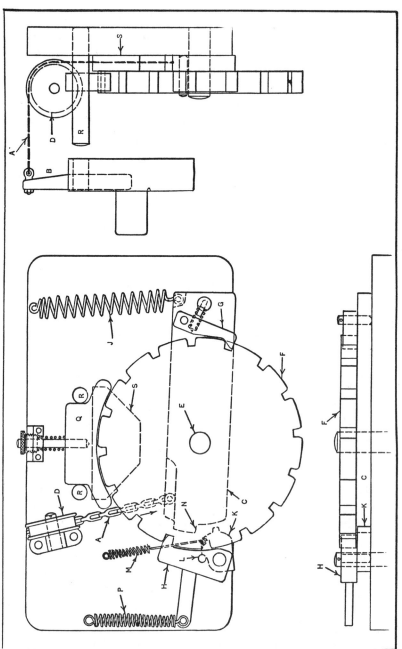

Fig. 11. Automatic Dial Feed for Press Work, Which is Self-contained and Readily Placed on or Removed from the Press

ment is to insure that the bushing will be pushed out close to the face of the chuck before the plunger forces the blank out of the spring fingers.

Self-contained Automatic Dial Feed. — In this design, which is for power presses, the necessary indexing and locking movements are obtained by very simple means through a self-contained mechanism which can be mounted on a press or removed from it quickly, without drilling holes in the press or attaching connecting-rods, levers, or other operating parts, to the crankshaft. A simple chain connection with the punch-holder provides the motions required for unlocking the dial and indexing it to the next station or working position.

The upper end of this chain A (see Fig. 11, which is partly diagrammatic) is attached to extension B on the punch-holder, and the lower end connects with indexing lever C. As the end view shows, the chain passes around a guide pulley D. The swinging movements of lever C about its pivot E are utilized in conjunction with spring controls, as described later, to unlock, index, and again lock dial F.

Pawl G is used for indexing dial F, and pawl H for locking the dial so that each successive die is accurately located relative to its punch and the dial is held securely during the working stroke. How the indexing and locking movements are derived will now be explained by describing the action of the different parts during, first, a downward and then an upward stroke of the ram.

The dial is shown in its normal or locked position. As the ram moves downward, the horizontal part of the chain moves in the direction indicated by the arrow (see also upper view, Fig. 12) and spring J turns lever C around its pivot E. The locking-pawl release-lever K is normally held against stop-pin L in locking pawl H, by a light spring M, as shown in Fig. 11, but when the projection N on lever C engages lever K, as shown by the upper view Fig. 12, lever K turns about its pin, allowing C to pass. Pawl H, however, is not disturbed, the dial remaining locked.

This turning movement of lever C also withdraws indexing

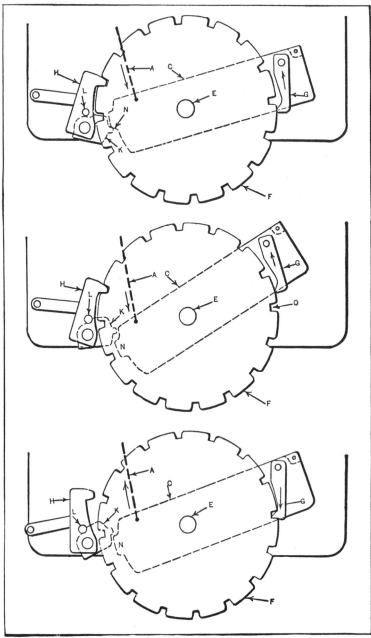

Fig. 12. Three Views Showing the Action of the Locking and Indexing Mechanism of the Automatic Dial Feed

pawl G preparatory to the next indexing movement. The central view in Fig. 12 shows the relative positions of the parts when the ram is near the bottom of its stroke. The projection N has passed lever K, thus allowing lever K to swing back to its position against stop-pin L. Meanwhile, ratchet G has withdrawn nearly a space and a half around dial F.

As the upward stroke of the ram begins, the movements are, of course, reversed, as indicated by the arrows in the lower view, Fig. 12. The chain connecting with the punch is now pulling lever C in the opposite direction. While pawl G is moving from the position shown in the central view around into engagement with slot O, the dial is unlocked by the engagement of lever C with K, which, in turn, acting against pin L, swings pawl H back to the position shown by the lower view. The continued movement of lever C, acting through pawl G, indexes the dial, and just before the ram reaches the top of its stroke, lever K clears projection N, thereby allowing the larger and more powerful spring P, Fig. 11, to swing pawl H into the locking position against the tension of the lighter spring M. This completes the cycle of movements.

It will be noted that the important motions required for unlocking and indexing are derived from the positive action or pull of the chain. This dial feed is used on a press that runs at 90 revolutions per minute. It is advisable to have a hard wood brake Q to assist pawl H in preventing the dial from over-running at the end of the indexing movement. Two guide pins R assist in aligning the punch and in keeping all parts together when the attachment is removed from the press. A hardened steel plate S takes the thrust of the punching, forming, or drawing operation. Spring P is ½ inch in diameter, 4 inches long, and made of 0.060-inch steel wire; spring M is 5/16 inch in diameter, 2½ inches long, and made of 0.035-inch wire; and spring J for the indexing lever is 1 inch in diameter, 4 inches long, and made of 0.080-inch wire.

The particular dial feed illustrated was designed for use

on a standard press having a 2-inch stroke, and it is used in assembling small locks requiring a number of operations, such as bending lugs, upsetting pins, etc., the work being indexed successively under the different punches (not shown) attached to the punch-holder. Locating gages or pockets are attached to plate F and the completed parts are ejected in front by air pressure. Plate F also serves as a bolster plate in order to provide ample die space.

The ease and rapidity with which this dial feed can be placed in position or removed from the press is an important feature of the design, as it can be applied or removed as quickly as an ordinary die having leader or guide pins. Owing to the simplicity of the design of this mechanism, it costs little to construct, so that it is practicable to have a number made for different operations or parts, the self-contained feeding mechanisms being interchanged on the press, the same as dies.

CHAPTER XVII

DESIGN OF AUTOMATIC FEEDING MECHANISMS

WHEN an automatic machine designer has solved a great many problems during the course of his experience, he reaches a point finally where he does not need to do so much experimenting before originating a plan or design for a certain operation. Two experienced men, however, working on the same problem will seldom decide upon the same method of handling, yet both solutions may be equally good.

Two jobs may be similar but they are not often exactly alike. A machine may have been designed and built for a certain operation, and the designer may be called upon to design another for a piece of almost the same shape. A small difference may make it necessary to use an entirely different method of handling. When experimenting with a model, if it is found that it works properly without a slip or failure under every test, it may be considered satisfactory, but when it fails to function *just once,* the trouble must either be overcome or a different scheme tried. It is best to develop an idea along lines that previously have been found successful, whenever this is possible, but for much of this work there is no precedent, and a new idea must be worked out to suit the case. Take nothing for granted, until it has been proved.

A customer may say, "I want to dump all these three sizes of pieces into a hopper and have them come out through three chutes separated according to their size. Can you do it?" Assuming that the designer is confronted with a case of this kind (which is by no means uncommon) it may be best to answer the question by asking another, for example: "Why don't you separate them first as they come from the manufacturing machines, and keep them separate?" There may be conditions which prevent this, but usually much can be done to simplify a design by adopting improved methods for previous operations. Many people think that one can do anything with an automatic machine, and while there is some truth in this, the cost may be prohibitive. One complicated machine can sometimes be designed for several operations on a difficult piece of work, but it is always expensive and likely to get out of order. It is nearly always better to use two or three simple machines than to make one complicated one, for in the latter case, when anything goes wrong with the machine the entire production stops until the trouble is overcome. The careful designer considers these points before he starts actual work.

A designer is frequently called upon to handle and arrange pieces that are dumped into a hopper and many times a much simpler arrangement could be used by starting at the beginning of the problem instead of in the middle. Any piece of work that is made in a machine is removed from it in a uniform manner, and if, after completion, these pieces are to be packed in a carton, package, or box, the feeding into the packing machine can be greatly simplified by devising a simple arrangement to apply to the machine that does the manufacturing. Such a device can often be made to stack the pieces into a removable carrier or magazine which can then be attached directly to the packing machine. The nature of the work sometimes prohibits the use of anything of this sort, but it is well to keep it in mind as a possibility.

In the selection of pieces from a mass, there are several principles frequently used: Gravity, vibration, oscillation, rotation, and centrifugal force. Some kinds of pieces will fall by gravity, perhaps assisted by vibration of inclined planes or hoppers. Others need to be oscillated in order to disturb the mass and change the arrangement continually. The principle of rotation is often applied in many ways. Centrifugal force can be applied successfully to separating devices, but its application in automatic machine design is seldom appreciated by the average designer. Gravity and vibration are probably most used, yet both of these methods must be applied properly to secure positive results. If we should place a single piece of work of a given shape in a fixed position on an inclined plane, we know that it will slide downward if the angle is great enough, but if other pieces are in contact with it, we cannot tell what may happen. Several examples will be shown to indicate the advantages of prearrangement of pieces before handling.

Feeding Shallow Boxes Top Side Up.— At A in Fig. 1 is shown a cup-shaped wooden disk which is to be passed through a machine and the depression filled with a composition. Our problem calls for feeding the pieces in such a way that they will always come through the machine cup side up. If they are arranged in a stack as at B, the matter is quite simple, as they will drop easily into the carrier C ready for any other operation.

If we are required to handle the pieces from a hopper, the first step is to flatten them out so that they will slide edgewise down a chute, as indicated at D. To do this, they may be placed on a vibrating inclined plate E so that they pass through the opening at F into the chute. An oscillating brush at G can be used to disturb the pieces and break up any fixed arrangement. When the pieces have passed through into the chute D, the cup sides will not face in the same direction— some will be one way and some another. Referring to the enlarged view at H, the end of the chute is seen at K with one piece L entering the selector, and another following it at M. The selector revolves intermittently in the direction shown by the arrow at N. There are six slots equidistantly placed in the edge of the selector, and a guard plate covers its face, as shown at O in the sectional view. As the selector revolves, it takes one piece at a time from the chute and carries it around to the point P. If the piece has entered in the position shown at Q, it drops down through the chute R, but if it has entered in the reverse position as at S, it is carried around to the next station T and drops through the other chute U.

The mechanism governing this is simple. The revolving

selector V has at each one of the six slots a spring plunger
W, at the end of which there is a hardened roll X which
travels around the circular plate Y. This plate is continuous
except at the point P, where it is interrupted. If the cups
face in the direction Q, the light spring behind plunger W

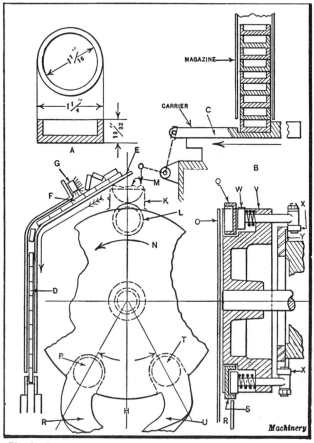

Fig. 1. Feeding Device for Cup-shaped Wooden Disk Shown at A

simply throws the piece out into the chute at R, but if the
cup lies as at S, the plunger enters the cup and prevents it
from dropping until the selector has passed that point. Before reaching T, roller X rides up again on plate Y, thus
withdrawing the plunger and allowing the piece to drop into

chute U. It is only necessary to twist the chutes R and U in opposite directions, making a quarter turn, to have both pieces come out facing the same way at the bottom of the chutes, at which point they can be easily handled as required. The spring behind the plunger must be very light in order to prevent friction when the pieces slide out.

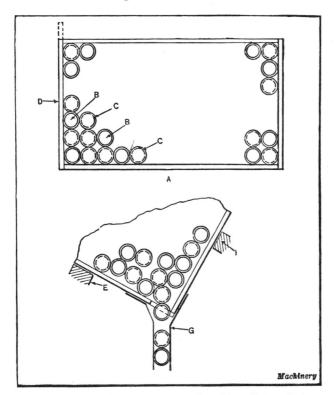

Fig. 2. Inclined Tray Method of Feeding Pieces Into a Chute

Considerable expense can be saved by using trays in which to arrange the pieces before feeding them into the machine, instead of using a hopper. Fig. 2 shows at A a form of tray that can be used, the pieces being rapidly spread out by hand so that only one layer is in the tray. Some of the pieces will be right side up, as shown at B, while others will be upside down, as at C. If the tray is made with one end D so that it can be pushed back far enough to allow one piece at a time to come through, this opening will serve as a gate, and the tray can be set up on an incline and against supports at E and F, so that the pieces will drop down one at a time into the chute G. From this they can go into the selector as in the preceding case.

Feeding Mechanism for Plain Flat Disks.— Let us take another example of a disk A (Fig. 3). This is just like the other, except that it has no depression and could be handled from the hopper or from a stack by means of a reciprocating slide, as shown at C in Fig. 1, or by a rotating selector without the plunger. If flat pieces are being handled which are to be packed in bottles or boxes, the preceding method would be much too slow. Assuming that the tablet shown at A is to be deposited in a bottle, thirty pieces by count, it would be out of the question to use any such method as that mentioned. The pieces are to be dumped into a large hopper, several thousand at a time, as the requirements for production are such that not more than one second can be allowed for counting and putting them into the bottle. It would be difficult to make a counting device to operate as rapidly as this and with certainty, but the pieces could be weighed, or they could be arranged thirty in a row approximately $11\frac{1}{4}$ inches long, as shown at B. If they could be arranged this way, it would still be difficult to keep the chute full and deliver the pieces as rapidly as required.

If the pieces are dumped into a hopper, the previously demonstrated methods can be employed for making the parts come through an opening flat side down on an incline great enough so that they will slide down it to a gateway shown at C. This gateway is open at D over a rubber-faced disk E which revolves slowly. A guard strip F runs across the revolving table at an angle, and there is also a guard at G, the lower end of which blends into the chute H. From point K to L there is just room enough for thirty pieces to lie. In operation, as the pieces drop through on the disk at D, the circular movement combined with gravity causes the pieces to roll rapidly down the incline and into the chute H until they have filled it completely. They will then back up against the guard G until relieved by the opening of valve M. This is timed so that, as it opens, another valve at N closes. As there are four or five pieces O still in contact with the rubber

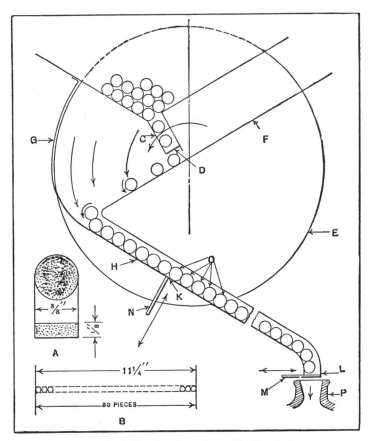

Fig. 3. Principle Applied in Counting a Large Number of Pieces

face of the disk, these will receive a rapid movement from it, pushing out those ahead so all the thirty pieces will shoot into the mouth of the bottle P.

Care must be taken in designing a mechanism of this sort to work out the important points by means of a model, as

otherwise it will not function satisfactorily. A slight change
in the angle of the guard plates, chutes, and the speed of the
rotating disk make a great difference in the operation, but if
the principle is understood, it is comparatively easy to make
up a simple model which can be used for demonstrating purposes and to obtain the right relation of the various guards
and chutes.

Filling Small Boxes with Tablets.— Fig. 4 shows at *A* a
pasteboard box containing twenty-four disk-shaped tablets.

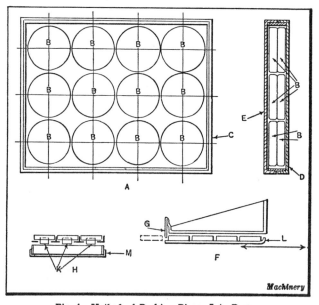

Fig. 4. Method of Packing Pieces Into Boxes

They are arranged in the box in layers, as shown at *B*. The
problem is to take the pieces from a hopper, arrange them in
the box as shown, and put on the cover *E*. The production
required is not less than twenty boxes per minute, and therefore not over one second must be consumed in putting in each
layer. Arranging the pieces and putting them in the box is
not particularly difficult, but we must also feed the boxes and
covers into position and put on a cover. It is advisable to
use two sets of chutes for the two layers in the box; this can

be done easily enough, although care must be taken to provide a sufficient supply so that both chutes will be kept full and still have a reserve. By using two chutes, one layer is put into the box which then moves over to the other chute and receives the second layer. If the chutes are arranged properly, the production time will be only that required for putting in one layer.

The arrangement for feeding is simple, and needs but a brief description. The tablets lie in a nesting device, a part of which is shown at F. They are held back by a rubber cowl G, and lie between guides shown in the end view H. These guides are open at the bottom and contain long steel fingers K, having a hook on the end at L. At the proper time these fingers move forward over the box M, pulling with them the tablets, which pass under the cowl G. The fingers K then move backward again, and as they do so, the cowl G restrains the pieces from following, and they drop into the box one after the other in their proper arrangement. It is advisable to make the fingers lie as close to the top of the box as possible, in order to minimize the amount of drop.

Let us now consider the handling of the box and cover. As both of these are rectangular and regular in shape, it is not difficult to arrange a feeding device for them, but the putting on of the cover is more difficult. If we take up one thing at a time, and decide to load the boxes from a stack, as shown at A in Fig. 5, we must realize that even with fifty boxes to a stack, a new magazine full would be required about every two minutes and a half. Therefore, it may be well to arrange for an indexing holder B containing six magazines, which can be quickly removed and replaced when empty. A device can be made which will index the table one station at every fifty strokes, either (1) by using a ratchet and pawl and a dog, which will drop into an index-plate at the proper time; (2) by putting exactly fifty holders on the conveyor belt and using a suitable dog on the side to index the table; or (3) by means of a cam. The first or third method is to be preferred.

The conveyor belt C removes the boxes one at a time by an

intermittent movement, and carries them along to the first loading station *D*, at which point the first layer of tablets is put in by the method previously described. The conveyor is arranged with cross-pieces *E* spaced only a trifle further apart than the width of the box. As it passes under the magazine

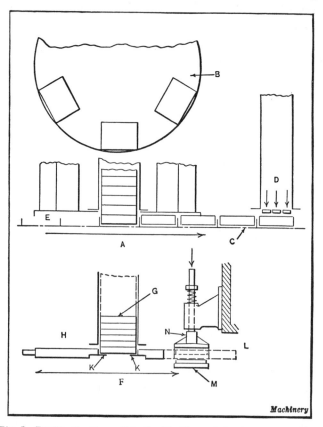

Fig. 5. Feeding the Boxes Into the Machine and Putting on the Covers

it stops, allows one box to drop, and continues to do this regularly. The next step is to put on the cover.

Placing Covers on Boxes. —By arranging the covers in a stack of similar form to that used for the boxes, a reciprocating slide can be used to remove them from the bottom of the stack, as shown by the diagram at *F*. Fig. 5. The cover *G* drops into the slide H and lies on top of two flexible rubber strips K which prevent it from falling through. At the proper time, the slide reciprocates and takes the position shown by the dotted lines at L, the cover then being directly over a box full of tablets shown at M. A pusher N lies directly over the

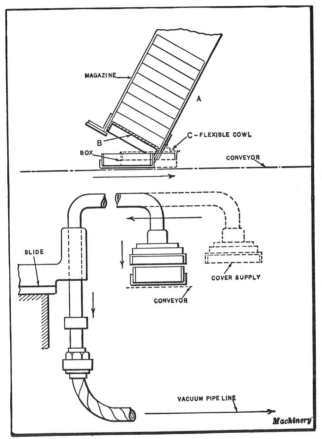

Fig. 6. Other Methods of Putting Covers on Boxes

cover, and is used to press the latter down on the box. The control of this mechanism is obviously very simple and does not need detailed description.

There are other ways of putting on a cover, one of these being indicated in Fig. 6 at A. Here the chute is set at such an angle that one side of the cover hangs low enough to be caught by the edge of the box B as it passes under it. A light rubber or cloth cowl at C drags over the cover as the conveyor pulls it through. This device might be satisfactory if it did not depend on gravity for the operation, but unless the covers are fairly heavy there is a possibility of a miss now and then.

The lower view shows still another method often used for picking up pieces of this sort. Here the covers fall down a curved chute and come to rest as indicated. A perforated suction plate picks up one cover by vacuum, moves it over to the box and presses it down by a vertical movement of the ram. This device is rather more complicated than the others, but could be made very satisfactory. The pneumatic method of handling can often be used for conditions of this kind, but care must be taken with any materials that lie close together and that are somewhat porous, as there is always a possibility of the suction going through more than one piece and picking up two, when only one is wanted.

It will be noted that in working out this problem it was considered in unit form, and attention has not yet been given to the power application. Often a great deal of unnecessary trouble is caused by the designer attempting to decide upon the method of driving and timing up a unit during the progress of the design. The units can be designed and located from the preliminary freehand sketch, and after their positions have been settled, the main driving shaft cams, levers, etc., can be placed most advantageously.

CHAPTER XVIII

HOPPER DESIGN FOR AUTOMATIC MACHINERY

WHEN materials are to be handled in bulk, they are often dumped into a hopper, from which they are removed through an opening in the bottom. It is difficult to classify the various kinds of hoppers and divide them rigidly into groups, but by considering the kinds of materials handled, we can make an arbitrary distinction sufficient for our purpose. The following list covers the types most commonly required:

1. Granular materials, which flow readily.
2. Liquids of all kinds, heavy or light.
3. Plastic materials, such as cement, asphaltum, and other mixtures, candy in certain forms, cream cheese, flour dough, etc., some of which are sticky or spongy.
4. Disk-shaped or oval pieces of uniform size and shape.
5. Cylindrical or conical pieces of uniform size or shape.
6. Irregular shapes of various kinds.

The first thing to consider in the design of a hopper for materials of any kind which are to be handled in bulk, is the general shape of the material; and the next, the manner in which the pieces or materials can be taken from the hopper— that is, their arrangement as they issue from it. Problems connected with these two points are some of the most important in automatic machine design, for upon their correct solution the functioning of the machine is largely dependent. A hopper may be cylindrical, conical, rectangular or any other shape best suited to the material. The different conditions encountered can best be understood by citing a few examples.

Hoppers for Different Substances. — Granular substances, unless they are very irregular (as might be the case if the material is formed in crystals) will fall down almost any kind of an incline and through almost any form of opening. Liquids, of course, will do the same, but for pasty or gummy substances a gravity feed is usually impractical, and the ma-

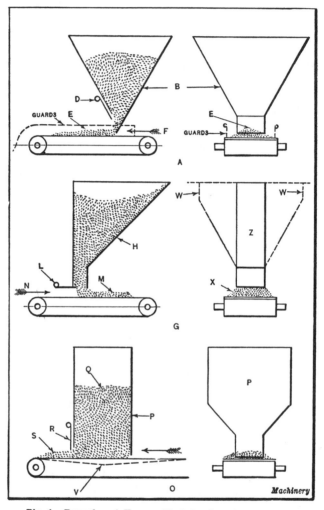

Fig. 1. Examples of Hoppers Used for Granular Materials

terial must be forced through the hopper by mechanical means. Disk-shaped or oval pieces are often fed by gravity and removed by inclined chutes. In such cases it is frequently necessary to use an agitator or some other form of mechanical feed to keep the flow constant. Cylindrical or conical pieces are very likely to clog in a hopper, and must be stirred up continually or agitated to prevent lodging. In some cases it is not even possible to use a regular form of hopper, and the pieces must be spread out on a large inclined plane. In analyzing a given condition, think first of the arrangement of pieces as applied to the method of handling after the work has passed through the hopper.

Examples of Hopper Design. — Fig. 1 shows several examples of different forms of hoppers, to illustrate the manner in which the material passes through the hopper and is discharged from it. In example A, the hopper B is of pyramidal form. The material has no tendency to wedge or crowd, but flows readily in a continuous stream to the bottom of the hopper at which point an adjustable gate or valve D controls the amount passing through. If handled on a belt moving in the direction F, the material will naturally spread out as shown at E, so that it can be carried and delivered in the quantities required. Side guards can be used (as shown by dotted lines) if desired. In example G, one side of the hopper H is straight and the other is angular. Often the space available makes it more convenient to have a hopper of this form. In the other direction, the sides may be straight, as at Z, or tapered, as at W. A valve at L permits adjustment according to requirements, and the material is deposited on the belt M. Care must be taken in proportioning the opening of the hopper so that too wide a mass is not distributed on the belt, as shown at X, for in a case of this kind there would doubtless be considerable loss as the belt moves along.

In example O a somewhat objectionable form is indicated at P. With certain materials there might be no great objection to this form, but when handling anything of a heavy nature, the weight of the rectangular mass at Q might be sufficient to deflect the belt, as shown at V, thus causing variable amounts to pass by the valve R. Assuming that the belt travels in the direction of the arrow, the material would be drawn out at S and carried away as needed. With a wide-mouthed hopper like this, a great deal of pressure is continually being exerted by the column of material above the belt, and this pressure varies according to the quantity of material in the hopper. Conveyors for this sort of material are often made concave and lie on three rollers, in which case no guards are required, as the concave form keeps the material from falling off.

In Fig. 2, the example shown at K is a pyramidal hopper L containing a granular material M which is to be removed,

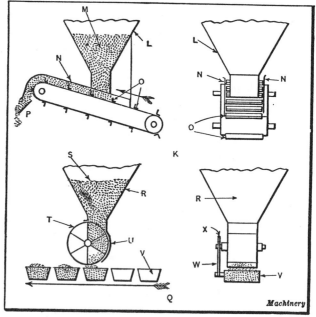

Fig. 2. Guarded Hoppers for Granular Substances

carried upward, and discharged at point P, intermittently. A plain belt conveyor is used, and on this are fastened at equal spaces, the clips O which hold the material and prevent its rolling down the incline. The guards on each side at N lie close to the cross strips and have leather or rubber ends which prevent loss of material. This type of hopper is useful when granular materials are to be fed in approximately uniform quantities and carried from one level to another. The example illustrated was used to remove bulk candy and transfer it to a higher level, where it was spread out into distributing chutes.

Valve Arrangements in Connection with Hoppers. — The example shown at Q, Fig. 2, is quite different in its general form, although the material is held in a hopper R of the ordinary form. The lower part of the hopper is narrowed so that the material passes down through it in the form of a rectangular column. A rotating valve T having six openings of equal size is located as shown, and one portion of the hopper U extends down around the valve. When in the position shown, the material fills the compartment shown uppermost and the guard U prevents the previously filled compartment from emptying until it is nearly upside down. The buckets V pass along under the valve, and a given quantity of the material is emptied into each bucket as it goes by. The movement of the buckets may be regular or intermittent, and the rotation of the valve can be easily controlled by means of projecting arms on the buckets, as shown in the side view at W. These arms can be made adjustable and placed so that they will strike a spider X and carry it around a given distance as they move along. In designing devices of this kind, it is a good plan to make the length of the arms adjustable and the positions of the striking portions on the buckets also adjustable, in order to compensate for slight variations in movement.

Another form of valve arrangement which is independent of the movement of the buckets, but is timed to suit their progress, is shown in Fig. 3. The usual form of hopper A carries the material B in the pyramidal portion, but the lower part is rectangular and quite long, being so proportioned that its contents will fill one of the buckets N. The valves C and D lie in guides at the top and bottom of the rectangular portion of the hopper. In the upper valve, there is an opening E, and in the lower an opening F, the two openings being so placed that when one is in line with the hopper the other is closed. The arm *G*, pivoted at *K*, connects the valves by sliding joints at *H*. Lever *L* connects with rod *M*, which pivots on block *O* in slotted disk *P*. A reciprocating movement is transmitted to lever *L*, as the disk revolves.

The kind of material that is being handled determines whether a small amount of leakage is permissible or not, but in normal cases the slide valves can be made of cast iron or brass working in guides of steel. Careful fitting is not usually

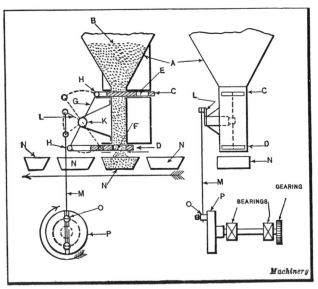

Fig. 3. Arrangement of Hopper Valves for Handling
Granular Substances

required on this kind of work, but the guides should always be long enough to permit a free sliding movement and good alignment.

Continuous Rotary Hopper Valves. — In the example shown in Fig. 4, a continuous rotary feeding device is shown. The hopper *O* and the valve *P* are similar to that shown at *Q* in Fig. 2, but the operation is by means of a ratchet gear *R* mounted as shown. The arm *T* has a pawl on it at *S* so arranged that it engages with ratchet *R*. The disk *X* is slotted at *W*, and the block *V* is adjustable in the slot to control the movement of lever T through the connecting-rod U, as in a preceding example. The disk is driven by a chain from the sprocket Y, and the buckets Q are controlled in their movement by suitable gears or sprockets and chain running from the same shaft. When two separate units are to be timed accurately, a chain drive or gearing should always be used.

Fig. 5 shows a form of rotating hopper valve which may occasionally be found useful when quantities are to be measured and deposited accurately in buckets or other receptacles. The hopper A contains the material B which flows downward until it enters the section H of the rotary valve. There are

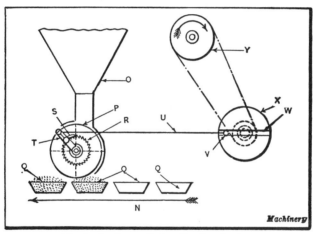

Fig. 4. Continuous Rotary Feeding Devices

six chambers H arranged radially about a center on which the valve rotates. The entire valve mechanism is supported by brackets C and D arranged according to the general design of the mechanism. The valve plate F fits closely against the under side of the hopper, and in each compartment there is a hole leading directly into the throat of the hopper. The plate G also has a series of corresponding holes, but there is only one opening in plate K from which the chute L leads. As the buckets M pass under the spout L the indexing mechanism operates at the proper point and discharges the contents of one of the compartments into the buckets, as indicated. If

small quantities only are delivered and if the conveyor movement is proportioned properly, the materials can be dumped
without stopping the conveyor, but for larger quantities it
may be necessary to use intermittent gears or some other convenient method to stop the movement for a moment when the
valve L is open to insure delivering the required amount.

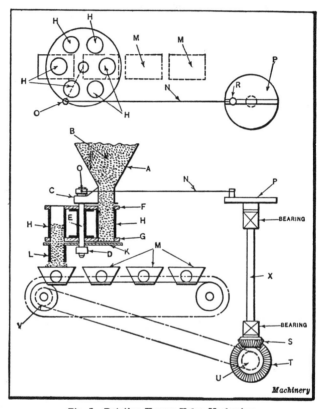

Fig. 5. Rotating Hopper Valve Mechanism

The operating mechanism of this device is very similar to
one previously described, the lever O being attached to a
pawl working on a ratchet gear, the movement of which is
controlled by the rod N which pivots in block R. The disk
P has a slot in which the block may be adjusted to govern the
movement of the pawl. By driving shaft X from the bevel

gear S by means of the ring gear T, the "take off" for the conveyor comes from the same shaft through sprockets U and V connected by a chain. Naturally, the location of the various driving units is dependent largely upon the general design of the machine and the other portions of the mechanism which must be driven. It is, of course, essential that all units which must function in a certain relation to each other must be controlled from a single shaft. When accurate timing is not necessary, intermediate gears and lever movements can sometimes be used to simplify the design.

Hoppers for Liquids. — So far we have considered only the hoppers used for materials that flow readily by their own weight, and as liquids of thin consistency are led through the hoppers in much the same manner, we will consider the problems arising in their handling. Although granular substances and liquids will both flow readily, there is considerable difference in the valve arrangements. In granular substances, a small amount of leakage around the valve is not serious, but when liquids are handled, more attention must be paid to this feature. Several points of importance should be noted:

1. The nature of the liquid, its value and its consistency are matters that affect the design of the valve mechanism. Waste and cleanliness do not go together; if you were to see any machine handling liquids which were dripping all over it, you would not think the machine thoroughly efficient. It is not a good plan to allow surplus material to run down or overflow into a tank or tray arranged to carry it off, unless for one reason or another the liquid must be flooded over the receptacle. Conditions of this kind are sometimes found in the making of molded products in order to be sure that the molds are thoroughly filled. A thin liquid will flow readily through a valve which can be shut off by suitable mechanism without difficulty, but a heavier fluid may drip all over the machine and the receptacles unless a suitable valve is used.

2. The distribution of the product into containers affects the design of the valve. Measuring valves are most commonly used for controlling the flow of liquid. When the valve

opening is very small, the pressure of the outside air may be sufficient to prevent the liquid from flowing unless a "vent" is provided. It is usually necessary to provide some means also for controlling the drip from the spout to prevent waste.

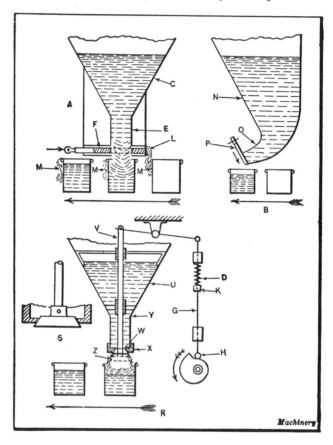

Fig. 6. Containers and Valves Designed for Thin Liquids

Valves Used for Thin Liquids. — Fig. 6 shows several forms of valves used for thin liquids. Some of these are good, while others are objectionable. At *A* the hopper *C* is of angular form. A point of importance is the variable pressures caused by small or large quantities of liquid in the hopper. The method of supply makes considerable difference in the design of the hopper, and it is quite common to find a cylindrical form used with a conical bottom leading to another cylindrical portion as shown at E. If the liquid is poured into the hopper from time to time, the pressure on the valve is variable, but if it flows into the hopper through a supply pipe, a float valve can be easily applied which will shut off and open the pipe as the liquid falls or rises in the container. By this means the pressure can be kept nearly uniform, and better results can be obtained. The type of valve shown in this example is unsuitable for use in handling liquid. The valve F slides backward and forward through guides of similar form to those used in handling granular substances, but it is almost impossible to make a mechanism of this kind tight, and the liquids will drip out at L and run down over the sides of the containers below. If the fluid is sticky, it may cling to the outside, as shown at M. If not, it will run down the sides and get all over the machine. In cases where the ends of the valve are covered, material is likely to collect there and cause trouble in cleaning.

The form shown at B is considerably better, as the hopper N terminates in a spout O, and the fluid is held back by a hinged shut-off valve P which fits tightly against the face of the opening. As this valve is operated it cuts off the supply in such a way that there is only a small amount of drip. This can be caught without difficulty by the containers, and although a few drops may remain with the heavier variety of liquids, those that are thin will not drip to any great extent. This form of shut-off valve is very frequently used and is recommended for its simplicity.

A form that is adapted to a variety of uses is shown at R in the same illustration. The hopper U contains a liquid which flows in the lower part and against the valve W which prevents it from passing through until the valve is opened. The valve seat X is fastened to the lower part of the hopper, and the angular surface of the valve keeps it tight. The rod V is guided as shown, and is connected by a sliding joint to an operating rod. This is pivoted in a bracket which may be attached in some convenient way, depending upon the design of the machine. The connecting-rod G has a roller H at the lower end which contacts with the cam shown. The roller is kept in contact with the cam by the spring D which thrusts against the collar K on the shaft. As the cam revolves in the direction shown by the arrow, the valve is opened and the liquid flows down over the conical portion Z and into the container. The spring D closes the valve quickly when the roller drops off the cam, but the spring must be fairly stiff in order to make a tight joint. The enlarged detail at S shows the construction of the valve seat and the valve. The conical surface should be made as narrow as possible, as it is easier to keep it properly fitted if this is the case.

When any sort of liquid is to be handled through a valve, it is very important that the construction permit quick and easy removal of the units for the purpose of cleaning. This point is of great importance when handling any sort of food product or any liquid that has a corrosive action on the metal. The selection of metals used in valve construction is also an important factor in the design. Stainless steel can often be used in valve work.

Experiments will usually show the effect of a fluid on a certain kind of material, but it is important to have all conditions that will obtain in using the machine fulfilled when making the test. Rubber is sometimes used for valves, and in such cases replacements should be made as easy as possible, because frequent renewals are necessary. A designer who does not consider such points as these fails in the essentials of good design. When considerable labor is necessary in replacing a certain part subject to wear, or when cleaning the machine after the day's work, this labor is just so much lost time that might have been avoided by greater forethought in designing.

CHAPTER XIX

MAGAZINE FEEDING ATTACHMENTS FOR MACHINE TOOLS

THE magazine attachment in any automatic machine must, first of all, deliver the work, even if it should be of irregular form, in a uniform manner. Some pieces are much more difficult to handle than others; for example, the bevel gear forging *A*, Fig. 1, is larger and therefore much heavier at one end than at the other. Also, as it is a forged piece, it may have fins, seams, or other small projections which will interfere with smooth movement. Obviously, a straight-guide magazine cannot be used for this piece, assuming that it is to be machined on centers or in a chuck. In any case it must be delivered from the magazine with the same side always in a given position. When irregular pieces are to be handled in a magazine, the greatest care must be used in planning the arrangement, in order to prevent wedging, interlockng, or cramping of any kind. Such conditions may develop from many causes, such as too much or too little clearance in the magazine guides, or a lack of forethought in arranging the pieces. Also, if forgings or castings are made from several dies or patterns, they will likely vary somewhat and cause trouble in feeding.

The first step in designing a magazine for an irregular piece is to look carefully at its general shape, the size of the fillets, and the location of seams, fins, or raised numbers which may prevent a smooth rolling movement. In considering the work shown at *A*, we know that it will undoubtedly have fins longitudinally where the dies part; that fillets will be in evidence at the junction of the head with the stem; and that it must be delivered from the magazine in about the position

495

shown. It is often necessary to find the approximate center of gravity of a piece in order to determine whether it will be likely to overbalance if handled in a certain way or in a particular position. Usually this can be determined by laying the work on a straight edge, such as a ruler or scale, or even by balancing it on the finger, and then marking the approximate center of gravity with a piece of chalk.

A sample piece is not always available in the preliminary stages of design, but an approximation can be obtained graphically from a drawing of the piece or from a wooden model

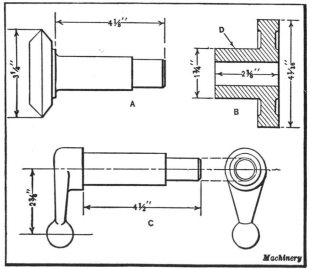

Fig. 1. Examples of Irregular Work for Which Feeding Magazines are to be Designed

which can usually be made in a short time. When gravity is used entirely for feeding the pieces, the possibilities of friction must not be overlooked, as much trouble is caused by neglecting this important factor, particularly when the pieces are arranged to slide down an inclined plane or when they rub against each other in the magazine. All surfaces which come into contact with the work or on which it slides, should be made narrow in order to reduce friction to a minimum.

In determining the shape of a magazine for a given piece,

several sketches should first be made of different schemes, and then a careful examination will usually show which idea is most practical. The use of paper, pasteboard, tin, and wooden models will often lead to an excellent development of an idea, the operation of which, on paper, might be problematical on account of unforeseen conditions. By this means it is possible to determine, without great expense, whether a certain mechanical contrivance will function properly.

One of the most important points in machine design is the elimination of the element of chance in the operation of some unit. A design may look all right on paper and may be mechanically correct, yet when it is built, some unforeseen condition may prevent it from functioning properly without more or less expensive changes in the construction. If the actual operation of a unit could be seen before it is built, many troublesome factors would be eliminated. The designer will do well to experiment with simple inexpensive models whenever there is an element of doubt regarding the working of a mechanism.

Designing a Magazine for a Bevel Gear.— Having considered the arrangement of pieces so that they will come to the carrier in the desired position, we have only to decide upon the best design of magazine. Gravity feeds are unsatisfactory for some kinds of work, and it is necessary to resort to mechanical operation. However, this is seldom required in magazines for feeding bar work, forgings, or castings of the sort being dealt with. There are several ways in which forging A, Fig. 1, can be fed. One form of magazine is shown in Fig. 2, the plan view clearly showing the general arrangement of the guide plates. In the right-hand elevation the magazine is shown emptying a piece into carrier A in which it is clamped by spring B and carried over to the centers at C on which it is to be turned. The clamp is withdrawn, as shown at D, after the chucking. In this magazine the upper pieces are held back by the shutter mechanism E. An adjustable stop for the carrier is provided at F.

The fillet at the point where the stem of the forging joins

the head might cause trouble unless clearance is provided. If there were only a small fillet, there would be no trouble, but a large one might result in wedging and stop the progress of the pieces through the magazine. If designed with sufficient clearance to take care of all possible variations, there would doubtless be no trouble with this design. The general

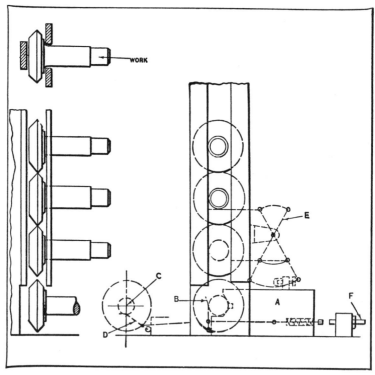

Fig. 2. Example of Design for Bevel Gear Magazine

construction of a magazine of this sort needs no particular comment, but it is advisable to make the back plate and one side plate adjustable so that the magazine can be set to suit varying conditions.

Magazine Used for Two Operations. — Example *B,* Fig. 1, is a small bronze casting which is to be fed from a magazine and faced, turned, and bored while held by stem *D.* In an-

other operation, the work is to be held on a plug and gripped
on the large finished diameter while the other side is finished
and the chucking stem cut off.   Assume that a magazine
adjustable to suit both operations is highly desirable, as the
work is to be done in a special chucking machine and the
device used in connection with the turret.   The magazine

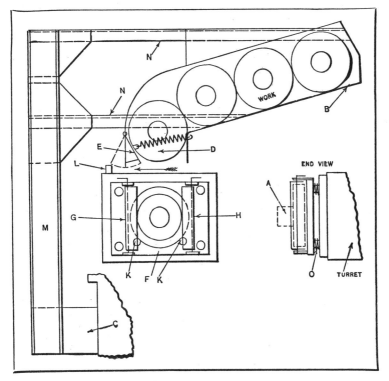

Fig. 3. Turret Lathe Magazine for Feeding Casting B, Fig. 1

should be set in such a position that at one complete revolution of the turret the chucking device would receive work
from the magazine.   The turret would then move forward
and allow the piece to be gripped by the chuck jaws.

Fig. 3 shows the general arrangement of a suitable magazine for this piece, and its position with respect to the turret.
It is loaded from the front of the machine by the operator,

who can obviously attend to several machines without difficulty. Sufficient details of construction are shown for the
designer to understand, but most important of all are the
principles involved, as these can frequently be applied to other
work of a similar kind. Magazine B should be supported in
some suitable manner, depending upon the type of machine
and the position in which the magazine is held. In this case
bracket C is fastened to a pad at the back of the machine and
on it is mounted a structural steel channel M, to the upper
part of which angle-irons N are fastened. These extend forward and are riveted to the sheet-metal guides of the magazine.
It is usually much better to arrange the supporting members
at the back of the machine rather than in front for several
reasons, one of these being that there is less likelihood of
interference with the working parts of the machine, and another, that, when so situated, it is not in the way of the
operator.

The magazine chute is constructed of sheet steel, bent to
shape, and is slightly inclined to allow the pieces to roll down
easily to point D, at which position the first piece is supported
by the hinged valve E. The turret chucking device F is
fastened to the face of the turret, and has metal guides G
and H at each side. The two locating pins K should be made
adjustable to take care of variations in the size of the pieces
handled. On the upper side of the turret, a trip L is located
in such a position that when the turret is indexed to the location shown, the trip strikes valve E and opens it, as indicated
by the dotted lines, thus allowing one piece to fall down into
nest F. After the piece has dropped, the turret moves forward and carries the work to the chuck, which should be air-
operated and have jaws of suitable form to grip the stem
ready to machine the part. After the work has been done,
the jaws should open automatically, and the piece drop down
into the bed of the machine, leaving the chuck ready to receive another piece. Coil springs O allow for variations in
placing the work in the chuck.

In designing a magazine and carriers of this sort, the designer must follow through the entire process carefully, always keeping in mind that when one piece is removed from the magazine another immediately takes its place, and so provision must be made to receive it and not permit it to follow the first piece into the carrier. When springs are applied for

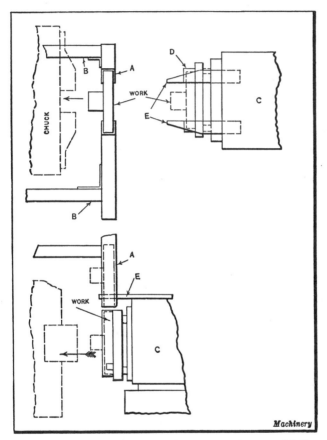

Fig. 4. Turret Lathe Magazine Placed Directly Over the Spindle

closing magazine valves, a little experimenting is sometimes necessary to obtain the proper form of spring and one that has the required strength. Also locations of pins on which the work rests may not always be correct, and then the piece will not be properly centered for chucking. Therefore, it is generally better to make such points adjustable so that suitable settings can be made without difficulty.

Magazine Located Above Spindle.— Another magazine for placing work in the chuck is indicated by the diagram Fig. 4. This magazine is for parts of the general form shown at *B*, Fig. 1. Here the chute *A* is arranged above the spindle, and is supported by brackets *B*. The design of the chute is much the same as in the previous example, but the work comes down and rests against a double valve instead of a single one. Turret *C* comes forward at the proper time with a carrier *D* having projecting fingers *E* above it which open the valve and allow the work to drop into the nest on the turret. A continuation of the turret movement carries the work forward to the chuck jaws. The dotted diagram in the lower part of the illustration shows the position of the work in the carrier after dropping and before chucking. In almost every form of magazine, the carrier or chucking device design is controlled by the shape of the work and the form of the chute. When mounted on the turret and used to push work into a chuck, such a device must be flexibly mounted if the nature of the work requires that it should be pushed back firmly into the chuck jaws. When this is not necessary, the movement can be controlled by the cam that operates the turret.

Magazine Placed at Rear of Machine.— A magazine for application to the rear of a machine is shown diagrammatically in Fig. 5. The mounting is similar to those used for bar or other work held on centers. The work lies in the chute with one piece over another, and the lowest one either resting on carrier *A* or in the chucking position *B*. With this arrangement the necessity for a spring valve is obviated, although the carrier must have suitable provision for keeping the pieces vertical, and also spring releasing members *C* which will allow the work to pass into the chuck when pushed by a device on the turret. The carrier is equipped with adjustable work supports *D*. The design of any magazine should nearly always be of open construction for several reasons: It is easy to see when the magazine needs refilling; it is not likely to become clogged with dirt; and it is cheaper to make than a closed form. There are cases when the closed form is preferable, but seldom on this kind of work. Sheet metal can be bent without difficulty into various forms and readily riveted or brazed together, which makes it a desirable material from which to construct

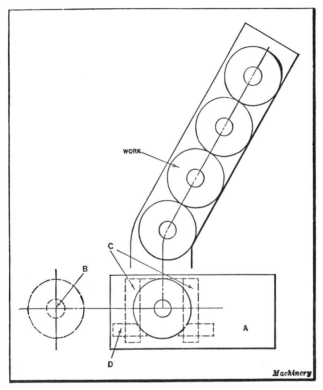

Fig. 5. Magazine Designed for Same Part as that in Fig. 4 but for Application to Rear of Machine

chutes. The inside surfaces should be kept smooth by using countersunk-head rivets, where necessary, or better still, by designing the magazine in such a way that rivets do not pass through walls where moving parts are carried.

Supports for the magazine should nearly always be made adjustable to provide for variations in castings or forgings.

More than one adjustment is often necessary for the size of the work and its relation to the carrier. When a gravity feed is used, it is not always necessary to support the magazine rigidly, but when mechanical feeding devices are applied, the supports must be rigid enough to carry out the operation without undue vibration.

Friction Between Rolling Parts.— Friction surfaces should be made as small as possible consistent with good design, always bearing in mind that line or point contacts with the work are to be preferred. Pieces that are to roll into

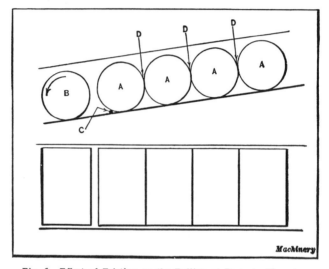

Fig. 6. Effect of Friction on the Rolling of Parts in Magazine

position should be fed in a chute having a sufficient incline so that they will roll freely. Care must always be taken with circular pieces having long surfaces that come into contact with each other, as shown in Fig. 6. Pieces *A,* considered one at a time, would roll down a slight incline unless they were badly distorted. However, when in a group, the first one starts to roll and gets away from the others, as shown at *B,* but a grain of sand or small chip in front of one of the others, as at *C,* may prevent it from starting because of the resistance at this point and the friction of the contact at points *D.* Of course, if the angle of the incline is great enough no trouble will be found, but while a 5-degree angle is sufficient to carry a single piece it is advisable to make the incline steep enough to prevent any possibility of sticking.

The importance of a smooth-working positive valve action for releasing pieces from a magazine should be emphasized,

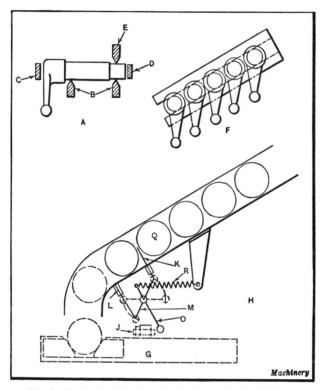

Fig. 7.   Design for Handling the Topheavy Forging C, Fig. 1

for on the correct admission of the pieces to the carrier depends the success of the entire mechanism.   Cases are found where springs, rubber bumpers, or sheet rubber "cowls" are necessary at the mouth of the magazine.

Feeding Forgings of Irregular Shape. — Piece *C*, Fig. 1 is to be machined on centers that have previously been drilled. Assume that a faceplate driver is used with an air-operated

device for driving the work by gripping the ball-end, a floating movement being provided to allow for variations in the
distance between the shaft center and the ball. There are
several interesting problems connected with designing an attachment for this piece. First, the faceplate must be stopped
at a specified position in relation to the work, in order that
the arm may enter the holding device properly.

The magazine for this work must be different from any
of the previous examples, as the pieces are so heavy at the
ball end that they cannot be balanced on the cylindrical portion. Therefore a special arrangement of guide rails and
guards is necessary, as shown at *A*, Fig. 7, in which the pieces
rest on knife-edges *B* with a guard *C* and *D* at each end spaced
far enough apart to allow clearance. Another guard *E* prevents the piece from tipping out of position. These pieces
cannot be held in a vertical chute because the overhanging
ends would strike each other and cause incorrect positioning.
However, they can be held in guides at an angle of 45 degrees,
as shown at *F*, and while they will not roll they will slide
easily down this incline without wedging or cramping.

There is a necessity for a stop arrangement at the lower
end of this device similar to a valve, but this can be of trigger
form to allow one piece to come through and stop the others,
as shown in the enlarged diagram at *H*. There are two sliding cut-off blades *K* and *L*, both of which are controlled by a
movement of lever *M* which is attached to a pivoted shaft.
To this shaft another lever *O* is connected in such a way that
when carrier *G* comes into position for a load, it strikes the
end of lever *O* and pulls it over, thus withdrawing blade *L*
and letting one piece of work fall into the carrier. At the same
time blade *K* is moved up to prevent the next piece from following. When the carrier returns again, spring *R* pulls lever
O forward returning blade *L* into place and withdrawing blade
K. There is an adjustable trip *J* on the carrier. Care must
always be taken to design shut-off mechanisms with sufficient adjustments so there will be no difficulty in timing up
the movement.

CHAPTER XX

DESIGN OF MAGAZINE CARRIERS AND SLIDES

IN transferring parts from a magazine chute or hopper to the cutting position in an automatic machine, a carrier or some other mechanical means is required, unless the work can be fed directly by gravity. The important factors in designing a carrier are shape of work, form of magazine, method of transferring work from carrier to holding units, and method of holding the work. If an automatic machine is to be designed for continuous work on the same piece day after day, it is not necessary to provide adjustments, except for taking up lost motion and wear. However, it is well to remember that a standard product may be redesigned and that even minor changes in size or shape may make it necessary to adapt the machine to new conditions. It is nearly always possible, without much extra expense, to provide for such contingencies in designing, and it is therefore advisable to make locating points, guides, stroke of carrier, etc., adjustable within reasonable limits.

Points Involved in Designing a Carrier.— Let us consider the conditions encountered in designing a carrier for delivering bars from a magazine to a chucking device in which the work is held on centers and gripped on the outside by floating jaws. The easiest way to determine the points of importance in developing any design is to consider the things that must be done. In this case, the work must fall from the magazine into the carrier, which must move over toward the supporting centers, and at the end of the travel, the work must be pushed out of the carrier and gripped by the chuck; the carrier must then move back into position for receiving

507

another piece from the magazine. It is evident that the work must be properly located in the carrier and perhaps held in a fixed position while the carrier moves forward to the chucking position. At this point any device used for holding the work must be positively released, leaving the work free to be pushed forward as required. Provision must also be made for returning the carrier without interfering with the piece that is left on the centers.

It will be assumed that the magazine delivers the work to the carrier in a position parallel to the axis of the centers on which it is to be placed. The problem, then, is to move the

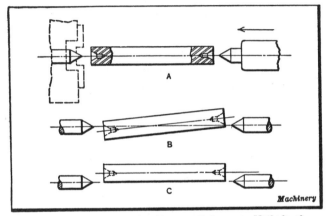

Fig. 1. Diagrams Showing Proper and Improper Methods of
Delivering Bar Work to Centers

work in the carrier from the magazine toward the front of the machine, until it reaches position *A*, Fig. 1, which represents a plan view. At this point it is pushed forward out of the carrier by a movement of the tailstock center, until it is located on the two centers and gripped by the floating jaws of the chuck.

The carrier must not interfere with the longitudinal movements of the work nor cause it to assume an incorrect position in relation to the centers or the chuck. At *B* is shown, exaggerated, a condition that might obtain if the work were located improperly in the carrier. In such a position, the centers would not enter the countersunk ends of the work and trouble will result. Of course, the centers will align work when it is not badly out of alignment, but they can only justify a certain amount of error. Then, again, the work may be parallel with the centers, but above, below, or at one side of them, as shown at *C*. Such a condition should exist only when the carrier supports are poorly designed, allowing chips or dirt to collect on the bearing seats, or when the adjustments for the carrier movement are loose or incorrectly set.

Carrier Supports. — Several examples of carrier supports for bar work are shown diagrammatically in Fig. 2, that at

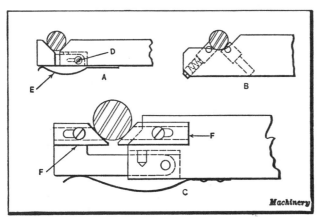

Fig. 2. Examples of Different Methods of Supporting Cylindrical
Work in Carriers

A being a plain V-form with one portion hinged at *D* to permit its withdrawal from the work after the latter has been placed on the centers. This design can easily be made adjustable by slotting the hole provided in the swinging piece for the hinge-pin, thus permitting a change in the relation of the two sides of the vee to suit different diameters of stock. A spring *E* is needed to hold the hinged portion in position, and this spring must be stiff enough so that it will not be depressed by the weight of the work. It must be so located as not to interfere with other parts of the mechanism.

In example *B*, the work is supported on two pins, one of which must either be of a spring-plunger type or held in a portion of the carrier that is hinged as in the preceding example. There are some advantages obtained in using a construction like this. It is simple and not likely to get out of order nor change its position in an operation. Another advantage is that the pins do not collect chips or dirt and can easily be replaced if worn or broken. Also, it is easy to provide an adjustment for various kinds of work.

In the final example *C*, the carrier is provided with two strips *F*, both of which may be made adjustable, or one fixed and the other adjustable, according to conditions. This question depends somewhat on the arrangement used to regulate the stroke of the carrier. If the stroke is not adjustable, both jaws of the vee must be adjustable, as otherwise they could not be arranged to receive and center the work properly. In this type of carrier also, one of the jaws must be hinged to allow easy withdrawal after chucking.

Examples of Carrier-slide Design. — Fig. 3 shows the arrangement of the magazine, carrier-slide, and carrier on an automatic machine handling bar stock. The work falls from the magazine to carriers *A* which are mounted in slides *B*. These slides, in turn, rest in saddles at *C*, which are adjustable lengthwise on a dovetail way extending longitudinally at the rear of the machine. The carriers and slides are moved back and forth across the machine by the bar *D* which passes through a boss at the rear of each slide. This construction permits longitudinal adjustment of the slides without disturbing the feeding mechanism.

Another construction is illustrated in Fig. 4, in which carrier *A* is a steel bar which is adjustable on slide *B*. The latter part is an iron casting, ribbed for stiffness, and dovetailed to fit saddle *C*. A taper gib *D* provides adjustment for wear. Carrier-slides should always have a long bearing in the saddles, extending out if necessary, as at *Y*, with an extension on the saddle serving as a support. The shaft used to move the carrier-slides backward and forward passes through the hole in boss E, and is a sliding fit in the hole so that the carriers may be readily adjusted for long or short work without disturbing any other mechanism.

Saddle C can be adjusted along dovetail F, after loosening gib G, which should again be tightened when the desired positions are reached. Adjustment of the carriers on the slides is possible by means of the slotted holes H. Portion J of the

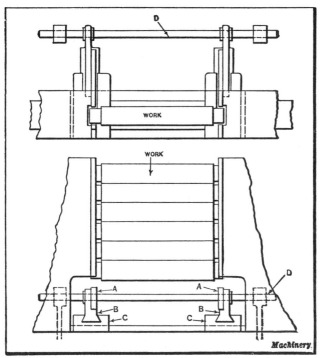

Fig. 3. Arrangement of the Magazine, Carrier-slide, and Carrier on an Automatic Machine

vee into which the work falls is hinged and supported by a flat spring as before, a slotted hole again permitting adjustments for diameter. Stud S limits the upward movement of jaw J. The dotted lines at K illustrate the operation of the swivel jaw when withdrawing from work placed on the machine centers.

In detail X, the magazine supports are dovetailed at L to

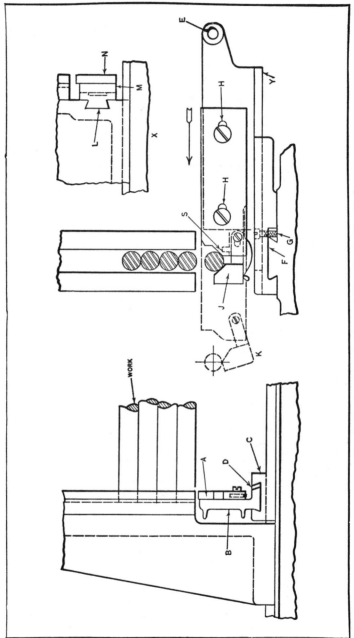

Fig. 4. Carrier-slides, Carrier, and Magazine for Bar Work

receive the carrier-slide M to which the carrier N may be fastened in any convenient way, so long as it is adjustable to some extent. There are both advantages and disadvantages in this construction, one advantage being that there are fewer pieces in the mechanism and adjustments are therefore simplified. The magazine supports must extend farther to the rear in order to give a long bearing to the slide. One disadvantage is that as the relation between the carriers and the magazine guides must always be the same, no adjustment is possible. A change in relation can only be obtained by substituting an offset carrier or adding special filler blocks between the slide and carrier to bring the latter into the required position. In designing this type of slide, the front face of slide M should be about in line with or a trifle back of the adjacent end of the work. In general, it will be found that the ideal construction is one in which simplicity is combined with a compact design requiring a minimum amount of space.

Control of Carrier-slides. — The movements of carrier-slides are usually controlled through levers actuated by a cam. The position of the camshaft, timing of the slide movements, and general method of operation are dependent to a large extent on the design of the machine. It is customary to use a single camshaft for controlling the movements in order to avoid the lost motion likely to occur if several shafts were used with intermediate gearing. With all movements taken from one shaft, it is much easier to regulate the timing and there is less possibility of a certain movement starting too late or too early on account of lost motion.

One method of operating a carrier-slide is illustrated in Fig. 5. The work A is shown just as it has dropped from the magazine into carrier B. Shaft D extends through both carrier-slides and enters slots in operating arms C, which control the movements. Both arms C may be fastened to pivot shaft E by means of keys, pins, or set-screws. Lever F has a roller G at one end, which engages with a cam H, so proportioned as to give the required movements at the proper time relative to other movements of the machine. Spring K keeps the roll in contact with the cam. The axis of the cam-shaft is in line with the work-holding centers.

Safety Devices for Slides. — With automatic machinery there is a possibility that the work may not lie properly in the slide, and this will cause considerable damage unless some

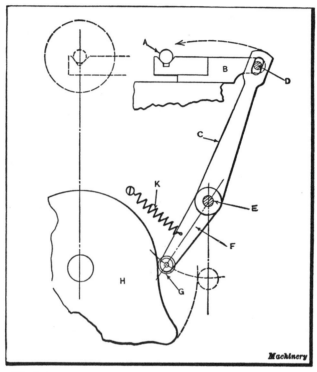

Fig. 5. Diagrammatic View Showing One Method of Imparting the Forward and Return Movements to the Carrier

provision is made to guard against such a contingency. It is well to anticipate such possibilities in designing by using the imagination and trying to conceive what the result would be if the piece did not fall properly into the carrier. One of the simplest safety devices that can be used for carrier-slides is a friction clutch, so placed that it will not drive the mecha-nism if an excessive strain occurs.

In Figs. 6 and 7 are shown examples of clutches suitable for such an application. In Fig. 6 the rocker-shaft is designated by *A*, and on it are mounted levers *B* which control the carrier-slide, only one lever being shown. A friction collar *C*, with a leather disk *D* secured to its face, is keyed or pinned

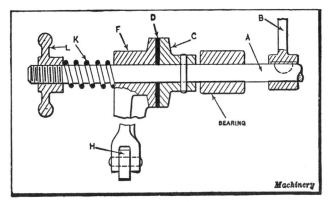

Fig. 6. Friction Clutch which may be Applied as a Safety Device to Carrier Mechanisms

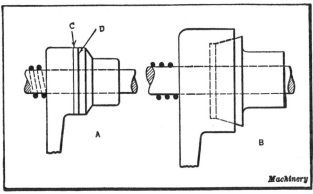

Fig. 7. Two Friction Clutch Designs which may be Used with Carriers

to the shaft. Lever *F* may be a steel or malleable-iron casting or a drop-forging, and is a running fit on the shaft. Roller *H* is held in contact with the cam in a manner similar to that shown in Fig. 5.

The hand-knob *L* is used to compress coil spring *K* and

produce the desired friction in the device by forcing the face of the hub on lever F against the leather face of collar C. The friction should be sufficient to permit the transmission of power to shaft A under normal conditions, but the clutch faces should slip when undue resistance is met. As the power required for this particular movement is small, an excessive pressure is not needed. It is an easy matter to regulate the friction by releasing or compressing the spring.

For mechanisms in which only a slight pressure is required, a steel disk C working against a cast-iron surface D, as shown at A in Fig. 7, is often sufficient, the steel disk being pinned to the lever. Another clutch that could be used for the same purpose is shown at B. This is of the conical type, possessing considerable pulling power which, of course, is dependent on the angle, diameter, width, etc., of the faces. This type is more expensive than the leather-disk form shown in Fig. 6, and it is not as well suited to the purpose.

Carriers for Irregular Pieces.— In handling symmetrical bar work, two parallel carriers of more or less fixed design are almost invariably used, but such forms are not often adapted to the handling of irregular work. Some of the important points that apply in designing a carrier for this kind of work are as follows:

1.   The shape of the piece to be held; it is essential to use a carrier that will hold the work invariably in the position required for its proper transfer to the chuck. If the work is fed to the carrier by gravity, and if it is heavier at one end than the other, provision must be made to prevent it from tipping or tilting to one side when entering the carrier.

2.   The release of the work as it is transferred; if there is more resistance on one portion than on another, the piece may be forced out of position so that it will not locate properly in the chuck. The pressure of any springs used for holding the work is often a troublesome matter. Experiments are frequently required before the most suitable form can be determined.

3.   The return of the carrier to the magazine for another

piece; the carrier should not "drag" unduly on the piece that has just been chucked, and it should close as soon as possible after leaving the piece so as to prevent the accumulation of chips or dirt. An air blast can be directed on the locating surfaces just before a new piece of work falls from the magazine, but this is necessary only when a great many chips are formed. Chips and dirt can also be eliminated by means of suitable guards so placed that they will not interfere with the work.

For certain kinds of irregular work, the carriers previously mentioned can be used, it being possible to make adjustments to suit conditions. When work of larger diameter is to be handled, however, another form is sometimes required, in which only one slide is used. Fig. 8 shows an example of this kind, which can be used when the magazine is arranged at the back of the machine. The work is contained in a vertical magazine A as indicated, and fed by gravity into carrier B. This piece of work has a chucking stem C, as illustrated clearly in the enlarged view Y, which is grasped in the jaws of a special chuck. As the work lies in the carrier, it is in front of the chuck, and so it must be pushed to the left to bring it into a position where the chuck jaws can seize it. This is accomplished by means of a pusher-rod on the turret of the machine and by a carrier design that readily releases the work.

Attached to the front side of parts B and D are three leaf springs E, mounted as clearly shown in the enlarged detail. The springs are made of sheet metal, bent over on each side to form a lug F adjacent to the walls of the carrier and assembled with a pin to form a hinge. Another spring G normally holds each spring E upright. On the opposite side of the carrier from the chuck, are three guard plates H which, in combination with springs E and the vee, form a more or less flexible "nest" in which the work is held upright while it is carried forward into the chucking device. When the carrier reaches this position, the turret pusher-rod advances and pushes the work past springs E into the chuck.

The carrier must make the forward motion without interference and must withdraw after the work has been chucked. In addition, it must hold the work approximately upright while moving from the magazine to the chucking position. A carrier having a wide range of adjustment is not always required for this kind of work, although it is advisable to provide a limited amount of adjustment to take care of normal

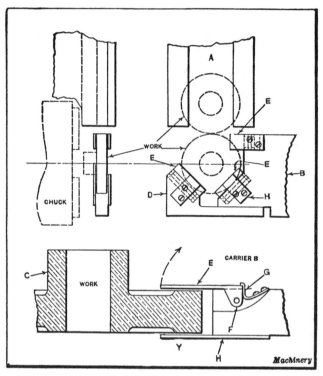

Fig. 8. Carrier in which the Work is Held by Means of Springs that Open Up in Placing the Work in the Chuck

variations in the size of work and to provide for minor changes in the design. This can be easily done by making member D separate from the carrier so that it can be moved backward or forward with relation to the other side of the vee.

Occasionally the same device may be used for several pieces, while in other cases it must be made to suit the particular piece. The method of holding the work in the carrier and the type of chuck employed affects the design to some extent. A similar spring valve arrangement can be applied to various forms of carriers, whether the magazine is at the rear of the machine, near the chuck, or above the turret. The requirements are that the carrier shall locate the work properly for chucking, hold it securely while moving into position, and release it readily when placed in the chuck.

Every form of carrier has its own peculiarities, and the designer must be continually on guard to avoid overlooking some small matter that would vitally affect the functioning of the mechanism. For example, when protruding chuck jaws are used for holding the work, if the chuck is stopped in a certain position when the piece is inserted, the position of the jaws may affect both the carrier and the pusher by means of which the work is inserted in the chuck. Also, when the work is picked up by a moving chuck, it may be necessary to so design either the carrier or the pusher that a certain portion can revolve while the work is being inserted in the chuck. Conditions like this tend to complicate the design and make it more difficult to avoid "hitches" in the operation.

It is not possible to give much information of value applying to conditions of this kind; the solution of such a problem is dependent largely upon the skill of the designer. Experiments with wooden or other models are almost always necessary in order to develop a scheme that will produce satisfactory results. The preliminary idea can often be developed on paper in proper proportion in order to make sure that it can be applied to the machine, but in perfecting the design, it is highly important to test it by using a working model.

INDEX

521